T0269270

Springer Undergraduate Mathematics Series

More information about this series at http://www.springer.com/series/3423

Shmuel Kantorovitz

Several Real Variables

Shmuel Kantorovitz
Bar-Ilan University
Ramat Gan
Israel

ISSN 1615-2085 ISSN 2197-4144 (electronic)
Springer Undergraduate Mathematics Series
ISBN 978-3-319-27955-8 ISBN 978-3-319-27956-5 (eBook)
DOI 10.1007/978-3-319-27956-5

Library of Congress Control Number: 2015959583

Mathematics Subject Classification: 26B05, 26B10, 26B12, 26B15, 26B20

Printed on acid-free paper

This Springer imprint is published by SpringerNature
The registered company is Springer International Publishing AG Switzerland

Preface

This book is based on the lecture notes of a course I gave half a dozen times to second-year undergraduates at Bar Ilan University. The prerequisites are two semesters on one variable Differential and Integral Calculus and a semester in Linear Algebra. Some familiarity with the language of elementary Abstract Algebra is assumed. Otherwise, the presentation is self-contained, but several of the more sophisticated proofs are omitted in order to keep the book within the appropriate constraints of width and depth. A description of the course follows.

Chapter 1 deals with the concept of continuity of functions of several real variables, or equivalently, functions on the k-dimensional space $\mathbb{R}^k$. This requires the introduction of a metric on $\mathbb{R}^k$. We discuss the Euclidean metric induced by the Euclidean norm, which is itself induced by the standard inner product on $\mathbb{R}^k$. More generally, we define inner product spaces and normed spaces, and study in particular the p-norms on $\mathbb{R}^k$ and the metric topology induced on $\mathbb{R}^k$ by any one of these equivalent norms. We then define compactness and show that compact sets are closed bounded sets in any metric space. The validity of the converse in $\mathbb{R}^k$ is then proved. Completeness is defined for metric spaces, and $\mathbb{R}^k$ is shown to be complete. These basic tools are then applied to the study of continuity of functions between metric spaces. The open sets characterization of continuity yields easily to the fundamental properties of continuous functions: the image of a compact set (connected set) by a continuous function is compact (connected, respectively), the inverse function of a bijective continuous function on a compact set is continuous, and the Intermediate Value Theorem is valid for any continuous real valued function on a connected metric space. The uniform continuity of continuous functions on a compact metric space and the equivalence of various definitions of connectedness of open sets are the closing subjects of Chap. 1.

The concept of differentiability for real or vector valued functions on $\mathbb{R}^k$ is introduced by means of linear approximation of the change of the function with respect to the change of the variable. We find sufficient conditions for differentiability and express the differential by means of the Jacobian matrix of the function with respect to the variable. We then prove the standard theorems on derivation of functions of several real variables: the chain rule, Taylor's theorem and sufficient conditions for local extrema.

In Chap. 3, we prove the Implicit Function Theorem for real (or vector) valued functions by applying the Banach Fixed Point Theorem for contractions on a complete metric space. The relevant space is a closed ball in the Banach space of all continuous real (or vector) valued functions on a cell in $\mathbb{R}^k$. This theory is applied to extrema with constraints (Lagrange multipliers). Some important properties of the Banach algebra of continuous functions on a compact subset of $\mathbb{R}^k$ (or on more general compact metric spaces) are studied in the last sections of the chapter: these include the Weierstrass Approximation Theorem and its Stone–Weierstrass generalization to compact metric spaces, and the Arzela–Ascoli Theorem on the Bolzano–Weierstrass Property.

Standard geometric applications in $\mathbb{R}^3$ are discussed at a slower pace, using intentionally some tools of the basic theory of systems of linear equations, in order to stress again the role played by Linear Algebra.

The Banach Fixed Point Theorem is also applied to obtain one of the versions of the Existence and Uniqueness Theorem for systems of ordinary differential equations. Consequences of the latter for linear systems are elaborated, with many details relegated to the Exercises section.

The chapter on integration begins with partial integrals (or integrals depending on parameters) of real functions of several real variables. We prove Leibniz's formula for derivation under the integral sign, and a theorem on the change of integration order.

The treatment of Riemann integration on a closed bounded domain in $\mathbb{R}^k$ does not seek greater generality because integration is more efficiently treated by means of the Lebesgue integral in more advanced texts. We define the (Jordan) content of bounded closed domains, and the Riemann integral of bounded real functions defined on them. The basic properties of Riemann integrable functions on such domains are proved. For normal domains in dimension 2 or 3, the integral is shown to reduce to an iterated (partial) integral. The change of variables formula is stated for any dimension, but the proof is omitted. The standard applications in dimension 2 or 3 follow. Integration on unbounded domains closes our general discussion of multiple integrals.

The last two sections are concerned with line and surface integrals. We define curve length, and obtain a formula for it in the (piecewise) smooth case. We then define line integrals of vector fields and conservative vector fields. We obtain necessary and sufficient conditions for a field to be (locally) conservative. Green's theorem is proved in two dimensions. The three-dimensional case (the Divergence Theorem) is stated but the proof is omitted. The generalization of Green's theorem to closed curves in $\mathbb{R}^3$ (Stokes' formula) is also stated without proof.

Most sections conclude with exercises. Many of the latter are routine, but some are rather sophisticated and complement the theory in certain important directions. Colleagues who taught parallel sections of the course at Bar Ilan University and my assistants in the past three or four years contributed some of the exercises and examples in the text, and I wish to thank them for it. In order to help students using this text for self-study, detailed solutions of a large portion of these exercises are given in the "Solutions" section at the end of the book. Last but not least, I wish to thank my colleague Prof. Jeremy Schiff, who helped me with the use of TEX in producing the manuscript.

Contents

Chapter 1
Continuity

1.1 The Normed Space $\mathbb{R}^k$

In this section, we shall introduce the space $\mathbb{R}^k$ on which the analysis of functions of k variables is done. We start with some notation and some basic algebraic notions.

The field of real numbers is denoted by $\mathbb{R}$. The Cartesian product $\mathbb{R}^k := \mathbb{R}\times\cdots\times\mathbb{R}$ (k factors) consists of all the ordered rows $x = (x_1, \ldots, x_k)$ with *components* $x_i \in \mathbb{R}$.

The study of functions of several real variables is the study of functions on $\mathbb{R}^k$. The analysis of these functions depends on a *distance function* on $\mathbb{R}^k$, which generalizes the standard distance $|x - y|$ between points $x, y \in \mathbb{R}$ to points $x, y \in \mathbb{R}^k$ for any $k \geq 1$.

The first step consists of generalizing the absolute value function

$$x \in \mathbb{R} \to |x| \in [0, \infty)$$

for $x \in \mathbb{R}^k$ with $k \geq 1$.

We recall first the elementary algebraic structure of $\mathbb{R}^k$. Addition in $\mathbb{R}^k$ is defined "componentwise": if $x = (x_1, \ldots, x_k)$ and $y = (y_1, \ldots, y_k)$ are elements of $\mathbb{R}^k$, then $x + y := (x_1 + y_1, \ldots, x_k + y_k)$. Similarly, one defines the multiplication of the *vector* $x \in \mathbb{R}^k$ by the *scalar* $\lambda \in \mathbb{R}$ by

$$\lambda x := (\lambda x_1, \ldots, \lambda x_k).$$

With these *operations* of *addition* and *multiplication by scalars*, $\mathbb{R}^k$ is a *vector space over the field* $\mathbb{R}$. The elements of $\mathbb{R}^k$ are called "points" of the space $\mathbb{R}^k$, or "vectors" in $\mathbb{R}^k$. The *standard basis* for the k-dimensional vector space $\mathbb{R}^k$ consists of the k vectors e^j, $j = 1, \ldots, k$, where e^j is the vector with j-th component 1 and all other

---

The original version of this chapter was revised. An erratum to this chapter can be found at DOI 10.1007/978-3-319-27956-5_5

S. Kantorovitz, *Several Real Variables*, Springer Undergraduate Mathematics Series, DOI 10.1007/978-3-319-27956-5_1

components 0 (that is, $e_i^j = \delta_{ij}$, the so-called *Kronecker delta*). Each $x \in \mathbb{R}^k$ has the unique representation as the linear combination $x = \sum_{j=1}^k x_j e^j$.

The j-th axis, or x_j-axis, of coordinates is the directed line

$$\mathbb{R}\, e^j := \{te^j;\ t \in \mathbb{R}\}.$$

In the above representation, each vector is uniquely decomposed into a sum of vectors belonging to distinct axis of coordinates.

The *inner product* (or *scalar product*) $x \cdot y$ of the vectors $x, y \in \mathbb{R}^k$ is defined by

$$x \cdot y = \sum_{i=1}^{k} x_i y_i.$$

In all the following sums, unless stated otherwise, the index runs from 1 to k.

The inner product is

(a) *bilinear*; that is, linear in each of the variables x and y.
(b) *commutative*: $x \cdot y = y \cdot x$ for all x, y.
(c) *positive definite*; that is, $x \cdot x \geq 0$ for all x, and $x \cdot x = 0$ iff $x = 0$, where 0 stands for the zero vector $(0, \dots, 0)$.

This brings us to the general concept of an "inner product space".

Inner Product Space

1.1.1 Definition Let X be a vector space over the field $\mathbb{R}$. An *inner product* on X is a function

$$(x, y) \in X \times X \to x \cdot y \in \mathbb{R}$$

with the properties (a)–(c).

If X is equipped with an inner product, it is called an *inner product space*.

With this terminology, we observed above that $\mathbb{R}^k$ is an inner product space (over the field $\mathbb{R}$).

A more general concept is that of a "normed space".

Normed Space

1.1.2 Definition A *norm* on a vector space X over the field $\mathbb{R}$ is a function

$$||\cdot|| : x \in X \to ||x|| \in [0, \infty)$$

with the following properties:

(a) *Definiteness*: $||x|| = 0$ iff $x = 0$.
(b) *Homogeneity*: $||\lambda x|| = |\lambda|\, ||x||$ for all $\lambda \in \mathbb{R}$ and $x \in X$.
(c) *The triangle inequality*: $||x + y|| \leq ||x|| + ||y||$ for all $x, y \in X$.

A *normed space* (over $\mathbb{R}$) is a vector space over the field $\mathbb{R}$ which is equipped with a norm. The defining properties of a norm are the essential properties of the absolute value $|\cdot|$ on the vector space $\mathbb{R}$.

We show below that every inner product space is a normed space for a norm induced by the inner product by an explicit formula.

1.1.3 Theorem *Let X be an inner product space over $\mathbb{R}$, and define $||x||$ as the non-negative square root of the non-negative number $x \cdot x$:*

$$||x|| = (x \cdot x)^{1/2} \qquad x \in X.$$

Then $||\cdot||$ is a norm on X and

$$|x \cdot y| \leq ||x||\,||y|| \qquad x, y \in X. \tag{1.1}$$

This norm on an inner product space is called *the norm induced by the inner product*. In particular, inner product spaces are normed spaces.

Unless stated otherwise, the norm on an inner product space is taken as the norm induced by the inner product.

The inequality (1.1) is referred to as the Cauchy-Schwarz inequality.

Proof Property (a) of $||\cdot||$ follows from its definition and the (positive) definiteness of the inner product. Property (b) follows from the bilinearity of the inner product: for all $x \in X$ and $\lambda \in \mathbb{R}$,

$$||\lambda x|| = [(\lambda x) \cdot (\lambda x)]^{1/2} = (\lambda^2)^{1/2}(x \cdot x)^{1/2} = |\lambda|\,||x||.$$

We proceed next to prove the Cauchy-Schwarz inequality (1.1).

By the bilinearity and positive definiteness of the inner product, we have for all $x, y \in X$ and $\lambda \in \mathbb{R}$

$$0 \leq (x - \lambda y) \cdot (x - \lambda y) = x \cdot x - 2\lambda\,(x \cdot y) + \lambda^2 (y \cdot y)$$

$$= ||x||^2 - 2\lambda\,(x \cdot y) + \lambda^2 ||y||^2.$$

In case $y \neq 0$, choose $\lambda = \frac{x \cdot y}{||y||^2}$. Then

$$0 \leq ||x||^2 - 2\frac{(x \cdot y)^2}{||y||^2} + \frac{(x \cdot y)^2}{||y||^2} = ||x||^2 - \frac{(x \cdot y)^2}{||y||^2}.$$

Hence $(x \cdot y)^2 \leq ||x||^2 ||y||^2$. Taking non-negative square roots, we obtain (1.1). In case $y = 0$, the inequality is trivially true (as an equality).

The triangle inequality for $||\cdot||$ follows from (1.1) and the bilinearity of the inner product:

$$||x+y||^2 = (x+y)\cdot(x+y) = ||x||^2 + 2(x\cdot y) + ||y||^2$$

$$\leq ||x||^2 + 2\,||x||\,||y|| + ||y||^2 = (||x|| + ||y||)^2. \qquad \square$$

We observed above that $\mathbb{R}^k$ is an inner product space under the inner product $x \cdot y = \sum x_i y_i$. The induced norm is the so-called Euclidean norm

$$||x||_2 := (x\cdot x)^{1/2} = \left(\sum x_i^2\right)^{1/2}.$$

For $k = 1$, we have clearly $||x||_2 = |x|$, so that $||\cdot||_2$ is a "possible" generalization of the absolute value on $\mathbb{R}$ to a norm on $\mathbb{R}^k$ for all $k \geq 1$.

In the geometric model for $\mathbb{R}^k$ when $k \leq 3$, the Euclidean norm of x is the usual distance from the point x to the "origin" 0, by Pythagores' theorem.

More generally, for any real number p, $1 \leq p < \infty$, we define the *p-norm* of x by

$$||x||_p := \left(\sum_i |x_i|^p\right)^{1/p}.$$

The norm properties (a) and (b) are trivial for $||\cdot||_p$ on the vector space $\mathbb{R}^k$. Property (c) is obvious as well in case $p = 1$. In order to prove Property (c) in case $1 < p < \infty$, we first prove *Hölder's inequality*, which generalizes the Cauchy-Schwarz inequality from the case $p = 2$ to all values of p, $1 < p < \infty$.

Hölder's Inequality

1.1.4 Theorem *Let $1 < p < \infty$, and let q, the conjugate exponent of p, be defined by the relation $(1/p) + (1/q) = 1$ (that is, $q = p/(p-1)$). Then for all $x, y \in \mathbb{R}^k$,*

$$|x\cdot y| \leq ||x||_p\,||y||_q.$$

Proof The inequality is trivial for $x = 0$ or $y = 0$. We then assume that both vectors are non-zero, and so $||x||_p > 0$ and $||y||_q > 0$.

It also suffices to prove the inequality for vectors with non-negative components, because then, in the general case, we have

$$|x\cdot y| \leq \sum |x_i|\,|y_i| \leq \left(\sum |x_i|^p\right)^{1/p}\left(\sum |y_i|^q\right)^{1/q} = ||x||_p\,||y||_q.$$

Since $||x||_p$ and $||y||_q$ are positive, the following are well-defined vectors

$$u = \frac{1}{||x||_p}x; \qquad v = \frac{1}{||y||_q}y,$$

with $u_i,\ v_i \geq 0$ and the additional property

$$||u||_p = ||v||_q = 1.$$

It suffices now to prove that $u \cdot v \le 1$, because substituting the expressions of u, v in terms of x, y, and using the bilinearity of the inner product, we then get the wanted inequality

$$\frac{x \cdot y}{||x||_p \, ||y||_q} \le 1.$$

The (natural) logarithm function $\log t$ is convex downward for $0 < t < \infty$, because $(\log t)'' = -1/t^2 < 0$. This means that the segment joining the points $(a, \log a)$ and $(b, \log b)$ on the graph of the function lies below the graph (for each $0 < a < b < \infty$). Since $(1/p) + (1/q) = 1$, we have $a < a/p + b/q < b$. The ordinates of the corresponding points on the segment and on the graph are $(\log a)/p + (\log b)/q$ and $\log(a/p + b/q)$ respectively. Since the first ordinate is less than or equal to the second (because the segment is below the graph), it follows from the properties of the logarithm that

$$a^{1/p} \, b^{1/q} \le \frac{a}{p} + \frac{b}{q}.$$

Write $\alpha := a^{1/p}$ and $\beta := b^{1/q}$ (clearly α, β are arbitrary positive numbers). Then

$$\alpha\beta \le \frac{\alpha^p}{p} + \frac{\beta^q}{q} \qquad (\alpha, \beta > 0).$$

Hence

$$u \cdot v = \sum u_i v_i \le \sum (\frac{u_i^p}{p} + \frac{v_i^q}{q})$$

$$= \frac{1}{p} \sum u_i^p + \frac{1}{q} \sum v_i^q = \frac{1}{p} ||u||_p^p + \frac{1}{q} ||v||_q^q = \frac{1}{p} + \frac{1}{q} = 1.$$ □

We may now prove the triangle inequality for $||\cdot||_p$, which is often called Minkowski's inequality.

Minkowski's Inequality

1.1.5 Theorem $||\cdot||_p$ *satisfies the triangle inequality for each* $p \in [1, \infty)$.

Proof As observed before, we may assume that $1 < p < \infty$. Let $q = p/(p-1)$ be the conjugate exponent. The triangle inequality is trivial if $||x + y||_p = 0$. Suppose then that $||x + y||_p > 0$. If the inequality is proved for x, y with non-negative components, the general case follows from the fact that the function t^α is increasing for $0 \le t < \infty$, for any $\alpha > 0$: since $|x_i + y_i| \le |x_i| + |y_i|$, one concludes that $|x_i + y_i|^p \le (|x_i| + |y_i|)^p$, and therefore, by the non-negative components case of the triangle inequality for $||\cdot||_p$,

$$||x + y||_p = \left(\sum |x_i + y_i|^p\right)^{1/p} \le \left(\sum (|x_i| + |y_i|)^p\right)^{1/p}$$

$$\leq \left(\sum |x_i|^p\right)^{1/p} + \left(\sum |y_i|^p\right)^{1/p} = ||x||_p + ||y||_p.$$

We then consider the case of vectors x, y with non-negative components such that $||x + y||_p > 0$, and $p > 1$. Write

$$||x + y||_p^p = \sum (x_i + y_i)(x_i + y_i)^{p-1} = \sum x_i (x_i + y_i)^{p-1} + \sum y_i (x_i + y_i)^{p-1}.$$

Therefore, by Hölder's inequality applied to the two inner products above,

$$||x + y||_p^p \leq (||x||_p + ||y||_p)\left(\sum (x_i + y_i)^{(p-1)q}\right)^{1/q}.$$

Since $(p-1)q = p$, the second factor above is equal to $||x + y||_p^{p/q}$. Divide the last inequality by this *positive* factor; since $p - p/q = p(1 - 1/q) = 1$, we obtain $||x + y||_p \leq ||x||_p + ||y||_p$, as wanted. □

The Norm $||\cdot||_\infty$ on $\mathbb{R}^k$

1.1.6 Definition

$$||x||_\infty = \sup_{1 \leq i \leq k} |x_i| \qquad (x \in \mathbb{R}^k).$$

(Since the supremum is taken over a *finite* set of numbers, it is equal to the greatest value of the numbers $|x_i|, i = 1, \ldots, k$.) Thus $|x_i| \leq ||x||_\infty$ for all i. The verification of the properties of the norm is trivial. We also have

$$|x \cdot y| = |\sum x_i y_i| \leq \sum |x_i|\,|y_i| \leq \sum ||x||_\infty |y_i| = ||x||_\infty ||y||_1,$$

and by changing the roles of x and y, also

$$|x \cdot y| \leq ||x||_1 ||y||_\infty.$$

This means that Hölder's inequality is valid also for $p = 1$ (with the conjugate exponent $q = \infty$) and $p = \infty$ (with conjugate $q = 1$).

Equivalence of the Norms $||\cdot||_p$

For all $p \in [1, \infty)$, since the function $t^{1/p}$ is increasing for $0 \leq t < \infty$, we have for all $i = 1, \ldots, k$

$$|x_i| = (|x_i|^p)^{1/p} \leq (\sum_j |x_j|^p)^{1/p} = ||x||_p.$$

and therefore

$$||x||_\infty \leq ||x||_p \qquad (x \in \mathbb{R}^k).$$

On the other hand,

$$||x||_p = (\sum_{i=1}^{k} |x_i|^p)^{1/p} \leq (\sum ||x||_\infty^p)^{1/p}$$

$$= (k||x||_\infty^p)^{1/p} = k^{1/p}||x||_\infty$$

for all $p \in [1, \infty)$ (and trivially for $p = \infty$). We conclude that, for all $p \in [1, \infty]$, there exist constants $A, B > 0$ (depending only on the dimension k and the exponent p) such that

$$A\,||x||_\infty \leq ||x||_p \leq B\,||x||_\infty \tag{1.2}$$

for all $x \in \mathbb{R}^k$. It follows that for any two exponents $p,\ p' \in [1, \infty]$, there exist constants $K, L > 0$ (depending only on $k,\ p,\ p'$) such that

$$K\,||x||_p \leq ||x||_{p'} \leq L\,||x||_p \tag{1.3}$$

for all $x \in \mathbb{R}^k$ (if $A',\ B'$ are the constants corresponding to p' as in (1.2), we may take $K = A'/B$ and $L = B'/A$).

Relation (1.3) is referred to as the *equivalence of the norms* $||\cdot||_p$ and $||\cdot||_{p'}$ for any two exponents $p,\ p' \in [1, \infty]$. We state this result formally:

1.1.7 Theorem *All the norms* $||\cdot||_p$ *(*$1 \leq p \leq \infty$*) are equivalent on* $\mathbb{R}^k$.

Actually, *any two norms on* $\mathbb{R}^k$ *are equivalent* (see Sect. 1.4.21, Exercise 10). The conceptual importance of this equivalence will be clarified in a later section.

1.1.8 Examples

Example 1. Fix $w \in \mathbb{R}^k$ with $w_i > 0$ for all $i = 1, \ldots, k$. If $p \in [1, \infty)$, define

$$||\cdot||_{p,w} : \mathbb{R}^k \to [0, \infty)$$

by

$$||x||_{p,w} = \Big(\sum_i w_i |x_i|^p\Big)^{1/p} \qquad (x \in \mathbb{R}^k).$$

Clearly $||\cdot||_{p,w}$ is positive definite and homogeneous. The triangle inequality for $||\cdot||_{p,w}$ follows from Minkowski's inequality for $||\cdot||_p$ (on $\mathbb{R}^k$) as follows. For all $x, y \in \mathbb{R}^k$,

$$||x+y||_{p,w} = \Big(\sum_i w_i|x_i + y_i|^p\Big)^{1/p} = \Big(\sum_i |w_i^{1/p}x_i + w_i^{1/p}y_i|^p\Big)^{1/p}$$

$$\leq \Big(\sum_i |w_i^{1/p}x_i|^p\Big)^{1/p} + \Big(\sum_i |w_i^{1/p}y_i|^p\Big)^{1/p} = ||x||_{p,w} + ||y||_{p,w}.$$

Thus each one of the functions $||\cdot||_{p,w}$ ($p \in [0, \infty)$, $w \in \mathbb{R}^k$ with $w_i > 0$ for all i) is a norm on $\mathbb{R}^k$. Let $m = \min_i w_i$ and $M = \max_i w_i$. Then for all $x \in X$,

$$m^{1/p} ||x||_p \leq ||x||_{p,w} \leq M^{1/p}||x||_p,$$

that is, the norms $||\cdot||_{p,w}$ and $||\cdot||_p$ are equivalent on $\mathbb{R}^k$. It follows then from the equivalence of the norms $||\cdot||_p$ on $\mathbb{R}^k$ that all the norms $||\cdot||_{p,w}$ are equivalent on $\mathbb{R}^k$.

Example 2. Let X be the set of all infinite rows $x = (x_1, x_2, \ldots)$ with components $x_i \in \mathbb{R}$, such that the series $\sum x_i^2$ converges. Let $x, y \in X$. Since $2ab \leq a^2+b^2$ for $a, b \in \mathbb{R}$, we have $(x_i + y_i)^2 = x_i^2 + 2x_i y_i + y_i^2 \leq 2x_i^2 + 2y_i^2$, and therefore $x + y \in X$ when addition of rows is defined componentwise. Defining $\lambda x = (\lambda x_1, \lambda x_2, \ldots)$ for $\lambda \in \mathbb{R}$, we have trivially $\lambda x \in X$. With the above operations of addition and multiplication by scalars, X is a vector space over $\mathbb{R}$. For $x, y \in X$, we define

$$x \cdot y = \sum_{i=1}^{\infty} x_i y_i.$$

The series converges absolutely since $|x_i y_i| \leq (1/2)(x_i^2 + y_i^2)$. It is clearly an inner product on X. Its norm is

$$||x||_2 = (\sum_{i=1}^{\infty} x_i^2)^{1/2} \qquad x \in X.$$

The inner product space X is usually denoted by l_2.

Example 3. Let $p \in [1, \infty)$, and let X be the set of all infinite rows $x = (x_1, x_2, \ldots)$ such that the series $\sum |x_i|^p$ converges (equivalently, such that the partial sums of this series are bounded). Define

$$||x||_p = (\sum_{i=1}^{\infty} |x_i|^p)^{1/p} \qquad (x \in X).$$

Let $x, y \in X$, and consider the row $x + y$ with components $x_i + y_i$. For each $k = 1, 2, \ldots$, we have by Minkowski's inequality on $\mathbb{R}^k$

$$\sum_{i=1}^{k} |x_i + y_i|^p \leq \Big[(\sum_{i=1}^{k} |x_i|^p)^{1/p} + (\sum_{i=1}^{k} |y_i|^p)^{1/p}\Big]^p \leq (||x||_p + ||y||_p)^p.$$

Therefore $x + y \in X$, and by letting $k \to \infty$, we also have the triangle inequality (or Minkowski's inequality) for $||\cdot||_p$ on X. Clearly $\lambda x := (\lambda x_1, \lambda x_2, \ldots) \in X$ for $\lambda \in \mathbb{R}$ and $x \in X$, and $||\lambda x||_p = |\lambda|\,||x||_p$. It follows that X is a vector space over the field $\mathbb{R}$, and $||\cdot||_p$ is a norm on X. Thus X is a normed space for the norm $||\cdot||_p$. It is usually denoted by l_p.

The infinite rows e^j with components $e_i^j = \delta_{ij}$ $(j = 1, 2, \ldots)$ are linearly independent, and therefore each space l_p is not finite dimensional. If $p, p' \in [1, \infty)$ are distinct, say $p < p'$, then for any $r \in (1/p', 1/p)$, we have $rp < 1 < rp'$, so that $\sum_i i^{-rp}$ diverges and $\sum_i i^{-rp'}$ converges. This means that the row (i^{-r}) belongs to $l^{p'}$ but not to l^p. In particular, this shows that the spaces l^p are distinct for distinct values of p.

Let X be the linear span of the rows e^j, that is, the set of all finite linear combinations of the vectors e^j. It is a normed space for each one of the norms $||\cdot||_p$, $1 \le p < \infty$. These norms are not equivalent on X. Indeed, if $1 \le p < p' < \infty$ and r is chosen as before, consider the vectors

$$x_n = (1, 2^{-r}, \ldots, n^{-r}, 0, 0, \ldots) \in X.$$

Then $||x_n||_p \to \infty$ while $||x_n||_{p'} \to ||(i^{-r})||_{p'} < \infty$. Therefore there does not exist a positive constant K such that $||x||_p \le K||x||_{p'}$ for all $x \in X$.

Example 4. Let $I = [a, b]$ be a closed real interval, and let $X = C(I)$ be the space of all continuous functions $f : I \to \mathbb{R}$, with vector space operations defined pointwise:

$$(f + g)(t) = f(t) + g(t), \quad (\lambda f)(t) = \lambda f(t)$$

$(f, g \in C(I), \lambda \in \mathbb{R})$. Clearly X is a vector space over the field $\mathbb{R}$. The zero vector of X is the function identically equal to zero in I. The functions $f_j(t) = t^j$ $(j = 0, 1, 2, \ldots)$ are linearly independent elements of X, so that X is not finite dimensional.

Let $p \in [1, \infty)$, and define

$$||f||_p = \Big(\int_I |f(t)|^p dt\Big)^{1/p} \qquad (f \in X),$$

where $\int_I := \int_a^b$ denotes the integration operation (from a to b) on $C(I)$.

The function $||\cdot||_p$ is non-negative on X and clearly $||\lambda f||_p = |\lambda|\,||f||_p$ for all $f \in X$ and $\lambda \in \mathbb{R}$. Suppose $||f||_p = 0$ for some $f \in X$, but f is *not* the zero function. The continuous funtion $|f|$ attains its maximum $M > 0$ in I at some point $s \in I$. By continuity of f, there exists a closed subinterval J of I containing s such that $0 \le |f(s)| - |f(t)| < M/2$ for all $t \in J$. Hence for all $t \in J$,

$$|f(t)| > |f(s)| - \frac{M}{2} = M - \frac{M}{2} = \frac{M}{2}.$$

Therefore

$$0 = ||f||_p^p = \int_I |f(t)|^p dt \ge \int_J |f(t)|^p dt \ge \int_J (\frac{M}{2})^p dt = (\frac{M}{2})^p |J| > 0,$$

where $|J|$ denotes the length of the interval J. This contradiction shows that $f = 0$, and therefore $||\cdot||_p$ is positive definite.

We can use the triangle inequality for the norm $||\cdot||_{p,w}$ on $\mathbb{R}^k$ (cf. Example 1) to prove the triangle inequality for $||\cdot||_p$ on $C(I)$. Let $P : a = t_0 < t_1 < t_2 < \cdots < t_n = b$ be a partition of I, and denote $w_i = t_i - t_{i-1}$. Let $f, g \in C(I)$. Define $x, y \in \mathbb{R}^n$ by $x_i = f(t_i)$ and $y_i = g(t_i)$. The triangle inequality

$$||x + y||_{p,w} \le ||x||_{p,w} + ||y||_{p,w}$$

translates as the following inequality between Riemann sums

$$\Big(\sum_i |f(t_i)+g(t_i)|^p(t_i - t_{i-1})\Big)^{1/p} \le \Big(\sum_i |f(t_i)|^p(t_i - t_{i-1})\Big)^{1/p} + \Big(\sum_i |g(t_i)|^p(t_i - t_{i-1})\Big)^{1/p}.$$

As the parameter of the partition tends to zero, the three Riemann sums above converge to the corresponding integrals, and we obtain the wanted inequality $||f + g||_p \le ||f||_p + ||g||_p$. We conclude that $||\cdot||_p$ is a norm on $C(I)$, so that $C(I)$ is a normed space for each one of these norms. The reader can easily verify that these norms are not equivalent.

The space $C(I)$ is also an inner product space for the inner product

$$f \cdot g := \int_a^b f(t)g(t)\,dt \qquad f, g \in C(I).$$

The norm $||\cdot||_2$ is the norm induced by this inner product.

Metric

Having generalized the absolute value concept on $\mathbb{R}$ to the norm concept on general vector spaces over the real field, we proceed to generalize the distance function $|x-y|$ on $\mathbb{R}$ to an appropriate concept on abstract sets, and in particular on normed spaces.

1.1.9 Definition A *metric* (or *distance function*) on a non-empty set X is a function

$$d : X \times X \to [0, \infty)$$

with the following properties:

(a) *definiteness*: $d(x, y) = 0$ iff $x = y$;
(b) *symmetry*: $d(y, x) = d(x, y)$ for all $x, y \in X$;
(c) *triangle inequality*: $d(x, y) \le d(x, z) + d(z, y)$ for all $x, y, z \in X$.

The set X with the metric d on it is called a *metric space*.

Several metrics may be defined on the same set X; if we wish to stress that we consider a particular metric d on X, we refer to the metric space as the ordered pair (X, d). Its points are the elements of the set X, and $d(x, y)$ is called the distance between the points x and y.

Any (non-empty) subset E of the metric space (X, d) is itself a metric space with the metric d restricted to $E \times E$. It will be understood that E is equipped with this metric, whenever $E \subset X$. We use the notation $E \subset X$ to state that E is either a proper subset of X or is equal to X.

Example 1. If X is a normed space (with the norm $||\cdot||$), the metric *induced by the norm* is defined as

$$d(x, y) := ||x - y|| \qquad (x, y \in X).$$

The verification of Properties (a)–(c) of a metric is trivial. This means that normed spaces are special cases of metric spaces. They will be usually equipped with the metric induced by their norm.

Example 2. We may specialize Example 1 to the normed space $\mathbb{R}^k$ with anyone of the equivalent norms $||\cdot||_p$, $1 \le p \le \infty$. Thus $\mathbb{R}^k$ is a metric space for anyone of the (induced) metrics

$$d_p(x, y) := ||x - y||_p$$

($p \in [1, \infty]$). These metrics are *equivalent*, that is, there exist constants $K, L > 0$ (depending only on k, p, p') such that

$$K\, d_p(x, y) \le d_{p'}(x, y) \le L\, d_p(x, y)$$

for all $x, y \in \mathbb{R}^k$.

The special metric d_2, called the *Euclidean metric*, and given by

$$d_2(x, y) = \left(\sum (x_i - y_i)^2\right)^{1/2},$$

is the usual distance function of elementary Euclidean Geometry for dimensions $k = 1, 2, 3$.

The Metric Topology

We shall now use the metric on a given metric space to define open sets, limit points, etc.

1.1.10 Definition Let (X, d) be a metric space.

(i) Given $x \in X$ and $r > 0$, the (open) *ball with centre x and radius r* is the set

$$B(x, r) := \{y \in X;\ d(y, x) < r\}.$$

The latter is also called a neighborhood of x, or the r-neighborhood of x.

In Figs. 1.1, 1.2, 1.3 and 1.4, the ball $B(0, 1)$ in $\mathbb{R}^2$ for the metrics d_1, d_2, d_3, and d_∞ is the set of all points "inside" the shown curve.

(ii) The point $x \in X$ is called an *interior point* of the subset E of X if it has an r-neighborhood contained in E.

(iii) The set E is *open* if every point of E is an interior point of E.

Fig. 1.1 $|x| + |y| = 1$

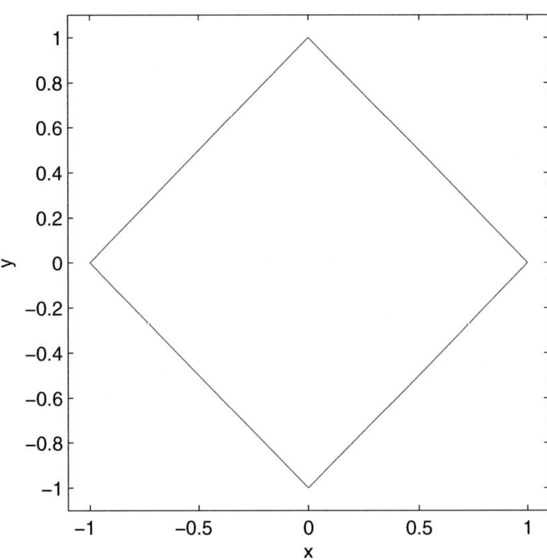

(iv) The set E is *closed* if its complement

$$E^c := \{x \in X;\ x \notin E\}$$

is open.

(v) The point $x \in X$ is a *limit point* of E if *every neighborhood of* x *contains a point* $y \in E$, $y \neq x$.

(vi) The point $x \in E$ is an *isolated point* of E if it is not a limit point of E, that is, if there exists a neighborhood of x which contains no point of E distinct from x.

Observe that if x is a limit point of E, then every neighborhood of x contains actually an *infinite* number of points of E. Indeed, if there is a neighborhood $B(x, r)$ containing only finitely many points of E, denote these points that are distinct from x by y_j, $j = 1, \ldots, n$. Necessarily $n \geq 1$, because x is a limit point of E. Define

$$\delta = \min_{j=1,\ldots,n} d(y_j, x).$$

Then $\delta > 0$, and $B(x, \delta)$ clearly contains no point of E distinct from x, contradicting the hypothesis that x is a limit point of E.

It follows in particular that a finite subset of X has no limit points.

Example 1. A ball $B(x, r)$ is an open set.

Indeed, if $y \in B(x, r)$, then $s := r - d(y, x) > 0$. If $z \in B(y, s)$, then by the triangle inequality for d,

$$d(z, x) \leq d(z, y) + d(y, x) < s + d(y, x) = r,$$

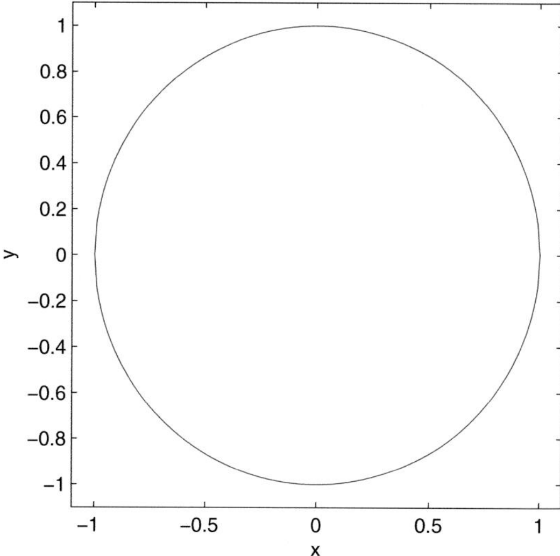

Fig. 1.2 $x^2 + y^2 = 1$

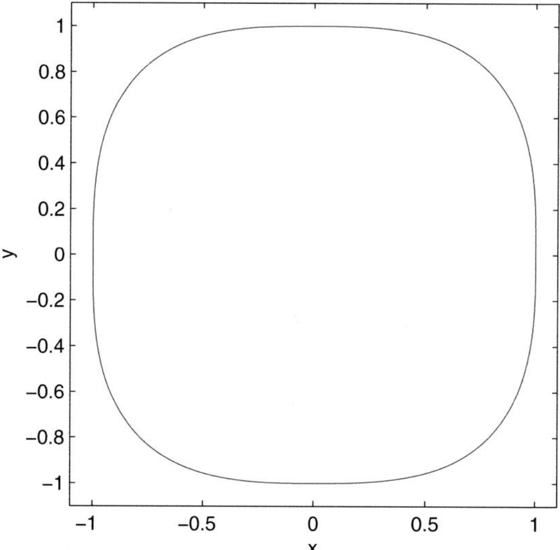

Fig. 1.3 $|x|^3 + |y|^3 = 1$

which proves that $B(y, s) \subset B(x, r)$, that is, y is an interior point of $B(x, r)$. Hence $B(x, r)$ is open.

Example 2. The set

$$\overline{B}(x, r) := \{y \in X;\ d(y, x) \le r\} \qquad (x \in X,\ r > 0)$$

is a closed set.

The set $\overline{B}(x, r)$ is called the closed ball with centre x and radius r.

Fig. 1.4 $\max(|x|, |y|) = 1$

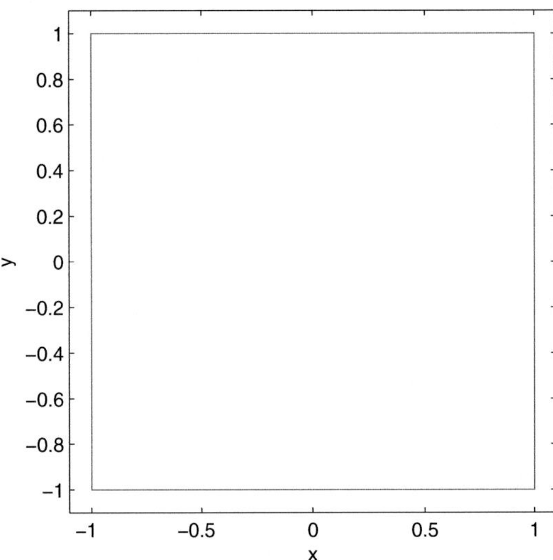

In order to verify the above statement, we we must show that the complement E of $\overline{B}(x, r)$ is open. We use an argument analogous to the argument in Example 1. We have

$$E = \{y \in X;\ d(y, x) > r\}.$$

Let $y \in E$ and set $s = d(y, x) - r$ (> 0). If $z \in B(y, s)$ is *not* in E, then

$$d(y, x) \le d(y, z) + d(z, x) < s + d(z, x) \le s + r = d(y, x),$$

contradiction. This shows that $B(y, s) \subset E$, so that y is an interior point of E, and we conclude that E is open.

Example 3. If E is a subset of the metric space X, denote by E' the set of all the limit points of E. Set $E'' = (E')'$. We shall verify that $E'' \subset E'$.

Let $x \in E''$ and let $B(x, r)$ be an arbitrary r-neighborhood of x. There exists $y \in E'$ such that $y \in B(x, r)$ and $y \neq x$. Since $B(x, r)$ is open, there exists $s > 0$ such that $s < d(y, x)$ and $B(y, s) \subset B(x, r)$. The neighborhood $B(y, s)$ of $y \in E'$ contains some $z \in E$, $z \neq y$. Then $z \in B(x, r)$, and $z \neq x$ (since $z \in B(y, s)$ but $x \notin B(y, s)$). We conclude that $x \in E'$, as desired.

Notation. Set Theory notation is used in the standard way, except that we simplify $\subseteq$ as $\subset$.

The family τ of all open subsets of X has the following obvious properties:

(a) $\emptyset,\ X \in \tau$;
(b) if $E_\alpha \in \tau$ for all α in an *arbitrary* index set I, then

$$\bigcup_{\alpha\in I} E_\alpha \in \tau;$$

(c) if $E_j \in \tau$ for $j = 1, \dots, n$, then

$$\bigcap_{j=1}^{n} E_j \in \tau.$$

We verify Property (c): if x belongs to the intersection, then $x \in E_j$ for all $j = 1, \dots, n$, and since E_j are open, there exist balls $B(x, r_j) \subset E_j$; the ball $B(x, r)$ with $r := \min_{j=1,\dots,n} r_j$ is then contained in $\bigcap_j E_j$.

Any family τ of subsets of a given set X which possesses Properties (a)–(c) is called a *topology* on X. The sets inluded in the family τ are called the *open sets* of the given topology. The set X with the specified topology τ is called a *topological space*.

In case X is a metric space, the open sets defined above determine the *metric topology* (induced by the given metric). Thus, metric spaces with their metric topology are special cases of topological spaces.

Example 4. Let X be a metric space. We show that $E \subset X$ is open iff it is a union of balls. By Property (b) and Example 1, a union of balls is an open set. On the other hand, if E is open, each $x \in E$ has a r_x-neighborhood contained in E, where $r_x > 0$ depends on x. Then

$$E \subset \bigcup_{x\in E} B(x, r_x) \subset E,$$

and we conclude that E is equal to the union of the balls $B(x, r_x)$, with the index x running in E.

Example 5. In the present example, we shall verify that equivalent metrics on a set X induce the same topology on X.

Let d and d' be equivalent metrics on X. Let then $K, L > 0$ be constants such that

$$K\, d(x, y) \le d'(x, y) \le L\, d(x, y) \qquad (x, y \in X).$$

Denote by $B(x, r)$ and $B'(x, r)$ the r-neighborhoods defined by means of the metrics d and d' respectively. Then clearly

$$B(x, \frac{r}{L}) \subset B'(x, r); \quad B'(x, Kr) \subset B(x, r).$$

Let $E \subset X$. If x is an interior point of E relative to the metric d, there exists a ball $B(x, r)$ contained in E. Then $B'(x, Kr) \subset E$, which shows that x is an interior point with respect to the metric d'. Similarly, in the latter case, we have $B'(x, r) \subset E$ for some $r > 0$, hence $B(x, r/L) \subset E$, and therefore x is an interior point with respect to d. Thus x is an interior point of E with respect to d iff it is an interior point of E

with respect to d'. It follows that E is open with respect to d iff it is open with respect to d'. In other words, equivalent metrics on X induce the same metric topology on X.

A similar argument shows that x is a limit point of E with respect to d iff it is a limit point of E with respect to d'.

Observe that we may use the De Morgan laws to translate Properties (a)–(c) of open sets into dual properties of closed sets, in any topological space X. We recall:

De Morgan's Laws. Let $\{E_\alpha; \alpha \in I\}$ be a family of subsets of a set X. Then

(a) $\left(\bigcup_{\alpha\in I} E_\alpha\right)^c = \bigcap_{\alpha\in I} E_\alpha^c$.
(b) $\left(\bigcap_{\alpha\in I} E_\alpha\right)^c = \bigcup_{\alpha\in I} E_\alpha^c$.

We obtain the following properties of closed sets:

(a) $\emptyset$ and X are closed sets;
(b) *Arbitrary intersections* of closed sets are closed;
(c) *Finite unions* of closed sets are closed.

Example 6. Let E be a subset of the topological space X. Denote by $\overline{E}$ the intersection of all the closed subsets of X that contain E. By Property (b) of closed sets, $\overline{E}$ is a closed set that contains E, and $\overline{E} \subset F$ for any closed set F containing E. We express this fact by saying that $\overline{E}$ is the *minimal* closed set containing E. The set $\overline{E}$ is called the *closure* of E. We say that E is *dense in* X if $\overline{E} = X$.

We discuss below some properties of the closure operation

$$A \to \overline{A}$$

operating on subsets A of X.

(a) If $A \subset B \subset X$, then $\overline{A} \subset \overline{B}$.
Indeed, if $A \subset B$, then $\overline{B}$ is a closed set containing A, and therefore $\overline{A} \subset \overline{B}$ by the above minimality property.
(b) If $E, F \subset X$, then
$$\overline{E \cup F} = \overline{E} \cup \overline{F}.$$
First, since both E and F are subsets of $E \cup F$, it follows from (a) that both $\overline{E}$ and $\overline{F}$ are subsets of $\overline{E \cup F}$. Hence $\overline{E} \cup \overline{F} \subset \overline{E \cup F}$. On the other hand, $\overline{E} \cup \overline{F}$ is closed by Property (c) of closed sets, and contains $E \cup F$. By the minimality property of the closure, it follows that $\overline{E \cup F} \subset \overline{E} \cup \overline{F}$, and (b) is verified.

Characterization of Closed Sets

We may use the concept of limit points to characterize closed subsets of a metric space.

1.1.11 Theorem *Let X be a metric space, and $E \subset X$. Then E is closed iff it contains its limit points.*

Proof Suppose E is closed (i.e., E^c is open), and x is a limit point not in E, that is, $x \in E^c$. Then x is an interior point of E^c. Hence there exists $r > 0$ such that $B(x, r) \subset E^c$, which means that $B(x, r)$ contains no points of E, and consequently x is *not* a limit point of E, contradiction! This shows that E contains all its limit points.

Conversely, suppose E contains all its limit points, but E is *not* closed. Then E^c is not open, that is, there exists $x \in E^c$ which is *not* an interior point of E^c. This means that every ball $B(x, r)$ contains points of E (necessarily distinct from x, since $x \in E^c$). Hence x is a limit point of E, but is not in E, contradiction! This proves that E is closed. □

Using the notation in Example 3, Theorem 1.1.11 states that the subset E of X is closed iff $E' \subset E$. In case $E' = E$, we say that E is a *perfect set*. By Theorem 1.1.11, a perfect set is a closed set with no isolated point.

If E is a finite subset of X, then $E' = \emptyset \subset E$, hence E is closed.

Looking at complements, we see that if we delete finitely many points from X, we get an open set.

The set $E = \{1/n; n = 1, 2, \ldots\}$ in $\mathbb{R}$ is not closed, since it does not contain its limit point 0. Clearly $E' = \{0\}$.

Example 7. We saw in Example 3, Sect. 1.1.10, that $E'' \subset E'$ for any subset E of a metric space. By Theorem 1.1.11, this implies that E' is a closed set.

We show that for any subsets E, F of X,

$$(E \cup F)' = E' \cup F'. \tag{1.4}$$

Observe first that if $A \subset B \subset X$, then $A' \subset B'$.

Indeed, if $x \in A'$, every ball $B(x, r)$ contains a point of A distinct from x; since this point is also in B, this shows that $x \in B'$, that is, $A' \subset B'$.

For any two subsets E, F of X, both E and F are subsets of $E \cup F$, hence $E' \subset (E \cup F)'$ and $F' \subset (E \cup F)'$ by our preceding observation. Consequently

$$E' \cup F' \subset (E \cup F)'. \tag{1.5}$$

On the other hand, suppose x belongs to the complement of $E' \cup F'$, that is,

$$x \in (E')^c \cap (F')^c.$$

Then there exist balls $B(x, r_i)$ $(i = 1, 2)$ containing no point of E and F (respectively) distinct from x. Let $r = \min(r_1, r_2)$. Then $B(x, r)$ contains no point of $E \cup F$ distinct from x, that is, x is in the complement of $(E \cup F)'$, Together with the inclusion (1.5), we obtain the desired identity (1.4).

Example 8. We show that a closed ball $E = \overline{B}(x, r)$ in any normed space X is a perfect set, that is, $E' = E$.

We saw in Example 2, Sect. 1.1.10, that E is closed, that is, $E' \subset E$. On the other hand, let $y \in E$, and consider any s-neighborhood of y. If $y = x$, $B(y, s) \cap E$

contains the ball $B(x, \delta)$ where $\delta = \min(r, s)$, which is an infinite set when X is a normed space (why?). Hence $y \in E'$ trivially. Suppose then that $y \neq x$.

The so-called segment $L := \overline{xy}$ defined by

$$L := \{z_t := (1-t)x + ty,\ t \in [0, 1]\}$$

is contained in E, because

$$d(z_t, x) = ||z_t - x|| = ||t(y-x)|| = t\,||y-x|| \le tr \le r.$$

For all $t \in (0, 1)$ such that $t > 1 - s/r$,

$$d(z_t, y) = ||z_t - y|| = ||(1-t)x - (1-t)y|| = (1-t)||x-y|| \le (1-t)r < s,$$

that is, $z_t \in B(y, s)$ and $z_t \neq y$ since $y \neq x$. This shows that $y \in E'$, and we conclude that $E' = E$.

We note that in an arbitrary metric space X, closed balls are not generally perfect. For example, if $X = \mathbb{Z}$ with the metric $d(x, y) = |x-y|$, the closed ball $E = \overline{B}(x, r)$ $(x \in X)$ is not perfect, because any s-neighborhood of $x \in E$ with $s < 1$ contains no point of $E \setminus \{x\}$, that is, $x \in E \setminus E'$.

Connected Sets

In the following, it will be convenient to say that two sets A, B *meet* if $A \cap B \neq \emptyset$ and that they are *disjoint* otherwise.

1.1.12 Definition Let X be a metric space. A subset E of X is *connected* if it is *not* contained in the union of two *disjoint open sets* $A, B \subset X$, *both of which meet* E.

The following trivial example illustrates the meaning of non-connectedness of a set E as being made up of intuitively "separated" pieces. If $x, y \in X$ and $r := d(x, y) > 0$, the set

$$E = B(x, r/4) \cup \overline{B}(y, r/4)$$

is not connected, because we may choose $A = B(x, r/2)$ and $B = B(y, r/2)$ in the above definition.

Observe that if the connected set E is contained in the disjoint union of two non-empty open sets A, B, then either $E \subset A$ or $E \subset B$. (otherwise E meets both A and B, hence is not connected, contradiction!)

Example 1. Let X be a normed space. If $a, b \in X$, define the *segment* $\overline{ab}$ by

$$\overline{ab} := \{x_t := (1-t)a + tb;\ t \in [0, 1]\}.$$

If $a = b$, the segment reduces to the singleton $\{a\}$. In the geometric model for $\mathbb{R}^k$ when $k \le 3$, $\overline{ab}$ is the usual closed line segment with the end points $x_0 = a$ and $x_1 = b$.

We show that $\overline{ab}$ is a connected subset of X. We may assume that $a \neq b$, since singletons are trivially connected.

Suppose A, B are disjoint open subsets of X, both meeting $\overline{ab}$, such that $\overline{ab} \subset A \cup B$. Let then $s, t \in [0, 1]$ be such that $x_s \in A$ and $x_t \in B$. Clearly $s \neq t$ because the sets A, B are disjoint. We may assume that $s < t$ (since the roles of A and B can be interchanged). In particular, $s < 1$ and $t > 0$. Note that

$$d(x_u, x_v) = ||[a + u(b - a)] - [a + v(b - a)]|| = |u - v|\, ||b - a||.$$

Let $s \in [0, 1)$ and $t \in (0, 1]$ be any pair of points as above (i.e., $x_s \in A$ and $x_t \in B$). Since A, B are open, there exists $r > 0$ such that

$$B(x_s, r) \subset A; \quad B(x_t, r) \subset B.$$

Therefore $s + \epsilon \in (0, 1)$ for $0 < \epsilon < \min\{\frac{r}{||b-a||}, 1 - s\}$ and

$$d(x_{s+\epsilon}, x_s) = \epsilon\, ||b - a|| < r,$$

that is, $x_{s+\epsilon} \in B(x_s, r) \subset A$.

Define

$$v = \sup\{u \in [0, 1];\ x_u \in A\}.$$

If $x_v \in A$, necessarily $v < t \leq 1$ because $x_t \in B$, and $x_{v+\epsilon} \in A$ for $\epsilon > 0$ small enough, as we saw above. This contradicts the definition of v. Hence $x_v \notin A$. A similar argument shows that $x_v \notin B$. Therefore $x_v \notin A \cup B$, contradicting the hypothesis that $\overline{ab} \subset A \cup B$. This contradiction proves that $\overline{ab}$ is connected.

Example 2. Let X be a normed space. A subset E of X is said to be *convex* if for any pair of distinct points $a, b \in E$, the segment $\overline{ab}$ is contained in E. In order to verify the convexity of E, it suffices to show that $x_t \in E$ for $t \in (0, 1)$ (notation as in Example 1), since $x_0 = a \in E$ and $x_1 = b \in E$ by hypothesis.

The empty set and singletons are the "trivial" convex sets. Balls (open or closed) in a normed space X are convex. Indeed, if a, b are distinct points of the ball $B(p, r)$ in X, then for all $t \in (0, 1)$,

$$d(x_t, p) = ||x_t - p|| = ||(1-t)(a-p) + t(b-p)|| \leq (1-t)||a-p|| + t\,||b-p||$$

$$< (1 - t)r + tr = r.$$

The convexity of $\overline{B}(p, r)$ is verified in the same way.

Intersections of convex sets are convex. However unions and differences of convex sets are not convex in general, as can be seen by simple planar examples.

We show that convex sets are connected. If $E = \emptyset$ or E is a singleton, E is trivially connected, since two disjoint open sets cannot both meet a singleton. Assume then that E contains at least two distinct points. Let A, B be disjoint open sets, both meeting E, say at the points a, b respectively, such that $E \subset A \cup B$. We must reach a contradiction.

Since E is convex, we have

$$\overline{ab} \subset E \subset A \cup B.$$

Since $\overline{ab}$ is connected (cf. Example 1), it follows from the observation preceding Example 1 that either $\overline{ab} \subset A$ or $\overline{ab} \subset B$, hence either $b \in A \cap B$ or $a \in A \cap B$, contradicting the disjointness of the sets A, B.

Union of Connected Sets

1.1.13 Theorem *Let $\{E_i;\ i \in I\}$ be a family of connected subsets of the metric space X such that*

$$F := \bigcap_{i \in I} E_i \neq \emptyset.$$

Then

$$E := \bigcup_{i \in I} E_i$$

is connected.

Proof If A, B are as in the above definition, then for each $i \in I$ the connected set E_i is contained in $A \cup B$ with A, B disjoint, open, and non-empty (both meet E!). By the above observation, either $E_i \subset A$ or $E_i \subset B$. Suppose $E_i \subset A$ for some i. Then for all $j \in I$,

$$\emptyset \neq F = E_i \cap F \subset A \cap F \subset A \cap E_j,$$

that is, A meets every E_j, and therefore the inclusion $E_i \subset B$ cannot occur for any i, and consequently the union E is contained in A. Similarly, if we assume that $E_i \subset B$ for some i, it follows that E is contained in B. In any case, one of the sets A, B does not meet E, contradiction. □

Example 3. A subset E of a normed space X is called a *starlike set* if there exists a point $p \in E$ such that $\overline{px} \subset E$ for all $x \in E$. We say also that E is starlike relative to p. Clearly every non-trivial convex subset E of X is starlike relative to any point $p \in E$: for all $x \in E$, $\overline{px} \subset E$ by the convexity of E. However there are starlike sets E that are not convex. For example, if $a, b, c \in \mathbb{R}^2$ are not colinear, the "polygonal path"

$$E = \overline{ab} \cup \overline{bc}$$

is starlike (relative to b) but is not convex, since $\overline{ac}$ does not contain the point $b \in E$. The set

$$E := \{(x, y) \in \mathbb{R}^2;\ x, y \geq 0,\ \sqrt{x} + \sqrt{y} \leq 1\}$$

is starlike but not convex (see Fig. 1.5).

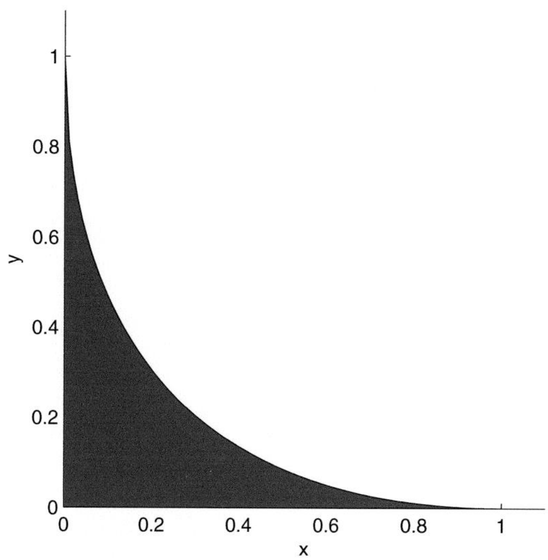

Fig. 1.5 $E = \{(x, y) \in \mathbb{R}^2;\ x, y \geq 0,\ \sqrt{x}+\sqrt{y} \leq 1\}$

If E is starlike relative to $p \in E$, we can write

$$E = \bigcup_{x \in E} \overline{px}.$$

The segments $\overline{px}$ are connected (cf. Example 1), and their intersection contains the point p. By Theorem 1.1.13, we conclude that E is connected.

Example 4. Let a_i, $i = 1, \ldots, n$ be $n \geq 2$ distinct points in the normed space X. The *polygonal path with vertices a_i* is the union of segments

$$E := \bigcup_{i=1}^{n-1} \overline{a_i a_{i+1}}.$$

We say that the polygonal path E *starts at the point a_1 and ends at the point a_n*, or that E *joins the points a_1 and a_n*.

We show that E is connected. If $n = 2$, E is a segment, and we saw in Example 1 that segments are connected. Proceeding by induction on $n \geq 2$, assume that polygonal paths with n vertices are connected, for some $n \geq 2$, and let E be a polygonal path with $n + 1$ vertices, say $a_1, \ldots, a_{n+1}$. Then $E = F \cup L$, where F is the polygonal path with the n vertices $a_1, \ldots, a_n$, and $L := \overline{a_n a_{n+1}}$. The set F is connected, by the induction hypothesis. We saw that L is connected in Example 1. Finally, we have $F \cap L \neq \emptyset$, since $a_n \in F \cap L$. We conclude from Theorem 1.1.13 that E is connected, and our claim follows by induction.

Sufficient Condition for Connectedness

1.1.14 Corollary *Suppose every two points of E belong to a connected subset of E. Then E is connected.*

Proof Fix $a \in E$. For any $x \in E$, there exists a connected subset $E_x \subset E$ such that $a, x \in E_x$. Then

$$a \in F := \bigcap_{x \in E} E_x,$$

so that $F \neq \emptyset$. By 1.1.13, the union E of all E_x $(x \in E)$, is connected. □

Example 5. Let X be a normed space. A set $E \subset X$ is *polygonally connected* if for every two distinct points $a, b \in E$, there exists a polygonal path joining them and lying in E. Convex sets are clearly polygonally connected. The same is true for starlike sets: if E is starlike relative to some $p \in E$, then for any distinct points $a, b \in E$ the polygonal path with vertices a, p, b (in this order) joins the points a, b, and lies in E by the starlike property.

An open ring in $\mathbb{R}^2$ is clearly polygonally connected, but is not starlike.

Every two points of a polygonally connected set E belong to a polygonal path joining them and lying in E; the path is a connected subset of E (cf. Example 4). By Corollary 1.1.14, it follows that E is connected. This shows that for subsets of a normed space, polygonal connectedness implies connectedness.

1.1.15 Exercises *In the following exercises, X is a metric space, with the metric d.*

1. Define
$$d'(x, y) = \frac{d(x, y)}{1 + d(x, y)} \qquad (x, y \in X).$$
Show that d' is a metric on X (Note that $d'(x, y) < 1$ for all $x, y \in X$ and $d'(x, y) \le d(x, y)$ for all x, y, with equality holding iff $x = y$.)
2. Let $Y \subset X$ (considered as a metric space with the metric d restricted to $Y \times Y$). Then $E \subset Y$ is open in Y if and only if there exists an open set V in X such that $E = V \cap Y$.
3. Let $E \subset X$. Recall that E' denotes the set of all limit points of E (cf. Examples 3 and 7, Sect. 1.1.10), and $\overline{E}$ denotes the closure of E (cf. Example 6, same section). Prove

   (a) $\overline{E} = E \cup E'$.
   (b) $x \in \overline{E}$ iff every ball $B(x, r)$ meets E.

   In particular, E is dense in X iff every ball in X meets E.
4. Let $E \subset X$. Denote by E^o the union of all the open subsets of X contained in E (the union may be empty!). Prove:

(a) E^o is the maximal open subset of X contained in E (explain in analogy with Example 6, Sect. 1.1.10). In particular, E is open iff $E = E^o$.
E^o is called the *interior* of E.
(b) $x \in E^o$ iff x is an interior point of E.
(c) How does the operation $E \to E^o$ on subsets of X relate to the union of two sets?
(d) Denote $\partial E = \overline{E} \setminus E^o$. Prove that $x \in \partial E$ iff every ball $B(x, r)$ meets *both* E and E^c. ∂E is called the *boundary* of E.

5. For $E,\ F \subset X$, define

$$d(E, F) := \inf_{x \in E,\ y \in F} d(x, y).$$

In particular, $d(x, F) := d(\{x\}, F)$ for any $x \in X$; $d(x, F)$ is called the *distance from* x *to* F, and $d(E, F)$ is called *the distance between the sets* $E,\ F$, although these are *not* "distance functions" in the sense used in the text. Prove:

(a) $x \in \overline{E}$ iff $d(x, E) = 0$.
(b) The subsets $E = \{(x, 0);\ x \geq 1\}$ and $F = \{(x,\ 1/x);\ x \geq 1\}$ of $\mathbb{R}^2$ are closed, disjoint, non-empty subsets with $d(E, F) = 0$.

6. Denote by $B(X)$ the set of all *bounded* functions $f : X \to \mathbb{R}$; the function f is said to be bounded if its range is a bounded subset of $\mathbb{R}$. Define $||f|| := \sup_{x \in X} |f(x)|$.

(a) Prove that $B(X)$ is a normed space for the pointwise operations and the above norm.
(b) Fix $p \in X$, and for each $x \in X$, define

$$f_x(v) = d(v, x) - d(v, p) \qquad (v \in X).$$

Prove that $f_x \in B(X)$ and $|f_x(v) - f_y(v)| \leq d(x, y)$ for all $x, y, v \in X$.
(c) Prove that $||f_x - f_y|| = d(x, y)$ for all $x, y \in X$. This identity means that the map $x \to f_x$ is an *isometry* of the metric space X onto the metric space $Y := \{f_x;\ x \in X\} \subset B(X)$ with the metric of $B(X)$.

7. Let $X^k := \{x = (x_1, \ldots, x_k);\ x_i \in X\}$. For any $p \in [1, \infty)$, define $d_p : X^k \times X^k \to [0, \infty)$ by

$$d_p(x, y) = \Big(\sum_{i=1}^{k} d(x_i, y_i)^p\Big)^{1/p} \qquad (x, y \in X^k),$$

where $x_i,\ y_i$ are the components of the rows $x,\ y$ (respectively).

(a) Use the Minkowski inequality in $\mathbb{R}^k$ to prove that d_p is a metric on X^k.

(b) Show that all the metrics d_p are equivalent to one another and to the metric

$$d_\infty(x, y) := \max_i d(x_i, y_i) \qquad (x, y \in X^k).$$

(c) If X is a normed space, then X^k is a normed space for the componentwise vector space operations and anyone of the equivalent norms

$$||x||_p := \Big(\sum_{i=1}^k ||x_i||^p\Big)^{1/p} \qquad (x \in X^k),$$

$1 \leq p < \infty$, and $||x||_\infty := \max_i ||x_i||$.

In particular, if X is an inner product space, then X^k is an inner product space for the inner product

$$x \cdot y := \sum_i x_i \cdot y_i \qquad (x, y \in X^k),$$

whose induced norm is precisely $|| \cdot ||_2$.

8. Prove the equivalence of the following statements about X:

   (a) X is connected.
   (b) X is not the union of two disjoint non-empty closed sets.
   (c) The only subsets of X that are both open and closed are $\emptyset$ and X.
   (d) The boundary of any *proper* subset of X is non-empty.

9. The subsets $A, B \subset X$ are *mutually separated* if

$$\overline{A} \cap B = A \cap \overline{B} = \emptyset.$$

   Prove that $E \subset X$ is connected if and only if it is *not* the union of two non-empty separated subsets.

1.2 Compact Sets

Some important properties of a real continuous function f of one real variable are valid when the domain of f is a *closed interval* $[a, b]$: for example, f is bounded, f attains its maximum and minimum values, the inverse function of f (when it exists) is continuous, f is *uniformly continuous*. One of the methods used in the proofs is the application of the Heine-Borel Theorem, stating that every cover of the closed interval $[a, b]$ by open sets has a finite sub-cover. In the generalization of these results to real functions f on $\mathbb{R}^k$ for $k > 1$, or for functions between metric spaces, we shall take this "Heine-Borel Property" of $[a, b]$ as the defining property of the domain set of f. Such sets will be called "compact sets". We shall then proceed to identify the compact sets in $\mathbb{R}^k$ as the closed bounded sets in $\mathbb{R}^k$.

Compact subsets of X may be defined and studied as we do below for a topological space X more general than a metric space. We shall however restrict our attention to the case of a metric space, which is all we need in these lectures.

Open Cover and Compactness

1.2.1 Definition Let X be a metric space, and $E \subset X$. An *open cover* of E is a family of *open* sets V_α $(\alpha \in I)$ such that

$$E \subset \bigcup_{\alpha \in I} V_\alpha. \tag{1.6}$$

A *subcover* of the above cover of E is a subfamily of the above family, whose union contains E. The subcover is said to be *finite* if it has finitely many sets in it.

The set E is said to be *compact* if *every open cover of E has a finite subcover.*

A finite subset of X is compact. Indeed, if $E = \{x_1, \ldots, x_n\}$ and $\{V_\alpha\}$ satisfies (1.6), then for each $j = 1, \ldots, n$, there exists $\alpha_j \in I$ such that $x_j \in V_{\alpha_j}$. The family $\{V_{\alpha_j};\ j = 1, \ldots, n\}$ is a finite subcover of the given cover of E. Also finite unions of compact subsets are compact.

In order to simplify statements, X (or(X, d)) denotes a metric space, and all subsets considered below are subsets of X.

Properties of Compact Sets

1.2.2 Theorem *Every infinite subset of a compact subset E has a limit point in E.*

Proof Suppose $F \subset E$ is infinite and has *no limit point in E*. Hence each $x \in E$ is not a limit point of F, and has therefore a neighborhood $B(x, r_x)$ containing no point of F distinct from x. The family $\{B(x, r_x);\ x \in E\}$ is an open cover of E (since each $x \in E$ belongs to the corresponding ball $B(x, r_x)$). The compactness of E implies the existence of a finite subcover $\{B(x_j, r_{x_j});\ j = 1, \ldots, n\}$ of E, hence of its subset F. Each $B(x_j, r_{x_j})$ contains at most one point of F, namely the centre x_j, which may or may not be in F. Therefore the union of these balls contains at most n points of F. Since F is contained in this union, F consists of at most n points, contradicting the hypothesis that F is an infinite set. □

The converse of Theorem 1.2.2 is true, but will not be proved here, since it will not be needed in these lectures. Observe that compactness of subsets of a metric space is thereby characterized by the property in Theorem 1.2.2.

1.2.3 Theorem *Closed subsets of compact sets are compact.*

Proof Let F be a closed subset of the compact subset E. Suppose $\{V_\alpha;\ \alpha \in I\}$ is an open cover of F. Since F^c is open, $\{V_\alpha\} \cup \{F^c\}$ is an open cover of E. The compactness of E implies the existence of indices $\alpha_1, \ldots, \alpha_n$ such that

$$E \subset \bigcup_{j=1}^{n} V_{\alpha_j} \cup F^c.$$

Since $F \subset E$, the union above contains F. The points of F are necessarily contained in the first union. □

1.2.4 Theorem *Let E be a compact subset of the metric space (X, d). Then E is closed.*

Proof Let $x \in E^c$. For each $y \in E$, the points x, y are distinct, and therefore $d(x, y) \neq 0$. Let $r_y := (1/2)d(x, y)$. The family $\{B(y, r_y);\ y \in E\}$ is an open cover of E, since $y \in B(y, r_y)$ for each $y \in E$, and balls are open. By compactness of E, there exist $y_1, \ldots, y_n \in E$ such that

$$E \subset \bigcup_{j=1}^{n} B(y_j, r_{y_j}). \tag{1.7}$$

Set

$$r = \min_{1 \le j \le n} r_{y_j} \ (>0).$$

Suppose $B(x, r)$ contains a point $z \in E$. By (1.7), there exists j $(1 \le j \le n)$ such that $z \in B(y_j, r_{y_j})$. Then

$$d(x, y_j) \le d(x, z) + d(z, y_j) < r + r_{y_j} \le 2r_{y_j} = d(x, y_j),$$

contradiction! It follows that $B(x, r) \subset E^c$, that is, x is an interior point of E^c. This proves that E^c is open. □

1.2.5 Corollary *The intersection of a compact set and a closed set is compact.*

Proof Let E be compact and F be closed. Since E is closed (by Theorem 1.2.4), the intersection $E \cap F$ is a *closed* subset of the compact set E, and is therefore compact by Theorem 1.2.3. □

The subset E of the metric space X is said to be *bounded* if it is contained in some ball $B(p, R)$ in X.

1.2.6 Theorem *Let E be a compact subset of the metric space (X, d). Then E is bounded.*

Proof The family $\{B(x, 1);\ x \in E\}$ is an open cover of E. Let then $\{B(x_j; 1);\ j = 1, \ldots, n\}$ be a finite subcover. Fix $p \in X$. For each $x \in E$, there exists $j \in \{1, \ldots, n\}$ such that $x \in B(x_j, 1)$. Then

$$d(x, p) \le d(x, x_j) + d(x_j, p) < 1 + \max_{1 \le j \le n} d(x_j, p) := R,$$

which proves that $E \subset B(p, R)$. □

Thus *compact sets are closed and bounded*. The converse is false in general metric spaces, as we shall see in the following example.

Example 1. The closed ball $\overline{B}(0, 1)$ in the normed space l_p (cf. Section 1.1.8, Example 3) is a closed bounded set. Consider the infinite subset

$$F := \{e^j;\ j = 1, 2, \ldots\}.$$

Since $||e^j||_p = 1$, we have indeed $F \subset \overline{B}(0, 1)$. We claim that F has no limit point. To see this, note that if $j, l \in \mathbb{N}$ are distinct, then $y = e^j - e^l$ is the infinite row with components $y_j = 1$, $y_l = -1$, and $y_i = 0$ for all indices i that are neither j nor l. Therefore

$$d(e^j, e^l) = ||e^j - e^l||_p = [1^p + |-1|^p]^{1/p} = 2^{1/p}.$$

For any $x \in l_p$, if the ball $B(x, 1/2)$ in l_p contains two distinct points of F, say e^j and e^l, then by the triangle inequality,

$$2^{1/p} = d(e^j, e^l) \le d(e^j, x) + d(x, e^l) < \frac{1}{2} + \frac{1}{2} = 1,$$

which is absurd. Hence $B(x, 1/2)$ contains at most one point of F. Therefore x is not a limit point of F, because every neighborhood of a limit point of F contains infinitely many points of F (cf. Section 1.1.10). Since x was an arbitrary point of l_p, we have shown that F has no limit point. Thus $\overline{B}(0, 1)$ contains the infinite subset F which has no limit point, and therefore the (closed bounded) set $\overline{B}(0, 1)$ is not compact, by Theorem 1.2.3.

Example 2. We may discuss in a similar way the closed ball $\overline{B}(0, 1)$ in $C(I)$ with the norm $||\cdot|| = ||\cdot||_2$ (cf. Section 1.1.8, Example 4). Take $I = [0, 2\pi]$ for convenience, and consider the set $F = \{e_n;\ n \in \mathbb{N}\}$, where $e_n \in C(I)$ is defined by

$$e_n(t) = \frac{1}{\sqrt{\pi}} \cos nt \qquad (t \in I).$$

We have

$$||e_n||^2 = e_n \cdot e_n = \frac{1}{\pi} \int_I \cos^2(nt)\, dt = \frac{1}{2\pi} \int_I (1 + \cos 2nt)\, dt$$

$$= 1 + \frac{1}{4n\pi} \sin 2nt \Big|_0^{2\pi} = 1,$$

and for $n \neq m$ in $\mathbb{N}$,

$$e_n \cdot e_m = \frac{1}{\pi}\int_I \cos nt\, \cos mt\, dt = \frac{1}{2\pi}\Big(\int_I \cos(n+m)t\, dt + \int_I \cos(n-m)t\, dt\Big)$$

$$= \frac{1}{2\pi}\Big(\frac{\sin(n+m)t}{n+m} + \frac{\sin(n-m)t}{n-m}\Big)\Big|_0^{2\pi} = 0.$$

Thus

$$e_n \cdot e_m = \delta_{nm} \qquad (n, m \in \mathbb{N}), \tag{1.8}$$

where δ_{nm} is Kronecker's delta.

For $n \neq m$, we have

$$||e_n - e_m||^2 = (e_n - e_m)\cdot(e_n - e_m) = ||e_n||^2 - 2\, e_n \cdot e_m + ||e_m||^2 = 2. \tag{1.9}$$

In particular, the points e_n of F are distinct (for distinct indices), and belong to $\overline{B}(0, 1)$, so that F is an infinite subset of $\overline{B}(0, 1)$. Starting with (1.9), the argument in Example 1 may be used to show that F has no limit point. Therefore the (closed bounded) set $\overline{B}(0, 1)$ in $C(I)$ is not compact, by Theorem 1.2.2.

Example 3. The argument in Example 2 is based on the existence of an infinite set $F = \{e_n;\ n \in \mathbb{N}\}$ in an inner product space X, such that (1.6) is satisfied. Such a set is called an *orthonormal sequence* in X. For such a sequence, $||e_n|| = (e_n \cdot e_n)^{1/2} = 1$, and the identity (1.9) is valid. In particular, $e_n \neq e_m$ for $n \neq m$, and therefore F is an infinite subset of the closed ball $\overline{B}(0, 1)$ of X. The argument in Example 1 shows that F has no limit point, and therefore the (closed bounded) set $\overline{B}(0, 1)$ is not compact.

We shall prove however that closed bounded sets are compact *in the normed space* $\mathbb{R}^k$, endowed with anyone of the (equivalent) norms $||\cdot||_p$ ($p \in [1, \infty]$). Actually, the compactness of the closed unit ball $\overline{B}(0, 1)$ in a normed space X *characterizes* its finite dimensionality. This fact however is beyond the scope of these lectures (cf. N. Dunford and S.T. Schwartz, Theorem IV.3.5).

Compactness of Closed Bounded Sets in $\mathbb{R}^k$

We start with some preliminaries.

A *cell* I in $\mathbb{R}^k$ is a Cartesian product of k closed real intervals

$$I := \prod_{i=1}^{k}[a_i, b_i] := \{x \in \mathbb{R}^k;\ a_i \le x_i \le b_i,\ i = 1, \ldots, k\}.$$

The interval $[a_i, b_i]$ is called the i-th projection of I. We recall *Cantor's lemma* about sequences of closed intervals: *Let* $\{I_n; n = 1, 2, 3, \ldots\}$ *be closed intervals in* $\mathbb{R}$ *such that*

$$I_1 \supset I_2 \supset I_3 \ldots.$$

Then $\bigcap_n I_n \neq \emptyset$.

Cantor's lemma generalizes easily to the case of cells in $\mathbb{R}^k$:

1.2.7 Lemma (Cantor's lemma for cells). *Let* I_j, $j = 1, 2, \ldots$ *be cells in* $\mathbb{R}^k$ *such that* $I_{j+1} \subset I_j$ *for all* j. *Then the intersection of all the cells* I_j *in non-empty.*

Proof For each fixed $i \in \{1, \ldots, k\}$, the i-th projections of I_j $(j = 1, 2, \ldots)$ are nested closed intervals as in the hypothesis of Cantor's lemma on the real line. By the latter, there exists a point x_i common to all these projections. The point $x := (x_1, \ldots, x_k)$ is then a common point of all the cells I_j, $j = 1, 2, \ldots$. □

1.2.8 Theorem *Let* I *be a cell in* $\mathbb{R}^k$. *Then* I *is compact.*

Proof Suppose I is not compact. There exists therefore an open cover $\{V_\alpha\}$ of I which has no finite subcover. Subdivide I into 2^k subcells by writing each projection $[a_i, b_i]$ of I as the union of the two intervals $[a_i, c_i]$ and $[c_i, b_i]$ with $c_i = (1/2)(a_i + b_i)$, $i = 1, \ldots, k$. At least one of the subcells, say I_1, cannot be covered by finitely many V_α-s, since otherwise, I itself would be covered by finitely many V_α-s. Repeating the process, we get a subcell I_2 of I_1 that cannot be covered by finitely many V_α-s. Inductively, we then obtain a sequence of subcells I_j of I such that

(a) $I_{j+1} \subset I_j$, $j = 1, 2, \ldots$;
(b) I_j cannot be covered by finitely many V_α-s, for each j; and
(c) the *diameter* of I_j is equal to $\delta/(2^j)$, where δ denotes the diameter of I, that is, the supremum of the distances between points of I.

To verify (c), observe that I_{j+1} is the Cartesian product of intervals whose length is half the length of the intervals in the corresponding product I_j. Consequently, the diameter of I_{j+1} is equal to half the diameter of I_j.

By Cantor's lemma (1.2.7), there exists a point x common to all the cells I_j. Since $x \in I$ and $\{V_\alpha\}$ is a cover of I, there exists an index α_1 such that $x \in V_{\alpha_1}$. Since V_{α_1} is open. there exists a ball $B(x, r) \subset V_{\alpha_1}$. Fix j such that $\delta/(2^j) < r$ (i.e., $j > \log_2 \delta - \log_2 r$). For all $y \in I_j$, since $x \in I_j$, $d(y, x)$ is not greater than the diameter of I_j, which is $\delta/(2^j) < r$, i.e., $y \in B(x, r) \subset V_{\alpha_1}$. This shows that $I_j \subset V_{\alpha_1}$, that is, finitely many (one!) V_α-s do cover I_j, contradicting Property (b). □

As immediate corollaries, we obtain the following important properties of $\mathbb{R}^k$ (1.2.9 and 1.2.10).

1.2.9 The Heine-Borel Theorem *Closed bounded sets in* $\mathbb{R}^k$ *are compact.*

Proof Let $E \subset \mathbb{R}^k$ be closed and bounded. Since E is bounded, it is contained in a ball $B(a, r)$ $(a \in \mathbb{R}^k$, $r > 0)$. Since $||x||_\infty \le ||x||_p$ for any given $p \in [1, \infty]$, if $y \in B(a, r)$, then

$$|y_i - a_i| \le ||y - a||_\infty \le ||y - a||_p = d_p(y, a) < r$$

for all $i = 1, \ldots, k$, that is,

$$y \in I := \prod_{i=1}^{k} [a_i - r, a_i + r].$$

Hence $B(a,r) \subset I$. Consequently, E is a *closed* subset of the *compact* set I (cf. Theorem 1.2.8), and is therefore compact by Theorem 1.2.3. □

1.2.10 The Bolzano-Weierstrass Theorem *Every infinite bounded set in* $\mathbb{R}^k$ *has a limit point.*

Proof Let E be an infinite bounded set in $\mathbb{R}^k$. Since E is bounded, it is contained in a cell I (cf. proof of 1.2.9). Thus E is an infinite subset of the compact set I (cf. Theorem 1.2.8), and has therefore a limit point in I, by Theorem 1.2.2. □

1.2.11 Exercises (In the following exercises, X is a metric space under the metric d.)

1. Suppose X is compact.
   (a) Let $F_i \subset X$ ($i \in I$, an arbitrary index set) be closed sets such that
   $$\bigcap_{j=1}^{n} F_{i_j} \neq \emptyset \tag{1.10}$$
   for any $i_j \in I$ and $n \in \mathbb{N}$. Prove that
   $$\bigcap_{i \in I} F_i \neq \emptyset. \tag{1.11}$$
   (b) In case $I = \mathbb{N}$, if F_i are non-empty closed subsets of X such that $F_{i+1} \subset F_i$ for all i, then (1.11) holds.
2. Prove that $E \subset X$ is compact if and only if every cover of E by open *balls* has a finite subcover.
3. A subset E of X is *totally bounded* if for every $r > 0$, there exist $x_1, \ldots, x_n \in E$ such that
   $$E \subset \bigcup_{i=1}^{n} B(x_i, r). \tag{1.12}$$
   Prove:
   (a) If E is compact, then it is totally bounded.
   (b) If $\overline{E}$ is compact, then E is totally bounded.
   (c) If $E \subset \mathbb{R}^k$ is totally bounded, then $\overline{E}$ is compact.

4. The metric space X is *separable* if it has a dense countable subset E (cf. Exercise 4, Sect. 1.1.12). Prove that every compact metric space is separable.
5. Prove that every bounded set in $\mathbb{R}^k$ has a compact closure.
6. The metric space X is said to be *locally compact* if every point $x \in X$ has a *compact neighborhood*, that is, a compact set K which contains an open neighborhood U of x. Let $Y \subset X$, considered as a metric space with the metric of X restricted to $Y \times Y$. Prove that if X is locally compact and Y is either open or closed in X, then Y is locally compact (Cf. Exercise 2(b) in Sect. 1.1.12.)
7. Let X be a compact metric space, and let $\{V_i;\ i \in I\}$ be an open cover of X (I is an arbitrary index set). Prove that there exists $r > 0$ such that, for every $x \in X$, $B(x, r)$ is contained in *some* V_i. A number r with this property is called a *Lebesgue number* for the given cover.

1.3 Sequences

Basics

Let (X, d) be a metric space. A sequence in X is a function $x(\cdot)$ from the set $\mathbb{N} := \{1, 2, 3, \ldots\}$ of natural numbers to X. It is customary to write x_n instead of $x(n)$, for $n \in \mathbb{N}$, and to identify the sequence with its ordered range, denoted by $\{x_n\}$. Note that if $X = \mathbb{R}^k$, we use superscripts rather than subscripts, and denote the sequence by $\{x^n\}$, because we agreed to use subscripts to denote components of vectors in $\mathbb{R}^k$. Accordingly, x_i^n is the i-th component of x^n.

We shall define convergence of a sequence by using the distance function $d(\cdot, \cdot)$ in the same way as it is done in the special case $X = \mathbb{R}$ with the metric $d(x, y) = |x - y|$: the sequence $\{x_n\} \subset \mathbb{R}$ converges if there exists $x \in \mathbb{R}$ with the following property: (P) For every $\epsilon > 0$, there exists an index $n(\epsilon) \in \mathbb{N}$ such that $|x_n - x| < \epsilon$ for all $n > n(\epsilon)$.

The general case in formulated below.

1.3.1 Definition The sequence $\{x_n\}$ in the metric space X *converges* if there exists $x \in X$ such that for every $\epsilon > 0$, there exists $n(\epsilon) \in \mathbb{N}$ such that $d(x_n, x) < \epsilon$ for all integers $n > n(\epsilon)$. A non-convergent sequence is said to be *divergent*.

If the sequence $\{x_n\}$ converges, the point x in the above definition is *uniquely determined*. Indeed, if x' has the same property as x (with $n'(\epsilon)$ replacing $n(\epsilon)$), then for any n greater than both $n(\epsilon)$ and $n'(\epsilon)$,

$$d(x, x') \le d(x, x_n) + d(x_n, x') < 2\epsilon.$$

Since $d(x, x')$ does not depend on the arbitrary number $\epsilon > 0$, it follows that $d(x, x') = 0$, and therefore $x = x'$.

This uniquely determined point x is called the *limit* of the (convergent) sequence $\{x_n\}$, and is denoted $\lim x_n$. We use also the notation $x_n \to x$ (read x_n *converges to* x).

A convergent sequence in X is necessarily *bounded*. Indeed, if $x_n \to x$, then $d(x_n, x) < 1$ for all $n > n(1)$, and therefore $x_n \in B(x, R)$ for all $n \in \mathbb{N}$, if $R := 1 + \max_{n \le n(1)} d(x_n, x)$.

A sequence which is eventually constant (that is, $x_n = x$ for all $n > n_0$) trivially converges to that constant x.

Note that if $x_n \to x$ and $y_n \to y$, then $d(x_n, y_n) \to d(x, y)$, because by the triangle inequality

$$|d(x_n, y_n) - d(x, y)| \le d(x_n, x) + d(y_n, y) \to 0.$$

In particular, if $x_n \to x$, then $d(x_n, a) \to d(x, a)$ for any $a \in X$ (take $y_n = a$, a constant sequence).

Exercise. The concepts of convergence and limit of a sequence are unchanged when we replace the metric on X by an equivalent metric.

1.3.2 Proposition *Let E be a subset of a metric space X. Then*

(a) If x is a limit point of E, then there exists a sequence $\{x_n\} \subset E$ such that $x_n \to x$.
(b) Conversely, if $\{x_n\} \subset E$ converges to x, then either $x \in E$ or x is a limit point of E. In particular, if E is closed, then $x \in E$.

Proof (a) For each $n = 1, 2, \ldots$, the ball $B(x, 1/n)$ contains a point $x_n \in E$. We have $d(x_n, x) < 1/n$, hence $d(x_n, x) \to 0$, i.e., $x_n \to x$.
(b) Let $\{x_n\} \subset E$ converge to x, and suppose x is not in E and is not a limit point of E. Then there exists a ball $B(x, \epsilon)$ containing no point of E. Let n_0 be an integer such that $d(x_{n_0}, x) < \epsilon$ (such an index n_0 exists, since $x_n \to x$). Then $x_{n_0} \in B(x, \epsilon) \cap E$, contradiction. □

If X is a *normed space*, the limit concept is consistent with the normed vector space structure in the usual way:

1.3.3 Proposition *Let X be a normed space, and let $\{x_n\}$, $\{y_n\}$ be sequences in X. Then*

(i) if $x_n \to x$ and $y_n \to y$, it follows that $x_n + y_n \to x + y$.
(ii) if $x_n \to x$ and $\lambda_n \in \mathbb{R} \to \lambda$, then $\lambda_n x_n \to \lambda x$.
(iii) if $x_n \to x$, then $||x_n|| \to ||x||$.
(iv) In case $X = \mathbb{R}^k$, if $x^n \to x$ and $y^n \to y$, then $x^n \cdot y^n \to x \cdot y$
(v) In case $X = \mathbb{R}^k$, $x^n \to x$ iff $x_i^n \to x_i$ for all $i = 1, \ldots, k$.

Proof The elementary proofs for real sequences apply to (i)–(ii), with absolute values replaced by norms; (iii) is a special case of a previous observation, since $||x_n|| = d(x_n, 0)$; (iv) can be proved by using Schwarz' inequality, since we may assume in (iv)–(v), without loss of generality, that the norm is the Euclidean norm, because the concepts involved are unchanged when the norm is replaced by an equivalent norm:

$$|x^n \cdot y^n - x \cdot y| \le |(x^n - x) \cdot y^n| + |x \cdot (y^n - y)|$$

$$\leq ||x^n - x|| \, ||y^n|| + ||x|| \, ||y^n - y|| \to 0,$$

because $||x^n - x|| \to 0$, $||y^n|| \to ||y||$, and $||y^n - y|| \to 0$.

In particular, with $y^n = y$ (a constant sequence), if $x^n \to x$, then $x^n \cdot y \to x \cdot y$. Taking $y = e^i$ for any $i = 1, \ldots, k$, we get

$$x_i^n = x^n \cdot e^i \to x \cdot e^i = x_i$$

for all $i = 1, \ldots, k$. Conversely, if $x_i^n \to x_i$ for all $i = 1, \ldots, k$, then

$$||x^n - x|| = \Big(\sum_{i=1}^{k} (x_i^n - x_i)^2\Big)^{1/2} \to 0,$$

by the elementary properties of limits of numerical sequences and the continuity of the square root function. Thus (v) is verified. □

Another way to see the necessity of the condition in (v) is to observe that for all $i = 1, \ldots, k$,

$$|x_i^n - x_i| \leq ||x^n - x||_\infty \leq ||x^n - x|| \to 0,$$

for any p-norm on $\mathbb{R}^k$.

Subsequences

1.3.4 Definition A *subsequence* of the sequence $\{x_n\}$ in the metric space X is a sequence $j \to x_{n_j}$, where

$$n_1 < n_2 < \cdots .$$

Note that we have necessarily $n_j \geq j$.

If $x_n \to x$, then $x_{n_j} \to x$ for any subsequence. Indeed, given $\epsilon > 0$, let $n(\epsilon)$ be as in the definition of convergence; then if $j > n(\epsilon)$, we have $n_j \geq j > n(\epsilon)$, hence $d(x_{n_j}, x) < \epsilon$. Consequently, a sequence possessing a divergent subsequence or two subsequences converging to distinct limits, is divergent.

We recall that every bounded sequence in $\mathbb{R}$ has a convergent subsequence. In any metric space X, we say that $E \subset X$ has the Bolzano-Weierstrass Property (BWP) if every sequence in E has a convergent subsequence. Such a set E is also called a *sequentially compact set* (in X). We generalize the above property of $\mathbb{R}$ to $\mathbb{R}^k$ for all $k \geq 1$.

1.3.5 Theorem *Every bounded sequence in* $\mathbb{R}^k$ *has a convergent subsequence.*

Proof Let $\{x^n\}$ be a bounded sequence in $\mathbb{R}^k$. If its range is a *finite set*, at least one element in the range, say x^{n_1}, is repeated for infinitely many indices $n_1 < n_2 < \cdots$. Then the subsequence $\{x^{n_j}\}$ is a *constant sequence*, hence converges trivially.

We may then assume that the range E of the sequence is an *infinite set*. As an infinite bounded subset of $\mathbb{R}^k$, E has a limit point x in $\mathbb{R}^k$ (by the Bolzano-Weierstrass theorem, cf. 1.2.10). Each ball centered at x contains infinitely many points of E (cf. 1.1.7). Thus $B(x, 1)$ contains a point of E, say x^{n_1}. Having chosen j indices $n_1 < n_2 < \cdots < n_j \in \mathbb{N}$ such that the corresponding points of the sequence satisfy $x^{n_r} \in B(x, 1/r)$ for $r = 1, \ldots, j$, we then consider the ball $B(x, 1/(j+1))$; since it contains *infinitely* many points of $E := \{x^n\}$, there exists an index $n_{j+1} > n_j$, such that $x^{n_{j+1}} \in B(x, 1/(j+1))$. By induction, we then obtain a subsequence $\{x^{n_j}\}$ such that $x^{n_j} \in B(x, 1/j)$ for all j, that is, $d(x^{n_j}, x) < 1/j \to 0$. Hence x^{n_j} converges to x. □

An alternative proof would use the case $k = 1$ on each real sequence $\{x_i^n\}$, $i = 1, \ldots, k$ (exercise).

The well-known Cauchy Criterion for the convergence of sequences in $\mathbb{R}$ motivates the following concept of "Cauchy sequences" in a general metric space.

Cauchy Sequences

1.3.6 Definition A sequence $\{x_n\}$ in the metric space X is called a *Cauchy sequence* if for every $\epsilon > 0$, there exists $n(\epsilon) \in \mathbb{N}$ such that $d(x_n, x_m) < \epsilon$ for all indices $m, n > n(\epsilon)$.
By symmetry of the metric, it suffices to consider indices $m > n$.

The condition in the above definition is called Cauchy's condition. It follows from the triangle inequality that every convergent sequence is Cauchy: if $x_n \to x$ and $n(\epsilon)$ is as in the definition of convergence, then for all $m > n > n(\epsilon/2)$,

$$d(x_n, x_m) \le d(x_n, x) + d(x, x_m) < \frac{\epsilon}{2} + \frac{\epsilon}{2} = \epsilon.$$

The converse is false in general : in the metric space $X = (0, 1]$ with the metric $d(x, y) = |x - y|$, the sequence $\{1/n\}$ is Cauchy, since it converges in $\mathbb{R}$, but it does *not* converge in X, because its uniquely determined limit in $\mathbb{R}$, namely 0, does not belong to the space X.

1.3.7 Definition 1. The metric space X is said to be *complete* if every Cauchy sequence in X converges in X.

2. A *complete normed space* is called a *Banach space*.

3. A complete inner product space is called a *Hilbert space*.

The completeness of a metric (or normed) space is unchanged when the metric (or the norm) is replaced by an equivalent metric (or norm, respectively).

Proposition *Let F be a subset of the complete metric space X. Then F (with the metric of X restricted to $F \times F$) is a complete metric space if and only if F is closed in X.*

Proof If F is closed and $\{x_n\} \subset F$ is Cauchy, it converges in X, and its limit belongs to F, because F is closed (cf. 1.3.2). Conversely, if F is complete and x is a limit point of F, there exists a sequence $\{x_n\} \subset F$ converging to x. The sequence is necessarily Cauchy, hence converges in F by completeness of F. The uniqueness of the limit of a convergent sequence implies that $x \in F$. This shows that F is closed (cf. Theorem 1.1.11). □

In particular, for any Banach space Y and $b > 0$, the closed ball

$$Y_b := \overline{B}_Y(0, b) := \{y \in Y;\ ||y|| \le b\}$$

is a complete metric space (cf. 1.1.7).

Example 1. We show that l_p ($1 \le p < \infty$) is a Banach space, and in particular, l_2 is a Hilbert space (cf. Section 1.1.8, Example 3). We need only to verify the completeness of l_p. Let $\{x^n\}$ be a Cauchy sequence in l_p. Given $\epsilon > 0$, let $n(\epsilon)$ be as in Definition 1.3.6. For each $i \in \mathbb{N}$, we have

$$|x_i^n - x_i^m| \le ||x^n - x^m||_p < \epsilon$$

for all $n, m > n(\epsilon)$. This means that $\{x_i^n;\ n \in \mathbb{N}\}$ is a Cauchy sequence in $\mathbb{R}$. By the well known completeness of $\mathbb{R}$, the sequence converges; define $x_i = \lim_n x_i^n$, and let $x = \{x_i\}$.

Fix $N \in \mathbb{N}$. For all integers $m > n > n(\epsilon)$, we have

$$\sum_{i=1}^{N} |x_i^n - x_i^m|^p \le \sum_{i=1}^{\infty} |x_i^n - x_i^m|^p < \epsilon^p.$$

Letting $m \to \infty$, we get

$$\sum_{i=1}^{N} |x_i^n - x_i|^p \le \epsilon^p$$

for all $n > n(\epsilon)$. Since this is true for all N, we get

$$\sum_{i=1}^{\infty} |x_i^n - x_i|^p \le \epsilon^p$$

for all $n > n(\epsilon)$. The convergence of the series means that $x^n - x \in l_p$, and therefore $x = x^n - (x^n - x) \in l_p$. Moreover, $||x^n - x||_p \le \epsilon (< 2\epsilon)$ for all $n > n(\epsilon)$, that is $x^n \to x$ in the l_p metric. In conclusion, the Cauchy sequence $\{x^n\}$ converges in l_p, and the completeness of l_p has been verified.

We prove below that $\mathbb{R}^k$ is a Banach space. We start with two simple lemmas on Cauchy sequences in general metric spaces.

1.3.8 Lemma *Cauchy sequences are bounded.*

Proof Let $\{x_n\}$ be a Cauchy sequence. By definition, $d(x_n, x_m) < 1$ for all $m, n > n(1)$. *Fix* $m > n(1)$, and let

$$R := 1 + \max_{1 \le n \le n(1)} d(x_n, x_m).$$

Then $d(x_n, x_m) < R$ *for all* n, that is $\{x_n\} \subset B(x_m, R)$. □

1.3.9 Lemma *If a Cauchy sequence has a convergent subsequence, then the sequence converges (necessarily to the subsequence limit).*

Proof Let $\epsilon > 0$ be given, and let $n(\epsilon)$ be as in Definition 1.3.6 for the given Cauchy sequence $\{x_n\}$ in the metric space (X, d). Let $\{x_{n_j}\}$ be a subsequence of $\{x_n\}$ which converges to the point $x \in X$. Let then $j(\epsilon) \in \mathbb{N}$ be such that $d(x_{n_j}, x) < \epsilon$ whenever $j > j(\epsilon)$. Define

$$n^*(\epsilon) := \max[n(\frac{\epsilon}{2}),\ j(\frac{\epsilon}{2})].$$

Fix $j > n^*(\epsilon)$. Then $j > j(\epsilon/2)$, and therefore $d(x_{n_j}, x) < \epsilon/2$. Also $n_j \geq j > n^*(\epsilon) \geq n(\epsilon/2)$, and therefore $d(x_n, x_{n_j}) < \epsilon/2$ for all $n > n^*(\epsilon)(\geq n(\epsilon/2))$, by the Cauchy condition on the sequence. Hence, for all $n > n^*(\epsilon)$,

$$d(x_n, x) \leq d(x_n, x_{n_j}) + d(x_{n_j}, x) < \frac{\epsilon}{2} + \frac{\epsilon}{2} = \epsilon.$$

This proves that the sequence $\{x_n\}$ converges to x. □

We are now ready to prove the completeness of $\mathbb{R}^k$, generalizing thereby the Cauchy Criterion to any dimension $k \in \mathbb{N}$.

1.3.10 Theorem $\mathbb{R}^k$ *is a Banach space.*

Proof We need to prove the completeness of the normed space $\mathbb{R}^k$. Any Cauchy sequence in $\mathbb{R}^k$ is bounded (by Lemma 1.3.8), hence it has a convergent subsequence, by Theorem 1.3.5. Therefore the sequence converges, by Lemma 1.3.9. □

Note that the special structure of $\mathbb{R}^k$ is only used in the application of Theorem 1.3.5. Therefore we have the following

1.3.11 Theorem *Let X be a normed space in which every bounded sequence has a convergent subsequence. Then X is a Banach space.*

An alternative way to prove the completeness of $\mathbb{R}^k$ is to rely on the well-known completeness of $\mathbb{R}$. Indeed, let $\{x^n\}$ be a Cauchy sequence in $\mathbb{R}^k$ normed by anyone of the p-norms. Given $\epsilon > 0$, let $n(\epsilon)$ be as in Definition 1.3.6. For each $i = 1, \ldots, k$, we have

$$|x_i^n - x_i^m| \leq ||x^n - x^m|| < \epsilon$$

whenever $n, m > n(\epsilon)$, that is, $\{x_i^n;\ n \in \mathbb{N}\}$ is a Cauchy sequence in $\mathbb{R}$, and therefore $x_i := \lim_n x_i^n$ exists, by the completeness of $\mathbb{R}$. Let $x := (x_1, \ldots, x_k)$. Then $x^n \to x$, by Proposition 1.3.3 (v), and the completeness of $\mathbb{R}^k$ is proved.

Example 2. As in Example 3, Sect. 1.1.8, consider the infinite rows e^j with components $e_i^j = \delta_{ij}$, and their linear span X. Fix $p \in [0, \infty)$. Obviously X is a proper subspace of l_p: a trivial element of $l_p \setminus X$ is $(2^{-i}; i \in \mathbb{N})$. For any $x \in l_p$, define

$$x^n = (x_1, \ldots, x_n, 0, 0, \ldots) \qquad (n \in \mathbb{N}).$$

Then $x^n \in X$, and

$$||x - x^n||_p^p = \sum_{i=n+1}^{\infty} |x_i|^p \to 0$$

as $n \to \infty$, by the convergence of the series $\sum_i |x_i|^p$. This shows that x is a limit point of X. If we choose $x \notin X$, we conclude that X is not closed in l_p, hence is not complete as a normed space with the p-norm. Another conclusion is that every $x \in l_p$ is either in X or in X', hence in the closure $\overline{X}$ of X (cf. Exercise 3(a), Sect. 1.1.15, to the effect that $\overline{X} = X \cup X'$). Thus X is dense in the complete normed space l_p. Consequently l_p is the completion of X with respect to the p-norm (cf. concluding comments of Exercise 8, Sect. 1.3.12).

1.3.12 Exercises 1. Let (X, d) be a metric space.

(a) Let $E \subset X$ be compact and $F \subset X$ be closed, both non-empty. Prove that $d(E, F) = 0$ if and only if $E \cap F \neq \emptyset$.
(b) Part (a) is false if we replace the assumption that E is compact by the assumption that E is (merely) closed (cf. Section 1.1.15, Exercise 5(b)).

2. Let (X, d) be a metric space, and denote by $\mathcal{K}$ the family of all compact subsets of X. Prove

(a) $\mathcal{K}$ is closed under finite unions and intersections.
(b) $\mathcal{K}$ is not closed in general under countable unions.
(c) If $E, F \in \mathcal{K}$, does it follow that $E \setminus F \in \mathcal{K}$?

3. Let X be a normed space. If $E,\ F \subset X$, define

$$E + F := \{x + y;\ x \in E,\ y \in F\}.$$

(a) Prove that if E is compact and F is closed, then $E + F$ is closed. Is it true that $E + F$ is compact?
(b) Prove or disprove that if $E,\ F$ are both compact, then $E + F$ is compact.

4. Let X be a normed space, and let $\{x_i\}$ be a sequence in X. Define $s_n = \sum_{i=1}^n x_i$, $n \in \mathbb{N}$. If $s_n \to s$ in X, we say that the series $\sum_{i=1}^\infty x_i$ converges in X, and s is its sum. If the numerical series $\sum ||x_i||$ converges, we say that the series $\sum x_i$ converges absolutely. Prove:

(a) If X is complete, then absolute convergence of a series in X implies its convergence in X.
(b) If $\{x_n\}$ is a Cauchy sequence in X, then it has a subsequence $\{x_{n_i}\}$ such that

$$||x_{n_{i+1}} - x_{n_i}|| < \frac{1}{2^i} \qquad (i \in \mathbb{N}).$$

(c) If absolute convergence of any series in X implies its convergence in X, then X is complete (cf. Lemma 1.3.9).

5. Let X, Y be vector spaces (over the real field). Recall that a map $T : X \to Y$ is said to be *linear* if

$$T(\alpha x + \beta y) = \alpha Tx + \beta Ty$$

for all $x, y \in X$ and $\alpha, \beta \in \mathbb{R}$.

In the special case $Y = \mathbb{R}$, the linear map T is called a *linear functional* on X.

Suppose X, Y are normed spaces; norms on both spaces are denoted by $||\cdot||$, except when $Y = \mathbb{R}$, in which case the Y-norm is denoted by $|\cdot|$ as usual. To avoid trivial situations, we assume that $X \neq \{0\}$.

For $T : X \to Y$ linear, denote

$$||T|| := \sup\{||Tx||;\ x \in X,\ ||x|| = 1\}. \tag{1.13}$$

The supremum can be infinite; T is said to be *bounded* if $||T||$ is *finite* ($||T|| < \infty$). Let $B(X, Y)$ denote the set of all bounded linear maps $T : X \to Y$.

(a) Prove

$$\begin{aligned} ||T|| &= \sup\{||Tx||;\ x \in X,\ ||x|| \le 1\} \\ &= \sup\{||Tx||;\ x \in X,\ ||x|| < 1\} \\ &= \sup\{\frac{||Tx||}{||x||};\ x \in X \setminus \{0\}\}. \end{aligned} \tag{1.14}$$

(b) $||Tx|| \le ||T||\,||x|| \quad (x \in X)$.

(c) $B(X, Y)$ with pointwise vector space operations and the norm (1.13) is a normed space.

(d) If Y is complete, so is $B(X, Y)$. In particular, $X^* := B(X, \mathbb{R})$ is complete for any normed space X.

(e) Take for granted that $X^* \neq \{0\}$. Therefore there is $h \in X^*$ such that $||h|| = 1$. Given $y \in Y$, define $T_y : X \to Y$

$$T_y x := h(x)y \qquad (y \in Y). \tag{1.15}$$

Prove that $||T_y|| \le ||y||$. Conclude from this that if $B(X, Y)$ is complete, so is Y.

6. Let (X, d) be a metric space. The *diameter* $\delta(E)$ of a subset E of X is defined by

$$\delta(E) := \sup_{x,y \in E} d(x, y). \tag{1.16}$$

Prove that X is complete if and only if it has the following property (P)

(P) if $\{F_i\}$ is a sequence of non-empty closed subsets of X such that $F_{i+1} \subset F_i$ and $\delta(F_i) \to 0$, then $\bigcap_i F_i \neq \emptyset$.

7. Let X be a metric space, and let $E \subset X$ be dense in X (i.e., $\overline{E} = X$). Suppose that every Cauchy sequence in E converges in X. Prove that X is complete.
8. Let (X, d) be a metric space, and fix $x_0 \in X$. A function $f : X \to \mathbb{R}$ is Lipschitz if there exists a constant $q > 0$ such that

$$|f(x) - f(y)| \leq q\, d(x, y) \qquad (x, y \in X).$$

Define Y as the set of all Lipschitz functions f on X such that $f(x_0) = 0$. For $f \in Y$, define

$$||f|| = \sup_{x,y \in X;\, x \neq y} \frac{|f(x) - f(y)|}{d(x, y)}. \tag{1.17}$$

(a) Prove that Y is a normed space for the pointwise vector space operations and the norm above. Given $x \in X$, show that the function

$$f_x(y) := d(x, y) - d(x, x_0) \tag{1.18}$$

belongs to Y and $||f_x|| = 1$.

(b) Given $x \in X$, define $h_x : Y \to \mathbb{R}$ by

$$h_x(f) := f(x) \qquad (f \in Y). \tag{1.19}$$

Prove that $h_x \in Y^*$, with $||h_x|| \leq d(x, x_0)$, where $||h_x||$ is the Y^*-norm of h_x (cf. Exercise 5).

(c) Prove that

$$|(h_x - h_y)(f)| \leq d(x, y)\, ||f|| \qquad (x, y \in X;\ f \in Y), \tag{1.20}$$

with equality holding in (1.20) for $f = f_x$. Conclude that

$$||h_y - h_x|| = d(x, y) \qquad (x, y \in X), \tag{1.21}$$

that is, *the map* $\pi : x \to h_x$ *is an isometry of* X *into* Y^*.

(d) Let $Z := \overline{\pi(X)}$ denote the closure in Y of the isometric image of X by the isometry π. Prove that Z is complete (cf. Exercise 5(d)).

This is a way to construct a complete metric space, namely Z, in which X is injected isometrically as a dense subset, namely $\pi(X)$. Such a space Z is unique up to an isometric equivalence, and is called *the completion of* X. A more classical way of doing this is by taking the set of all equivalence classes of Cauchy sequences under some natural equivalence relation, with a metric induced by the metric of X (cf. [Kuller], proof of Theorem 3.3.4).

9. Let (X, d) be a metric space, and consider the metric space X^k of Exercise 7, Sect. 1.1.15, with anyone of the metrics d_p $(1 \leq p \leq \infty)$. Prove
   (a) A sequence $\{x^n\} \subset X^k$ converges to $x \in X^k$ iff $x_i^n \to x_i$ in X for all $i = 1, \ldots, k$.
   (b) X^k is complete iff X is complete.
10. Let X be a compact metric space. Prove:
   (a) Every sequence in X has a convergent subsequence.
   (b) X is complete.
11. Let E be a subset of the metric space X. Prove that $x \in \overline{E}$ iff there exists a sequence $\{x_n\} \subset E$ such that $x_n \to x$.

1.4 Functions

Basics

1.4.1 Notation Let X, Y be (non-empty) sets. A function (or map) f from X to Y (notation: $f : X \to Y$) assigns a unique element $y = f(x)$ in Y to each element x in the *domain* X of f.

The *restriction of f to the subset $E \subset X$*, denoted $f|_E$, is the function $f|_E : E \to Y$ such that $f|_E(x) = f(x)$ for all $x \in E$.

If $f : X \to Y$ and $g : Y \to Z$, the *composition* $h = g \circ f$ is the function $h : X \to Z$ defined by $h(x) = g(f(x))$ for $x \in X$.

The function $f : X \to Y$ is *injective* (or "one-to-one") if $f(x) \neq f(x')$ whenever $x, x' \in X$ are distinct; f is *surjective* (or "onto") if for each $y \in Y$ there exists $x \in X$ such that $y = f(x)$; f is *bijective* if it is *both injective and surjective*. In the latter case, for each $y \in Y$, there exists a *unique* $x \in X$ such that $y = f(x)$. This defines a function $g : Y \to X$, that assigns the unique element $x = g(y)$ for which $y = f(x)$, to each given $y \in Y$. The function g is called the *inverse function* of the (bijective) function f. From the definition, *the relations $x = g(y)$ and $y = f(x)$ are equivalent, $g \circ f$ is the identity map on X, and $f \circ g$ is the identity map on Y.*

If Y has an algebraic structure, algebraic operations between functions $f, g : X \to Y$ are defined *pointwise*; for example, $(f + g)(x) := f(x) + g(x)$ (if addition is defined on Y).

If $A \subset X$, the *image of A by $f : X \to Y$* is the set

$$f(A) := \{f(x);\ x \in A\}.$$

In particular, $f(X)$ is called the *range* of f.

If $B \subset Y$, the *inverse image of B by $f : X \to Y$* is the set

$$f^{-1}(B) := \{x \in X;\ f(x) \in B\}.$$

Observe that if f is bijective and g denotes its inverse function, then for all $A \subset X$.

$$g^{-1}(A) = f(A).$$

Indeed, $y \in g^{-1}(A)$ iff $x := g(y) \in A$, which is equivalent to $y = f(x)$ with $x \in A$, that is, to $y \in f(A)$.

The basic operations of set theory are clearly stable under inverse images, that is

$$f^{-1}(\bigcup_\alpha B_\alpha) = \bigcup_\alpha f^{-1}(B_\alpha);$$

$$f^{-1}(\bigcap_\alpha B_\alpha) = \bigcap_\alpha f^{-1}(B_\alpha);$$

$$f^{-1}(B^c) = [f^{-1}(B)]^c.$$

In the above union and intersection, the index α runs in an arbitrary index set. Stability under arbitrary unions is also true for the image by f, but for intersection we only have in general the inclusion

$$f(\bigcap A_\alpha) \subset \bigcap f(A_\alpha).$$

For the image of the complement A^c, we have the following relations:
(a) if f is *injective*, then

$$f(A^c) \subset [f(A)]^c;$$

(b) if f is *surjective*, then

$$[f(A)]^c \subset f(A^c).$$

Hence $f(A^c) = [f(A)]^c$ for a *bijective* function f.

For any $A \subset X$, we have

$$A \subset f^{-1}(f(A)),$$

with equality holding when f is injective.

For any $B \subset Y$, we have

$$f(f^{-1}(B)) \subset B,$$

with equality holding when f is surjective.

The verifications of the above statements are simple exercises.

Limits

1.4.2 Definition Let X, Y be metric spaces (with the respective metrics d_X and d_Y). Let $E \subset X$, $f : E \to Y$, and suppose p is a limit point of E. We say that $f(x)$ converges to $q \in Y$ (in symbols, $f(x) \to q$) as $x \to p$ if for every $\epsilon > 0$, there exists $\delta > 0$ such that

$$d_Y(f(x), q) < \epsilon \tag{1.22}$$

for every $x \in E$ for which $d_X(x, p) < \delta$.

In the above definition, the distance functions in X and Y are used to measure the "closeness" of $f(x)$ to q and of x to p; the condition means that $f(x)$ is arbirarily close to q, provided that x is sufficiently close to p.

If $f(x) \to q$ as $x \to p$, a slight modification of the argument given in the case of sequences shows that q is uniquely determined. It is called the limit of $f(x)$ as x tends to p, and is denoted by $\lim_{x \to p} f(x)$.

Proposition (Notation as in Definition 1.4.3.)
The following statements are equivalent:

(a) $f(x) \to q$ *as* $x \to p$;
(b) the sequence $\{f(x_n)\}$ *converges to* q *for every sequence* $\{x_n\} \subset E$ *converging to* p.
(c) $\lim_{x \to p} f|_F(x) = q$ *for every* $F \subset E$ *such that* p *is a limit point of* F.
(d) $\lim_{x \to p} d_Y(f(x), q) = 0$.

Proof Apply the definitions. □

In the above definition, we assumed that p is a limit point of E, because otherwise there is a ball $B(p, \delta)$ containing no point of E distinct from p. With this δ, Condition (1) is trivially true in case $p \in E$ (take then $q = f(p)$); when $p \notin E$, no point in $B(p, \delta)$ needs to be tested, since the ball does not meet E.

Example 1. Let

$$E = \{(x, y) \in \mathbb{R}^2;\ (x, y) \neq (0, 0)\}$$

(say, with the Euclidean metric), and let $f : E \to \mathbb{R}$ be defined by

$$f(x, y) = \frac{2xy}{x^2 + y^2} \qquad (x, y) \in E.$$

The point $(0, 0)$ is a limit point of E, and also of the subsets $F_1 = \{(x, 0);\ x > 0\}$ and $F_2 = \{(x, x);\ x > 0\}$. As $(x, y) \to (0, 0)$, we have

$$f|_{F_1}(x, y) = 0 \to 0,$$

and

$$f|_{F_2}(x, y) = 1 \to 1.$$

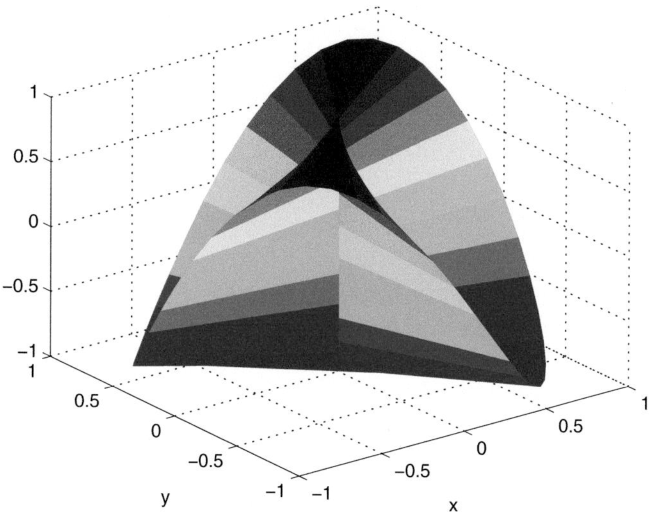

Fig. 1.6 $z = \frac{2xy}{x^2+y^2}$, $0 < x^2 + y^2 \leq 1$

Since these two limits are distinct, it follows from the above observations that the limit of f as $(x, y) \to (0, 0)$ *does not exist.*

The graph of f over the set $\{(x, y);\ 0 < x^2 + y^2 \leq 1\}$, that is, the set $\{(x, y, z) \in \mathbb{R}^3;\ z = f(x, y),\ 0 < x^2 + y^2 \leq 1\}$, is sketched in Fig. 1.6.

In the following proposition, we collect the basic properties of limits in the presence of an algebraic structure on the range space Y.

Proposition *Let X be a metric space, Y a normed space, and $f, g : E \subset X \to Y$. Let $p \in X$ be a limit point of E, and suppose $f(x) \to q$ and $g(x) \to r$ as $x \to p$. Then as $x \to p$,*

(i) $(f + g)(x) \to q + r$ and $cf(x) \to cq$ (for any real constant c);
(ii) if $Y = \mathbb{R}^k$, $(f \cdot g)(x) \to q \cdot r$;
(iii) when $Y = \mathbb{R}$, if $g \neq 0$ on E and $r \neq 0$, then $(f/g)(x) \to q/r$;
(iv) when $Y = \mathbb{R}^k$, $f(x) \to q$ iff $f_i(x) := f(x)_i \to q_i$ for all $i = 1, \ldots, k$.

The meaning of (iv) is that convergence of a vector valued function f to the vector q is equivalent to the convergence of its components to the corresponding components of q.

The above properties of limits follow easily from the definition.

Example 2. Let $\mathbb{R}^k$ be normed by the Euclidean norm $||\cdot|| = ||\cdot||_2$, let $E = \mathbb{R}^k \setminus \{0\}$, and let $f : E \to \mathbb{R}$ be defined by

$$f(x) = \frac{\sqrt{a^2 + ||x||^2} - a}{||x||^2} \qquad (x \in E),$$

where a is a positive constant.

Multiplying numerator and denominator by the non-zero factor $\sqrt{a^2 + ||x||^2} + a$, we obtain

$$\lim_{x \in E, x \to 0} f(x) = \lim \frac{1}{\sqrt{a^2 + ||x||^2} + a} = \frac{1}{2a}.$$

Continuous Functions

1.4.3 Definition Let X, Y be metric spaces, and let $E \subset X$. A function $f : E \to Y$ is *continuous* at the point $p \in E$ if for every $\epsilon > 0$, there exists $\delta > 0$ such that

$$d_Y(f(x), f(p)) < \epsilon \tag{1.23}$$

for all $x \in E$ such that $d_X(x, p) < \delta$.

If f is continuous at *every* point p of E, we say that f is *continuous on* E.

Comparing this definition to the definition of the limit, we see that if p is a limit point of E, then f is continuous at p iff

$$\lim_{x \to p} f(x) = f(p). \tag{1.24}$$

This is precisely the usual condition for a real valued function of a real variable to be continuous at the point p of its domain.

If p is not a limit point of $E \subset X$, that is, if p is an isolated point of E, there exists a neighborhood $B(p, \delta)$ in X containing no point of E distinct from p, and therefore (1.23) is satisfied trivially with this δ for any ϵ, because if $x \in E$ is such that $d(x, p) < \delta$, i.e., $x \in B(x, \delta)$, necessarily $x = p$, hence $d(f(x), f(p)) = 0 < \epsilon$. This means that every function is trivially continuous at the isolated points of its domain.

Example 1. Let X be a normed space, and let $f(x) = ||x||$. If $x, y \in X$, then

$$||x|| = ||(x - y) + y|| \leq ||x - y|| + ||y||.$$

Hence

$$||x|| - ||y|| \leq ||x - y||. \tag{1.25}$$

Interchanging the roles of x and y, we have

$$-(||x|| - ||y||) = ||y|| - ||x|| \leq ||y - x|| = ||x - y||. \tag{1.26}$$

Since $\Big| ||x|| - ||y|| \Big| = \pm(||x|| - ||y||)$, it follows from (1.25) and (1.26) that

$$\Big| ||x|| - ||y|| \Big| \leq ||x - y||. \tag{1.27}$$

Consider the norm function $f : x \in X \to ||x|| \in \mathbb{R}$. By (1.27),

$$d_{\mathbb{R}}(f(x), f(y)) := |f(x) - f(y)| \leq ||x - y|| := d_X(x, y).$$

Given $\epsilon > 0$, we choose $\delta = \epsilon$ in the definition of continuity; thus, whenever $d_X(x, y) < \delta$, $d_{\mathbb{R}}(f(x), f(y)) < \delta = \epsilon$. This shows that the norm function is continuous on X.

Example 2. The function f defined on E in Example 1 of Sect. 1.4.2 cannot be extended to a function continuous on the whole plane $\mathbb{R}^2$. Indeed, $(0, 0)$ is a limit point of E, and we saw that the limit of $f(x, y)$ as $(x, y) \to (0, 0)$ does not exist (so (2) is not satisfied).

Example 3. The function $f : E = \mathbb{R}^k \setminus \{0\} \to \mathbb{R}$ in Example 2 of Sect. 1.4.2 extends as a continuous function on $\mathbb{R}^k$, since $\lim_{x\to 0} f(x)$ exists. The value of the extended function at the point 0 is uniquely given by this limit, namely $1/(2a)$.

1.4.4 Theorem *Let X, Y, Z be metric spaces, $E \subset X$, $f : E \to Y$, and $g : f(E) \to Z$. Suppose f is continuous at the point $p \in E$, and g is continuous at the point $q := f(p)$. Then $g \circ f$ is continuous at p.*

Proof Let $\epsilon > 0$. By the continuity of g at q, there exists $\delta_1 > 0$ such that $d_Z(g(y), g(q)) < \epsilon$ for all $y \in f(E)$ such that $d_Y(y, q) < \delta_1$. By continuity of f at p, there exists $\delta > 0$ such that

$$d_Y(f(x), q) = d_Y(f(x), f(p)) < \delta_1$$

for all $x \in E$ such that $d_X(x, p) < \delta$. Therefore, *for these x*,

$$d_Z(h(x), h(p)) = d_Z(g(f(x)), g(q)) < \epsilon.$$

□

In the next proposition, we show that continuity is stable with respect to algebraic operations in the range space, when the latter has an algebraic structure.

Proposition *Let X be a metric space, $E \subset X$, and let Y be a normed space. Suppose $f, g : E \to Y$ are continuous at a point $p \in E$. Then*

(a) $f + g$ is continuous at p;
(b) cf is continuous at p for any real constant c;
(c) If $Y = \mathbb{R}^k$, then $f \cdot g$ is continuous at p;
(d) $f = (f_1, \ldots, f_k) : E \to \mathbb{R}^k$ is continuous at p iff its component functions $f_i : E \to \mathbb{R}$ are continuous at p for all $i = 1, \ldots, k$;
(e) In case $Y = \mathbb{R}$, if $g(p) \neq 0$, then f/g is defined for all $x \in E$ in a neighborhood of p and is continuous at p.

Proof Apply the properties of limits (cf. Section 1.4.2.) We remark only that in the proof of (e), we use the fact often applied in the sequel that if $g : E \to Y$ is continuous at a point $p \in E$, Y is a normed space, and $g(p) \neq 0$, then $g(x) \neq 0$ for all $x \in E$ in some neighborhood of p. Indeed, take $\epsilon = ||g(p)||$ (> 0), and let δ be as in the definition of continuity for g at p. Then for all $x \in E$ in the neighborhood $B(p, \delta)$ of p,

$$||g(x)|| = ||g(p) - [g(p) - g(x)]|| \geq ||g(p)|| - ||g(x) - g(p)|| > 0.$$

(cf. Section 1.4.3, (3)). □

Note that the projections $x \to x_i$ of $\mathbb{R}^k$ onto $\mathbb{R}$ are continuous on $\mathbb{R}^k$ (for each $i = 1, \ldots, k$), because

$$|x_i - y_i| \leq ||x - y|| \qquad (x, y \in \mathbb{R}^k)$$

for any norm $|| \cdot || = || \cdot ||_p$ on $\mathbb{R}^k$. Therefore, by Properties (a)–(c), polynomials in k real variables, that is, functions f defined by

$$f(x) = \sum a_{n_1,\ldots,n_k} x_1^{n_1} \ldots x_k^{n_k},$$

where the sum is finite and the coefficients $a_{n_1,\ldots,n_k}$ are real constants, are continuous on $\mathbb{R}^k$.

By Property (e), a rational function, that is, the ratio f/g of such polynomials, is continuous on the complement in $\mathbb{R}^k$ of the zero set of the polynomial g.

Example 4. Let $f : \mathbb{R}^2 \to \mathbb{R}$ be equal to $x \sin(y^2/x)$ on $E = \{(x, y) \in \mathbb{R}^2;\ x \neq 0\}$, and $f = 0$ on E^c. By the continuity of the sine function on $\mathbb{R}$ and of rational functions on their domain of definition, it follows from Theorem 1.4.4 that f is continuous on E. For any point $(0, y_0)$ (of E^c), we have $|f(x, y)| \leq |x| \to 0 = f(0, y_0)$ as $(x, y) \to (0, y_0)$. Hence f is continuous on $\mathbb{R}^2$.

Example 5. Let E be as in Example 4, and let $f : E \to \mathbb{R}$ be defined by $f(x, y) = \arctan(y/x)$. By Theorem 1.4.4, the function f is continuous and bounded by $\pi/2$ on the set E. Is it possible to extend f as a continuous function $\tilde{f}$ beyond the set E? The points of $\mathbb{R}^2 \setminus E$ are the points $(0, y)$ of the y-axis, and are limit points of E. A continuous extension $\tilde{f}$ of f exists at a point $(0, y_0)$ of the y-axis iff the following limit as $(x, y) \in E$ tends to $(0, y_0)$ exists:

$$\lim \tilde{f}(x, y) = \lim f(x, y) = \lim \arctan(\frac{y}{x}). \tag{1.28}$$

In that case, $\tilde{f}(0, y_0)$ is uniquely given by the limit (1.28).

If $y_0 > 0$ $(y_0 < 0)$, then as $x \to 0\pm$ and $y \to y_0$, we have $f(x, y) \to \pm\pi/2$ $(f(x, y) \to \mp\pi/2$, respectively), and therefore the limit (1.28) does not exist.

The restrictions of f to $F := \{(x,0);\ 0 \neq x \in \mathbb{R}\} \subset E$ ($F' = \{(x,x); 0 \neq x \in \mathbb{R}\} \subset E$) are the constant functions with value 0 ($\pi/4$), hence have the limits 0 ($\pi/4$, respectively) as $x \to 0$. Therefore the limit (1.28) does not exist for $y_0 = 0$.

We conclude from the above discussion that the limit (1.28) does not exist for all $y_0 \in \mathbb{R}$, and consequently f cannot be extended as a continuous function beyond the set E.

Properties of Continuous Functions

In this section, we shall generalize the well-known properties of real valued continuous functions on closed intervals to the general setting. We begin with an important characterization of global continuity by means of open sets.

1.4.5 Theorem *Let X, Y be metric spaces, and $f : X \to Y$. Then f is continuous on X iff $f^{-1}(B)$ is open (in X) for each open subset B of Y.*

Proof Suppose f continuous on X, and let $B \subset Y$ be open. Let $x \in f^{-1}(B)$. Since $f(x) \in B$ and B is open, there exists a ball $B_Y(f(x), \epsilon)$ contained in B. By the continuity of f at the point x, there exists $\delta > 0$ such that $d_Y(f(y), f(x)) < \epsilon$ for all $y \in X$ such that $d_X(y,x) < \delta$. This means that if $y \in B_X(x,\delta)$, then $f(y) \in B_Y(f(x), \epsilon) \subset B$, i.e., $y \in f^{-1}(B)$. This shows that $B_X(x,\delta) \subset f^{-1}(B)$, that is, every point x of $f^{-1}(B)$ is an interior point of that set. We thus proved that $f^{-1}(B)$ is open for each open $B \subset Y$.

Conversely, suppose the latter property is valid. Given $x \in X$ and $\epsilon > 0$, consider the open set $B := B_Y(f(x), \epsilon)$. Then by hypothesis, $f^{-1}(B)$ is open, and clearly contains the point x. Therefore there exists a ball $B_X(x,\delta)$ contained in $f^{-1}(B)$. This means that if $d_X(y,x) < \delta$, then $f(y) \in B$, i.e., $d_Y(f(y), f(x)) < \epsilon$. This proves that f is continuous at the arbitrary point x of X. □

Note that the above characterization of continuity uses only *open sets* in the spaces X and Y. It is then used as the *definition* of continuity for functions $f : X \to Y$ when X, Y are general topological spaces.

Using the above characterization of continuity, we show that a continuous function $f : X \to Y$ maps compact subsets of X onto compact subsets of Y.

1.4.6 Theorem *Let X, Y be metric spaces, and let $f : X \to Y$ be continuous. Then $f(E)$ is a compact subset of Y for every compact subset E of X.*

Proof Let $E \subset X$ be compact, and let $\{V_\alpha\}$ be an open cover of $f(E)$. By Theorem 1.4.5, the sets $f^{-1}(V_\alpha)$ are open for all indices α, and (cf. 1.4.1)

$$E \subset f^{-1}(f(E)) \subset f^{-1}(\bigcup_\alpha V_\alpha) = \bigcup_\alpha f^{-1}(V_\alpha). \tag{1.29}$$

Thus $\{f^{-1}(V_\alpha)\}$ is an open cover of the *compact* set E. Therefore there exist indices $\alpha_1, \ldots, \alpha_n$ such that

$$E \subset \bigcup_{i=1}^n f^{-1}(V_{\alpha_i}).$$

Hence (cf. 1.4.1)

$$f(E) \subset f(\bigcup_{i=1}^{n} f^{-1}(V_{\alpha_i}) = \bigcup_{i=1}^{n} f(f^{-1}(V_{\alpha_i})) \subset \bigcup_{i=1}^{n} V_{\alpha_i},$$

that is, $\{V_{\alpha_i};\ i = 1, \ldots, n\}$ is a finite subcover of the given open cover of $f(E)$. This proves that $f(E)$ is compact. □

Since the only concept involved in the above proof and in the definition of compactness is the concept of an open set, Theorem 1.4.6 is valid for general topological spaces X, Y.

The following properties of continuous functions are immediate consequences of Theorem 1.4.6.

1.4.7 Corollary *Let X, Y be metric spaces and $f : X \to Y$ be continuous. Then f is bounded on compact subsets of X.*

The meaning of the statement is that the set $f(E)$ is bounded whenever $E \subset X$ is compact. In particular, if Y is a *normed space*, then for each compact subset $E \subset X$, the set of (non-negative) numbers $\{||f(x)||;\ x \in E\}$ is bounded above, or equivalently, it has a *finite supremum*

$$||f||_E := \sup_{x \in E} ||f(x)|| < \infty. \tag{1.30}$$

Proof If $E \subset X$ is compact, then $f(E)$ is compact, by Theorem 1.4.6, hence bounded, by Theorem 1.2.6. □

In connection with (1.30), we recall that if $E \subset \mathbb{R}$ is bounded above (below), it has a supremum $q = \sup E \in \mathbb{R}$ (infimum $r = \inf E \in \mathbb{R}$, respectively), where q (r) is uniquely characterized by the two properties:

1. $x \le q$ $(x \ge r)$, for all $x \in E$;
2. for every $\epsilon > 0$, there exists $x \in E$ such that $x > q - \epsilon$ $(x < r + \epsilon$, respectively).

Note that every ϵ-neighborhood of q contains a point of the bounded above set E, so that either $q \in E$ or q is a limit point of E (or both). Hence $q \in E$ if E is *closed.* Similarly, $r \in E$ for any closed bounded below set E. In particular, *every closed bounded (i.e., compact) subset E of $\mathbb{R}$ contains* $\sup E$ *and* $\inf E$.

1.4.8 Corollary *Let X be a metric space and let $f : X \to \mathbb{R}$ be continuous. Then for each compact subset $E \subset X$, there exist points $a, b \in E$ such that $f(a) = \sup f(E)$ and $f(b) = \inf f(E)$.*

The supremum and infimum of the values of f on E are *values of the function at certain points of E*; it is then customary to call them *maximum* and *minimum* of f on E, respectively, and to say that f *assumes its extrema, that is, its maximum and minimum, on E.*

Proof By Theorem 1.4.6, $f(E)$ is a compact subset of $\mathbb{R}$ for each compact $E \subset X$; therefore $\sup f(E) \in f(E)$ and $\inf f(E) \in f(E)$. □

We prove next the "automatic" continuity of the inverse function of a continuous function defined on a compact metric space, whenever the inverse function is well-defined.

1.4.9 Corollary *Let X, Y be metric spaces, X compact. Suppose $f : X \to Y$ is continuous and bijective. Then f is an open map, that is, $f(E)$ is open in Y for all open subset E of X. Equivalently, the inverse function g of f is continuous.*

Proof Let $E \subset X$ be open. Then E^c is a closed subset of the compact set X, and is therefore compact, by Theorem 1.2.3. Since f is continuous, $f(E^c)$ is compact by Theorem 1.4.6. But $f(E^c) = f(E)^c$, because f is bijective (cf. 1.4.1). Hence we proved that $f(E)^c$ is compact, and thus closed (cf. Theorem 1.2.4), that is, $f(E)$ is open. Equivalently (cf. 1.4.1), $g^{-1}(E)$ is open for every open subset E of X, hence g is continuous, by Theorem 1.4.5. □

Uniform Continuity

1.4.10 Definition Let X, Y be metric spaces. The function $f : X \to Y$ is *uniformly continuous on* X if for each $\epsilon > 0$ there exists $\delta > 0$ such that

$$d_Y(f(x),\ f(y)) < \epsilon$$

for all $x, y \in X$ such that $d_X(x, y) < \delta$.

A comparison with the definition of continuity at every point $x \in X$ shows that the uniformity requirement consists of the possibility of choosing δ *independently of* x. For example, the function $f(x) = 1/x$ is continuous on $X := (0, \infty)$, but not uniformly continuous on X. The concept of uniform continuity will be important in the study of the Riemann integration. The following theorem will be the usual tool.

1.4.11 Theorem *Let X, Y be metric spaces. If X is compact and $f : X \to Y$ is continuous on X, then f is uniformly continuous on X.*

Proof Let $\epsilon > 0$ be given. For each $x \in X$, by continuity of f at x, there exists $\delta_x > 0$ such that

$$d_Y(f(y),\ f(x)) < \frac{\epsilon}{2} \tag{1.31}$$

for all $y \in B_X(x, \delta_x)$. The family of balls $B(x, \delta_x/2)$ $(x \in X)$ is an open cover of X. Since X is compact, there exists a finite subcover consisting of the balls $B(x_i, \delta_{x_i}/2)$, $i = 1, \ldots, n$, that is

$$X \subset \bigcup_{i=1}^{n} B(x_i,\ \frac{\delta_{x_i}}{2}). \tag{1.32}$$

Set

$$\delta := \min_{i=1,\ldots,n} \frac{\delta_{x_i}}{2}. \tag{1.33}$$

Let $x, y \in X$ be such that $d_X(x, y) < \delta$. By (1.32), there exists $i \in \{1, \ldots, n\}$ such that

$$x \in B(x_i, \frac{\delta_{x_i}}{2}).$$

For such an index i, we have by the triangle inequality and (1.33)

$$d_X(y, x_i) \le d_X(y, x) + d_X(x, x_i) < \delta + \frac{\delta_{x_i}}{2} \le \delta_{x_i}.$$

Hence by (1.31)

$$d_Y(f(y), f(x_i)) < \frac{\epsilon}{2}. \tag{1.34}$$

Since $d_X(x, x_i) < \delta_{x_i}/2 < \delta_{x_i}$, we also have by (1.31)

$$d_Y(f(x), f(x_i)) < \frac{\epsilon}{2}. \tag{1.35}$$

Therefore, by (1.34) and (1.35), and the triangle inequality,

$$d_Y(f(x), f(y)) \le d_Y(f(x), f(x_i)) + d_Y(f(x_i), f(y)) < \epsilon,$$

as desired. □

We generalize now the Intermediate Value Theorem (IVT). Recall that if I is an interval (open, closed, or half-closed), and $f : I \to \mathbb{R}$ is continuous, then f "assumes its intermediate values", that is, if $a, b \in I$, $a < b$, and γ is between $f(a)$ and $f(b)$, then there exists $c \in [a, b]$ such that $f(c) = \gamma$. The generalization of the IVT goes through the concept of connectedness. We show first that connectedness is preserved by continuous mappings.

1.4.12 Theorem *Let X, Y be metric spaces, and let $f : X \to Y$ be continuous. Then $f(E)$ is a connected subset of Y for any connected subset E of X.*

Proof Let $E \subset X$ be connected, and suppose $f(E) \subset A \cup B$, where A, B are disjoint open subsets of Y, both meeting $f(E)$.

By Theorem 1.4.5, $f^{-1}(A)$ and $f^{-1}(B)$ are open subsets of X, clearly disjoint, both meeting E, and

$$E \subset f^{-1}(f(E)) \subset f^{-1}(A \cup B) = f^{-1}(A) \cup f^{-1}(B).$$

This contradicts the connectedness of E. □

We consider now connectedness of subsets of $\mathbb{R}$.

1.4.13 Theorem *The set $E \subset \mathbb{R}$ is connected iff*

$$a, b \in E \ (a < b) \Longrightarrow [a, b] \subset E. \tag{1.36}$$

It is clear from the criterion above that the only non-empty connected subsets of $\mathbb{R}$ are the finite or infinite, open, closed, or half-closed intervals, and the singletons.

Proof Suppose $a, b \in E$, but $[a, b]$ is not contained in E. Then there exists $c \in (a, b)$ such that $c \notin E$, and therefore

$$E \subset (-\infty, c) \cup (c, \infty).$$

The sets on the right hand side are disjoint open sets, which meet E at a and b respectively. Hence E is not connected.

Conversely, suppose (1.36) is satisfied, but E is *not* connected. Let then A, B be disjoint open sets in $\mathbb{R}$, both meeting E, such that $E \subset A \cup B$. Pick $a \in E \cap A$ and $b \in E \cap B$. Without loss of generality, we assume that $a < b$. Let

$$c := \sup\, [a, b] \cap A. \tag{1.37}$$

Since $b \in B$ and B is open, there exists $r > 0$ such that $a < b-r$ and $(b-r, b] \subset B$. Hence $c \le b-r < b$. Also $c > a$ because A is open. If $c \in A$, there exists $s > 0$ such that $[c, c+s) \subset A$ and $c+s < b$, hence $\sup[a, b] \cap A \ge c+s > c$, contradicting (1.37). This shows that $c \notin A$.

An analogous argument shows that $c \notin B$: since $a \in A$ and A is open, there exists $s > 0$ such that $[a, a+s) \subset A$ and $a+s < b$. Then $c = \sup[a, b] \cap A \ge a+s > a$. If $c \in B$, then since B is open, there exists $r > 0$ such that $(c-r, c] \in B$, hence $(c-r, c]$ is not contained in A, and therefore $\sup[a, b] \cap A \le c-r$, contradicting (1.37). This shows that $c \notin B$.

In conclusion, $c \notin A \cup B$, and therefore $c \notin E$, because $E \subset A \cup B$. This contradicts (1.36), since $c \in (a, b)$. □

We can state now the wanted generalization of the IVT to continuous real valued functions on a metric space.

1.4.14 Theorem (Intermediate Value Theorem) *Let E be a connected subset of a metric space X, and let $f : E \to \mathbb{R}$ be continuous. Suppose $a < b$ are in $f(E)$. Then $[a, b] \subset f(E)$.*

The conclusion means that any "value" c beween the values a and b of f is indeed a value of the continuous function f.

Proof By Theorem 1.4.12, $f(E)$ is a connected subset of $\mathbb{R}$. Since $a, b \in f(E)$, it follows that $[a, b] \subset f(E)$ by Theorem 1.4.13. □

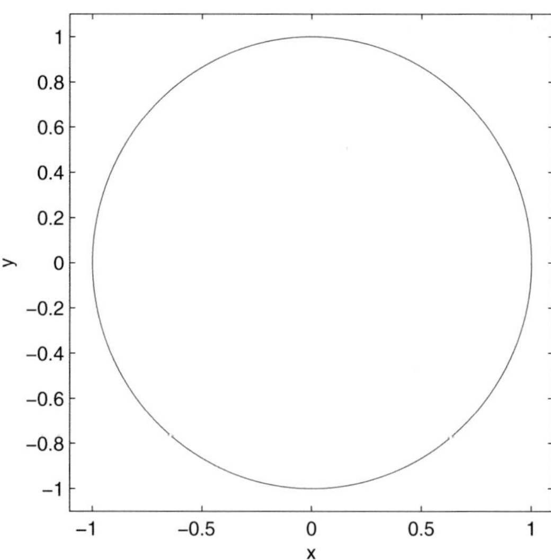

Fig. 1.7 $f(t) = (\cos t,\ \sin t), t \in [0, 2\pi]$

Paths

In this section, we consider a connectedness concept defined by means of "paths", and we show that it coincides with the connectedness concept on *open subsets of a normed space*.

1.4.15 Definition A *path* in the metric space X is the image $f(I)$ of a closed interval $I = [a, b]$ by a continuous function $f : I \to X$. The points $f(a)$ and $f(b)$ are called the *initial point* and the *final point* of the path, respectively. The path is said to connect its *end points* $f(a)$ and $f(b)$. If $f(a) = f(b)$, the path is a *closed path*.

For example, the functions $f : [0, 2\pi] \to \mathbb{R}^2$ given below define closed paths:

(a) $f(t) = (\cos t,\ \sin t)$ (the unit circle).
(b) $f(t) = (1 - \cos t)\,(\cos t,\ \sin t)$ (a *cardioid*).

(Cf. Figs. 1.7 and 1.8.)

The *cycloid*, defined by the function

$$f(t) = (t - \sin t,\ 1 - \cos t) \qquad (t \in [0, 2\pi]),$$

is not closed (cf. Fig. 1.9).

As an immediate consequence of Theorems 1.4.12 and 1.4.13, we obtain

1.4.16 Corollary *In any metric space, paths are connected subsets.*

Consider the special case of a normed space X. If $p, q \in X$, the (line) segment $\overline{pq}$ is the set of all points $x(t) = p + t(q - p)$, $t \in [0, 1]$. It is the image of $[0, 1]$ by the continuous function $x(\cdot)$, i.e., it is a path, hence a connected subset of X. More

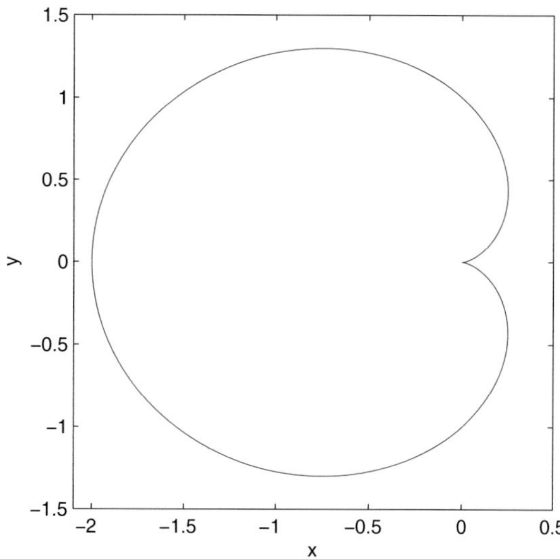

Fig. 1.8 $f(t) = (1 - \cos t)\,(\cos t,\ \sin t)$, $t \in [0, 2\pi]$

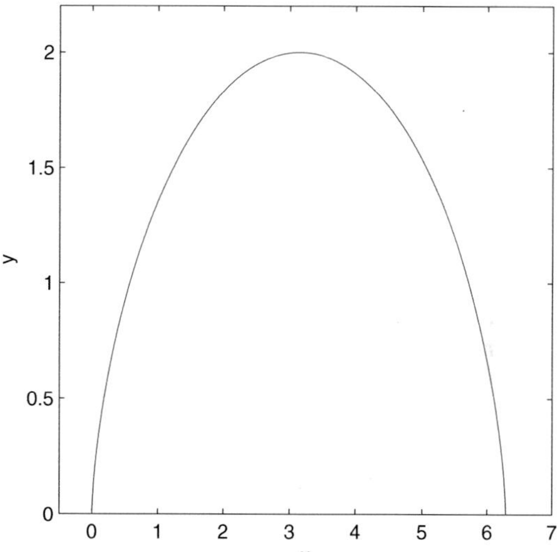

Fig. 1.9 $f(t) = (t - \sin t,\ 1 - \cos t)$, $t \in [0, 2\pi]$

generally, a *polygonal path* γ from p to q is the union of segments $\overline{p^i q^i}$, $p_i,\ q_i \in X$, $i = 1, \ldots, n$, such that $p^1 = p$, $q^n = q$, and $q^i = p^{i+1}$ for all $i = 1, \ldots, n-1$. One sees easily that γ is a path (called the "sum" or "composition" of its segments), hence is connected.

1.4.17 Definition Let X be a metric (normed) space. The set $E \subset X$ is *pathwise* (*polygonally*) connected if for every $p, q \in E$, there exists a path (a polygonal path, respectively) lying in E with end points p, q.

1.4.18 Theorem *(i) In any metric space X, pathwise connected subsets are connected.*

(ii) Connected open subsets of a normed space are polygonally (hence pathwise) connected.

Proof (i) follows from Corollaries 1.4.16 and 1.1.11.
Proof of (ii). Let E be a (non-empty) connected open set in the metric space X. Given $p \in E$, let A be the set of all $q \in E$ that can be connected to p by a polygonal path γ_{pq} lying in E. We show that A is open. If $q \in A$, then $q \in E$, and since E is open, there exists a ball $B(q,r) \subset E$. For any point $x \in B(q,r)$, the segment $\overline{qx}$ is contained in $B(q,r)$, since $||(q + t(x-q)) - q|| = t||x-q|| < r$ for all $t \in [0,1]$. Adding this segment to γ_{pq}, we get a polygonal path from p to x which lies in E. Hence $x \in A$, and we proved that $B(q,r) \subset A$. This shows that A is open.

Let $B = E - A$. We prove that B is open. If $q \in B$, then $q \in E$, and since E is open, there is a ball $B(q,r) \subset E$. If some point $x \in B(q,r)$ belongs to A, there exists a polygonal path $\gamma_{px} \subset E$ connecting p to x, and as before, the segment $\overline{xq}$ lies in $B(q,r)$, hence in E. Adding this segment to γ_{px}, we get a polygonal path in E connecting p and q, that is, $q \in A$, contradiction. This shows that $B(q,r) \subset B$, and we conclude that B is open. Thus A, B are disjoint open sets in X, such that $E = A \cup B$. Since E is connected, it cannot meet both A and B. However $p \in E \cap A$. Therefore $E \cap B = \emptyset$, and we conclude that $E = A$. Since $p \in E$ is arbitrary, this shows that *every* two points $p, q \in E$ can be connected by a polygonal path lying in E. □

Domains in $\mathbb{R}^k$

1.4.19 Definition A *domain* in $\mathbb{R}^k$ is a connected open set in $\mathbb{R}^k$.

By Theorem 1.4.18, for open subsets of $\mathbb{R}^k$, the concepts of *connected, pathwise connected*, and *polygonally connected* are equivalent. However, in general, the converse of Theorem 1.1.18 (i) is false. See Exercise 10, Sect. 1.4.21.

If $D \subset \mathbb{R}^k$ is a domain, a point $p \in \mathbb{R}^k$ is a *boundary point* of D if every ball $B(p,r)$ meets both D and D^c. The set of all boundary points of D is called the *boundary of* D, denoted ∂D. The set $\overline{D} := D \cup \partial D$ will be called a *closed domain* (with some abuse of logic).

1.4.20 Theorem *Let D be a domain in $\mathbb{R}^k$, and let $D \subset E \subset \overline{D}$. Then E is connected.*

This means that the connectedness of the domain is preserved when we add to it any part of its boundary.

Proof Suppose E is not connected. Let then A, B be disjoint open sets in $\mathbb{R}^k$, both meeting E, such that $E \subset A \cup B$. Let then $a \in A \cap E$ and $b \in B \cap E$. Since A, B are open, there exists $r > 0$ such that $B(a,r) \subset A$ and $B(b,r) \subset B$. Since

$a, b \in E$ and $E \subset \overline{D}$, these balls contain points $p, q \in D$, respectively. Therefore $D \subset E \subset A \cup B$, where A, B are disjoint open sets, both meeting D:

$$A \cap D \supset B(a, r) \cap D \supset \{p\}; \qquad B \cap D \supset B(b, r) \cap D \supset \{q\}.$$

This shows that D is not connected, contradiction. □

1.4.21 Exercises 1. Let X, Y be metric spaces and $f : X \to Y$. Prove that f is continuous if and only if $f^{-1}(E)$ is closed in X for every closed subset E of Y.

2. Let X, Y be metric spaces and $f : X \to Y$. Prove:

   (a) f is continuous if and only if $f(\overline{E}) \subset \overline{f(E)}$ for every $E \subset X$.
   (b) If X is compact and f is continuous, then $f(\overline{E}) = \overline{f(E)}$.

3. Let (X, d) be a metric space, and let

$$\emptyset \neq F \subset G \subset X,$$

   with F closed and G open. Define

$$h(x) = \frac{d(x, F)}{d(x, F) + d(x, G^c)}.$$

   Prove that $h : X \to [0, 1]$ is continuous, and

$$F = h^{-1}(0); \quad G^c = h^{-1}(1).$$

   We wrote $h^{-1}(0)$ instead of $h^{-1}(\{0\})$, etc.

4. The following real valued functions f are defined on $\mathbb{R}^2 \setminus \{(0, 0)\}$. For each function, determine its extendability as a continuous function $\tilde{f}$ on $\mathbb{R}^2$. In case of a positive answer, find $\tilde{f}(0, 0)$.

   (a) $f(x, y) = \frac{x^2 y}{x^2+y^2}$.
   (b) $f(x, y) = \frac{xy}{\sqrt{x^2+ay^2}}$, $a \geq 1$.
   (c) $f(x, y) = (x^{2m} + y^{2m}) \log(x^2 + y^2)$, $m \in \mathbb{N}$.

5. Let X be a metric space and let $f : X \to \mathbb{R}$ be continuous. Suppose $\gamma_j : [0, 1] \to X$ $(j = 1, 2)$ are paths from p to q and from q to p, respectively, where p, q are given points in X. Prove

   (a) The functions $f \circ \gamma_j$ coincide at some point of $[0, 1]$.
   (b) If $f(p) \neq f(q)$, the above functions coincide at some point of $(0, 1)$.

6. Let D be the domain $\{x \in \mathbb{R}^k;\ 2 < ||x||^2 < 3\}$, where $|| \cdot ||$ is the Euclidean norm. Let

$$f(x) = \cos \frac{1}{||x||^2 - 1} \qquad (x \in \mathbb{R}^k,\ ||x|| \neq 1).$$

Prove

(a) f is uniformly continuous in D.
(b) Let X, Y be metric spaces, Y complete, and $E \subset X$. Suppose $f : E \to Y$ is uniformly continuous. Then f is extendable to a continuous function on the closure $\overline{E}$ of E.
(c) The function f of Part (a) is *not* uniformly continuous in the unit ball $B(0, 1)$ in $\mathbb{R}^k$.

7. Let X, Y be normed spaces, and let $T : X \to Y$ be a linear map. Prove the equivalence of the following statements (a)–(d):

(a) T is continuous at $x = 0$.
(b) $T \in B(X, Y)$ (cf. 1.3.7 (5)).
(c) T is Lipschitz on X, that is, there exists a constant $q \geq 0$ such that

$$||Tx - Tx'|| \leq q\, ||x - x'||$$

for all $x, x' \in X$.
(d) T is uniformly continuous on X.

8. Let S be the unit sphere in $X := (\mathbb{R}^k, \ ||\cdot||_1)$, that is,

$$S := \{x \in \mathbb{R}^k;\ ||x||_1 = 1\}.$$

(a) Prove that S is compact in X.
(b) Let $||\cdot||$ be an arbitrary norm on $\mathbb{R}^k$. Denote

$$K := \max_{1 \leq i \leq k} ||e^i||.$$

Prove that

$$||x|| \leq K\, ||x||_1,$$

and conclude that $||\cdot||$ is continuous on X. Therefore $||\cdot||$ assumes its minimum value L on S, and $L > 0$ because $0 \notin S$.
(c) Prove that $L\, ||x||_1 \leq ||x||$ for all $x \in \mathbb{R}^k$.
By Parts (b) and (c), the norm $||\cdot||$ is equivalent to $||\cdot||_1$, and therefore *every two norms on* $\mathbb{R}^k$ *are equivalent*. Compare with Theorem 1.1.7.

9. A subset E of a vector space X is *convex* if

$$x, y \in E \Longrightarrow \overline{xy} \subset E,$$

where $\overline{xy}$ denotes the line segment from x to y, i.e.,

$$\overline{xy} := \{(1 - t)x + ty;\ t \in [0, 1]\}.$$

Suppose E is a compact convex subset of a normed space X and $0 \in E^o$, where E^o denotes the interior of E, cf. Exercise 1.1.15 (4). Define

$$h(x) = \inf\{t > 0;\ \frac{x}{t} \in E\} \qquad (x \in X).$$

Prove:

(a) $h(x+y) \le h(x)+h(y)$ and $h(cx) = c\,h(x)$ for all $x,\ y \in X$ and $c \in [0, \infty)$. Furthermore, $h(x) = 0$ iff $x = 0$.
(b) h is uniformly continuous on X.
(c) $E = h^{-1}([0, 1])$.
(d) $E^o = h^{-1}([0, 1))$.

10. Let E be any non-empty subset of $\{0\} \times [-1, 1]$ and let

$$Y = \{(x,\ \sin\frac{1}{x});\ x > 0\} \subset \mathbb{R}^2.$$

Prove

(a) $Y \cup E$ is connected (cf. Theorem 1.4.20.)
(b) $Y \cup \{(0, 0)\}$ is not pathwise connected.

This example shows that the converse of Theorem 1.4.18(i) is false even in $\mathbb{R}^2$.

Chapter 2
Derivation

In this chapter, we shall study the concept of differentiability of real or vector valued functions on $\mathbb{R}^k$. To fix the ideas, we assume that $\mathbb{R}^k$ is normed by the Euclidean norm

$$||x|| := ||x||_2 := (\sum_{i=1}^{k} x_i^2)^{1/2}.$$

The induced metric is the Euclidean metric

$$d(x, y) := d_2(x, y) := ||x - y||.$$

2.1 Differentiability

Directional Derivatives

If f is a real valued function of one real variable x, the derivative $f'(x)$ is a measure of the "rate of change" of f at the point x:

$$f'(x) := \lim_{h \to 0} \frac{f(x+h) - f(x)}{h}.$$

This expression has no meaning when f is a function of $x \in \mathbb{R}^k$ for $k \geq 3$, since $\mathbb{R}^k$ in not a field, and division by h is not defined. Note by the way that in case $k = 2$, a field structure is available: $\mathbb{R}^2$ is identified with the field of complex numbers $\mathbb{C}$. The above limit makes sense for f defined in a neighborhood of $x \in \mathbb{C}$ with values in $\mathbb{C}$, and is called the derivative of f at x, denoted $f'(x)$, when it exists. In general, we can still consider the rate of change of f in a given "direction", determined by a

---

The original version of this chapter was revised. An erratum to this chapter can be found at DOI 10.1007/978-3-319-27956-5_5

S. Kantorovitz, *Several Real Variables*, Springer Undergraduate Mathematics Series, DOI 10.1007/978-3-319-27956-5_2

"unit vector" u. Any vector $u \in \mathbb{R}^k$ with norm equal to 1 is called a *unit vector*. In particular, the unit vectors e^j $(j = 1, \dots, k)$ are the vectors with

$$e_i^j = \delta_{ij},$$

the Kronecker delta. As before, the subscript index is the component index. The unit vector e^j is called the unit vector in the direction of the axis x_j. In general, for any unit vector $u \in \mathbb{R}^k$ and $x \in \mathbb{R}^k$, the *axis through x in the direction u* is the directed line with the parametric representation

$$\gamma : t \in \mathbb{R} \to \gamma(t) := x + tu \in \mathbb{R}^k.$$

Let f be a real valued function defined in a neighborhood $B(x, r)$ of x in $\mathbb{R}^k$. The function $F := f \circ \gamma$, that is,

$$F(t) := f(x + tu), \tag{2.1}$$

is well defined for $|t| < r$, because for such t,

$$d(x + tu, x) = ||x + tu - x|| = ||tu|| = |t| < r$$

so that $x + tu \in B(x, r)$. Clearly, F is the restriction of f to the given line. Since the parameter value $t = 0$ corresponds to the point x, $F'(0)$ (when it exists) is a measure of the rate of change of f at the point x along the line γ, that is, "in the direction" u. This motivates the following formal definition.

2.1.1 Definition (*Notation as above*) If the derivative $F'(0)$ exists, it is called *the directional derivative of f at x in the direction u*, denoted $D_u f(x)$ or $\frac{\partial f}{\partial u}(x)$:

$$\begin{aligned} \frac{\partial f}{\partial u}(x) := F'(0) &= \lim_{t \to 0} \frac{F(t) - F(0)}{t} \\ &= \lim_{t \to 0} \frac{f(x + tu) - f(x)}{t}. \end{aligned}$$

In particular, if we take $u = e^j$, we shall denote the directional derivative at x of f in the direction e^j by $\frac{\partial f}{\partial x_j}(x)$, instead of $\frac{\partial f}{\partial e^j}(x)$. We use also the notation $f_{x_j}(x)$ or $D_j f(x)$ (instead of $D_{e^j} f(x)$). This directional derivative is called *the partial derivative at x of f with respect to x_j*. Thus

$$\frac{\partial f}{\partial x_j}(x) := \lim_{t \to 0} t^{-1}[f(x_1, \dots, x_j + t, \dots, x_k) - f(x_1, \dots, x_k)].$$

This is the ordinary derivative of f as a function of the single variable x_j at the point x_j, with all the other variables x_i $(i \neq j)$ kept fixed. If the partial derivatives of f at x exist for all $j = 1, \dots, k$, we define the *gradient of f at x* as the *vector*

$$\nabla f(x) := (\frac{\partial f}{\partial x_1}(x), \ldots, \frac{\partial f}{\partial x_k}(x)).$$

(∇f is read "nabla f", or "the gradient of f".)

For $k = 1$, the existence of the derivative of f at x implies the continuity of f at x. This elementary fact does not extend to the case $k > 1$, with the derivative replaced by the partial derivatives for all $j = 1, \ldots, k$. This is illustrated by the following example.

Example. Let $f : \mathbb{R}^2 \to \mathbb{R}$ be defined by $f(0, 0) = 0$ and $f(x, y) = 2xy/(x^2 + y^2)$ for $(x, y) \neq (0, 0)$. Then f is not continuous at $(0, 0)$ (cf. Example 2, Sect. 1.4.3). However for $h \neq 0$,

$$\frac{f(h, 0) - f(0, 0)}{h} = 0,$$

and therefore $\frac{\partial f}{\partial x}(0, 0)$ exists and is equal to zero. By symmetry, the same is true for the partial derivative with respect to y.

In this context, we need some hypothesis in addition to the existence of the partial derivatives at x, in order to insure the continuity of the function at x. The following proposition states a result of this type. Its proof, as well as other proofs in the sequel, apply the Mean Value Theorem (MVT) for functions of one real variable. We recall its statement for the reader's convenience.

Mean Value Theorem (MVT) for functions of one variable. *Let $f : [a, a+h] \to \mathbb{R}$ be continuous, and suppose f' exists in $(a, a + h)$.Then there exists $\theta \in (0, 1)$ such that*

$$f(a + h) - f(a) = h\, f'(a + \theta h).$$

The MVT will be usually applied under the stronger hypothesis that f is differentiable in $[a, a + h]$.

Proposition *Let f be a real valued function defined in a neighborhood $B(x, r) \subset \mathbb{R}^k$, $k > 1$. Suppose all the partial derivatives of f exist and are bounded in $B(x, r)$. Then f is continuous at x.*

Proof Let m_i be a bound for $|\frac{\partial f}{\partial x_i}|$ in the neighborhood $B(x, r)$, $i = 1, \ldots, k$, and let $m = (m_1, \ldots, m_k)$. Let $h \in \mathbb{R}^k$, $||h|| < r$. Define $h^0 = 0$, $h^k = h$, and

$$h^j = (h_1, \ldots, h_j, 0, \ldots, 0), \qquad 1 \leq j < k.$$

For all $j = 0, \ldots, k$, $||h^j|| \leq ||h|| < r$, hence $d(x + h^j, x) = ||h^j|| < r$, that is, $x + h^j \in B(x, r)$. Therefore $f(x + h^j)$ is well-defined, and

$$f(x + h) - f(x) = \sum_{j=1}^{k} [f(x + h^j) - f(x + h^{j-1})]. \tag{2.2}$$

Fix j, $1 \leq j \leq k$. Suppose $h_j \neq 0$, and define

$$F_j(t) := f(x + h^{j-1} + te^j)$$

for $t \in [0, h_j]$ (or $t \in [h_j, 0]$). By hypothesis, the derivative F_j' exists in the closed interval, and

$$F_j'(t) = \frac{\partial f}{\partial x_j}(x + h^{j-1} + te^j). \tag{2.3}$$

By the MVT for functions of one variable, there exists $\theta_j \in (0, 1)$ such that

$$F_j(h_j) - F_j(0) = h_j F_j'(\theta_j h_j). \tag{2.4}$$

Formula (2.4) is trivially true in case $h_j = 0$, and by (2.2)–(2.4),

$$f(x+h) - f(x) = \sum_{j=1}^{k} [F_j(h_j) - F_j(0)]$$
$$= \sum_j h_j \frac{\partial f}{\partial x_j}(x + h^{j-1} + \theta_j h_j e^j).$$

The partial derivatives are evaluated at points belonging to $B(x, r)$. Therefore their absolute value is $\leq m_j$. Hence by the Cauchy-Schwarz inequality,

$$|f(x+h) - f(x)| \leq \sum_j |h_j| m_j \leq ||m|| \, ||h||.$$

Given $\epsilon > 0$, let $0 < \delta < \min[r, \epsilon/(1 + ||m||)]$. Then if $||h|| < \delta$, we have $x + h \in B(x, r)$, and

$$|f(x+h) - f(x)| \leq \frac{||m||}{1 + ||m||}\epsilon < \epsilon.$$

This proves that f is continuous at the point x. □

Notation. We shall also use the symbolic vector notation

$$\nabla := (\frac{\partial}{\partial x_1}, \ldots, \frac{\partial}{\partial x_k}),$$

which operates on f as the multiplication of the formal vector ∇ by the scalar f, where $(\frac{\partial}{\partial x_j}) f := \frac{\partial f}{\partial x_j}$.

Example. Let $f(r, \phi, \theta) = r \cos\phi \sin\theta$, $r \geq 0$, $\phi \in [0, 2\pi]$, $\theta \in [0, \pi]$. Then

$$\nabla f(r, \phi, \theta) = (\cos\phi \sin\theta, \ -r\sin\phi \sin\theta, \ r\cos\phi \cos\theta)$$

at all points $(r, \phi, \theta) \in [0, \infty) \times [0, 2\pi] \times [0, \pi]$.

The Differential

Let f be a real valued function defined in a neighborhood $B(x, r)$ of the point $x \in \mathbb{R}^k$. The simplest non-constant functions on $\mathbb{R}^k$ are the *linear functions*. We wish to approximate the change of f at the given point $x \in \mathbb{R}^k$

$$\Delta_x f(h) := f(x+h) - f(x) \qquad (||h|| < r)$$

by a linear function of h.

Let us start by recalling some facts from Linear Algebra.

2.1.2 Definition The function $L : \mathbb{R}^k \to \mathbb{R}$ is a *linear function* (more often called a *linear functional*) if $L(\alpha x + \beta y) = \alpha L(x) + \beta L(y)$ for all scalars $\alpha, \beta \in \mathbb{R}$ and all vectors $x, y \in \mathbb{R}^k$.

We shall write Lx instead of $L(x)$ when L is a linear function.

Representation of linear functions. *The general form of a linear function L on $\mathbb{R}^k$ is*

$$Lh = h \cdot a,$$

where $a \in \mathbb{R}^k$ is uniquely determined.

The vector a is called the *representing vector of L*.

Proof Let $L : \mathbb{R}^k \to \mathbb{R}$ be a linear function. Denote $a_j = Le^j$ and define $a := (a_1, \ldots, a_k)$. Then for all $h \in \mathbb{R}^k$, we have by the linearity of L

$$Lh = L(\sum_j h_j e^j) = \sum_j h_j Le^j = \sum_j h_j a_j = h \cdot a.$$

On the other hand, for any $a \in \mathbb{R}^k$, the function $L : \mathbb{R}^k \to \mathbb{R}$ defined by $Lh = h \cdot a$ is linear, by the bilinearity property of the inner product. This means that $L : h \to h \cdot a$ is the general representation of linear functionals on $\mathbb{R}^k$.

The vector a in the representation $Lh = a \cdot h$ is uniquely determined, since if we also have $Lh = h \cdot b$ for some $b \in \mathbb{R}^k$ (for all $h \subset \mathbb{R}^k$), then

$$0 = Lh - Lh = h \cdot a - h \cdot b = h \cdot (a - b)$$

for all $h \in \mathbb{R}^k$. Taking in particular $h = a - b$, we obtain $(a - b) \cdot (a - b) = 0$, hence $a - b = 0$ by the positive definiteness of the inner product. □

A linear functional $L : h \in \mathbb{R}^k \to Lh \in \mathbb{R}$ is $O(||h||)$, that is, the ratios $\frac{|Lh|}{||h||}$ $(h \neq 0)$ are bounded. Indeed, if a is the representing vector of L, then by the Cauchy-Schwarz inequality,

$$\frac{|Lh|}{||h||} = \frac{|h \cdot a|}{||h||} \leq ||a||$$

for all $0 \neq h \in \mathbb{R}^k$.
This $O(||h||)$ property implies that L *is uniformly continuous on* $\mathbb{R}^k$. Indeed, fix $M > ||a||$, and let $\epsilon > 0$ be given. Choose $\delta = \epsilon/M$. Then whenever $h, h' \in \mathbb{R}^k$ satisfy $||h - h'|| < \delta$, we have

$$|Lh - Lh'| = |L(h - h')| \leq M\,||h - h'|| < M\delta = \epsilon.$$

Given f as above, we are interested in a linear function L (depending on x) which approximates the change $\Delta_x f$ for $||h|| < r$ in such a way that the "error"

$$\phi_x(h) := \Delta_x f(h) - Lh$$

is $o(||h||)$ when $h \to 0$. The latter property means that

$$\lim_{0 \neq h \to 0} \frac{\phi_x(h)}{||h||} = 0.$$

While the approximation Lh is $O(||h||)$, we want the error to be $o(||h||)$, which is a smaller order of magnitude in the sense that the relevant ratios tend to zero rather than being merely bounded.

When such a linear approximation exists, we shall say that f is *differentiable at the point* x. In this case, the above linear functional L is uniquely determined. Indeed, suppose $L' \neq L$ is also a linear functional satisfying the above requirement, let a, a' be the representing vectors of L, L' respectively, and let ϕ_x, ϕ'_x be the respective error functions. Then $a' \neq a$, and for $0 < t < r/||a - a'||$ and $h = t(a - a')$, we have $0 < ||h|| < r$, and

$$\begin{aligned} ||a - a'|| &= \frac{t(a - a') \cdot (a - a')}{t\,||a - a'||} = \frac{h \cdot (a - a')}{||h||} \\ &= \frac{Lh - L'h}{||h||} = \frac{\phi'_x(h) - \phi_x(h)}{||h||} \to 0 \end{aligned}$$

as $h \to 0$ (i.e., as $t \to 0$). Since the left hand side is a positive constant, we have reached a contradiction. This shows that $L' = L$.

The above discussion is formalized in the following

2.1.3 Definition 1. Let f be a real valued function defined in a neighborhood of the point x in $\mathbb{R}^k$. The function f is said to be *differentiable at the point* x if there exists a linear functional L on $\mathbb{R}^k$ and $r > 0$ such that

$$\phi_x(h) := f(x + h) - f(x) - Lh = o(||h||) \tag{2.5}$$

for all $h \in \mathbb{R}^k$ with norm smaller than r.

2. If f is differentiable at the point $x \in \mathbb{R}^k$, the linear functional L satisfying (2.5) is uniquely determined, and is called *the differential of f at the point x*, denoted $df(x)$ or $df|_x$.

The defining Eq. (2.5) of the differential may be written in the form

$$f(x+h) - f(x) = df|_x h + o(||h||) \qquad (||h|| < r). \tag{2.6}$$

Example 1. The case $k = 1$. By (2.5), the function $f : \mathbb{R} \to \mathbb{R}$ is differentiable at the point $x \in \mathbb{R}$ iff there exists a linear functional L on $\mathbb{R}$, say $Lh = ah$ for some $a \in \mathbb{R}$, such that

$$\frac{|f(x+h) - f(x) - ah|}{|h|} \to 0$$

as $h \to 0$. Equivalently,

$$\left|\frac{f(x+h) - f(x)}{h} - a\right| \to 0$$

as $h \to 0$. Thus f is differentiable at x iff the limit

$$\lim_{h \to 0} \frac{f(x+h) - f(x)}{h}$$

exists, i.e., iff the derivative $f'(x)$ exists, and in that case $a = f'(x)$, that is,

$$df|_x h = f'(x)h \qquad (h \in \mathbb{R}).$$

Example 2. Let $f = L + c$, where $L : \mathbb{R}^k \to \mathbb{R}$ is linear, and c is a real constant. Then for any $x \in \mathbb{R}^k$, $f(x+h) - f(x) = L(h)$. Therefore L is the wanted linear functional in the definition of differentiability, because

$$\phi_x(h) = f(x+h) - f(x) - L(h) = 0 = o(||h||).$$

Thus f is differentiable at all points $x \in \mathbb{R}^k$ and $df|_x = L$.

In particular, if $f = c$ (a constant function), then f is differentiable at all points $x \in \mathbb{R}^k$ and $df|_x = 0$ (the zero linear functional) for all x.

Example 3. Let $f : \mathbb{R}^3 \to \mathbb{R}$ be equal to 0 at 0 and

$$f(x) = \frac{x_1 \sin(x_2 x_3)}{||x||^a} \qquad x = (x_1, x_2, x_3) \in \mathbb{R}^3 \setminus \{0\},$$

where a is a real constant, $a < 2$.

We shall verify that f is differentiable at the point $x = 0$ and $df|_0$ is the zero linear functional (denoted 0), that is, $df|_0 h = 0$ for all $h \in \mathbb{R}^3$.

Since $|\sin t| \leq |t|$ for all $t \in \mathbb{R}$ and $|h_i| \leq ||h||$ for all $h = (h_1, h_2, h_3) \in \mathbb{R}^3$, we have (with $L = 0$)

$$\frac{|\phi_0(h)|}{||h||} = \frac{|h_1 \sin(h_2 h_3)|}{||h||^{a+1}}$$

$$\leq \frac{|h_1|\,|h_2|\,|h_3|}{||h||^{a+1}} \leq \frac{||h||^3}{||h||^{a+1}} = ||h||^{2-a} \to 0$$

as $h \to 0$, since $a < 2$. Thus Condition (2.5) is satisfied by f at the point $x = 0$, with L taken as the zero linear functional. This means that f is differentiable at 0 and $df|_0 = 0$.

Example 4. Let $f : \mathbb{R}^k \to \mathbb{R}$ be equal to $1/2$ at 0 and

$$f(x) = \frac{\sqrt{1 + ||x||^2} - 1}{||x||^2} \qquad x \in \mathbb{R}^k \setminus \{0\}.$$

We show that f is differentiable at 0 and $df|_0 = 0$, the zero linear functional on $\mathbb{R}^k$. Taking $L = 0$ in (2.5), we have

$$\phi_0(h) = \frac{\sqrt{1 + ||h||^2} - 1}{||h||^2} - \frac{1}{2} = \frac{1}{\sqrt{1 + ||h||^2} + 1} - \frac{1}{2}$$

$$= \frac{1 - \sqrt{1 + ||h||^2}}{1 + \sqrt{1 + ||h||^2}} = -\frac{||h||^2}{(1 + \sqrt{1 + ||h||^2})^2}.$$

Hence $\phi_0(h)/||h|| \to 0$ as $h \to 0$, that is, $\phi_0(h) = o(||h||)$ as desired.

The following theorem gives *sufficient conditions* for differentiability and a formula for the differential when the conditions are satisfied.

2.1.4 Theorem *Let f be a real valued function defined in the ball $B(x, r)$ in $\mathbb{R}^k$. Suppose $\nabla f : B(x, r) \to \mathbb{R}^k$ exists and is continuous at x. Then f is differentiable at x and*

$$df|_x h = h \cdot \nabla f|_x \qquad (h \in \mathbb{R}^k).$$

Note that under the conditions of the theorem, the representing vector of the linear functional $df|_x$ is $\nabla f|_x$.

Proof As in the proof of the proposition in Sect. 2.1.1, we consider the vectors $h^i \in \mathbb{R}^k$, $i = 0, \ldots, k$, given by $h^0 = 0$, $h^i = (h_1, \ldots, h_i, 0, \ldots, 0)$ for $i < k$, and $h^k = h := (h_1, \ldots, h_k)$. Then $||h^i|| \leq ||h||$, and therefore $f(x + h^i)$ is well defined for $||h|| < r$, and we have

$$f(x + h) - f(x) = \sum_{i=1}^{k} [f(x + h^i) - f(x + h^{i-1})]$$

$$= \sum_{i=1}^{k} [f(x + h^{i-1} + h_i e^i) - f(x + h^{i-1})]. \tag{2.7}$$

Suppose $h_i \neq 0$. The hypothesis of the theorem imply that for each $i = 1, \ldots, k$, the function $F_i(t) := f(x + h^{i-1} + te^i)$ is differentiable for $0 \leq t \leq h_i$ (or $h_i \leq t \leq 0$), and

$$F_i'(t) = \frac{\partial f}{\partial x_i}(x + h^{i-1} + te^i).$$

We apply to F_i the MVT for functions of one real variable. There exist $\theta_i \in (0, 1)$ such that

$$F_i(h_i) - F_i(0) = h_i F'(\theta_i h_i),$$

that is, the i-th summand in (2.7) is equal to

$$h_i \frac{\partial f}{\partial x_i}(x + h^{i-1} + \theta_i h_i e^i).$$

The same is trivially true if $h_i = 0$.

Denote

$$q_i(h) := \frac{\partial f}{\partial x_i}(x + h^{i-1} + \theta_i h_i e^i) - \frac{\partial f}{\partial x_i}(x),$$

$i = 1, \ldots, k$, and let $q(h) := (q_1(h), \ldots, q_k(h))$. Then by (2.7)

$$f(x + h) - f(x) = h \cdot \nabla f(x) + h \cdot q(h).$$

It remains to show that $\phi(h) := h \cdot q(h)$ is $o(||h||)$. By the Cauchy-Schwarz inequality, $|\phi(h)|/||h|| \leq ||q(h)||$ for $h \neq 0$. Since $h^{i-1} + \theta_i h_i e^i \to 0$ as $h \to 0$, the continuity of ∇f at x implies that $q_i(h) \to 0$ for all i, hence $||q(h)|| \to 0$ as $h \to 0$. □

On the other hand, if f is differentiable at x, we prove below that

2.1.5 Theorem *Let f be a real valued function defined in a neighborhood of the point x in $\mathbb{R}^k$ and differentiable at x. Then:*

(a) f is continuous at x.
(b) The directional derivatives $\frac{\partial f}{\partial u}(x)$ exist for all unit vectors u, and

$$\frac{\partial f}{\partial u}(x) = df|_x u. \tag{2.8}$$

In particular (for the unit vectors $u = e^i$), the partial derivatives of f exist at x,

$$\frac{\partial f}{\partial x_i}(x) = df|_x e^i, \tag{2.9}$$

and

$$df|_x h = h \cdot \nabla f|_x \qquad (h \in \mathbb{R}^k). \tag{2.10}$$

Proof (a) Using the notation of preceding sections, we have by the Cauchy-Schwarz inequality,

$$|f(x+h) - f(x)| \leq ||h||\,||a|| + |\phi(h)| \to 0$$

as $h \to 0$, because for $h \neq 0$,

$$|\phi(h)| = \frac{|\phi(h)|}{||h||}\,||h|| \to 0,$$

as $||h|| \to 0$, since $\phi(h) = o(||h||)$ by definition of differentiability.

(b) Let u be a unit vector in $\mathbb{R}^k$, and choose $h = tu$ with t real, $0 < |t| < r$. Then, with the notation of Sect. 2.1.1,

$$\left|\frac{F(t) - F(0)}{t} - df|_x u\right| = \left|\frac{f(x+tu) - f(x) - df|_x(tu)}{t}\right|$$
$$= \frac{|\phi(tu)|}{||tu||} \to 0$$

as $t \to 0$, since $\phi(h) = o(||h||)$. This proves (b). □

Examples.

1. Let f be as in the example in Sect. 2.1.1., and let u be the unit vector $u = (3/5,\ 0,\ 4/5)$. Then at all points (r, ϕ, θ), the directional derivative $\frac{\partial f}{\partial u}$ is given by

$$\frac{\partial f}{\partial u} = u \cdot \nabla f = \frac{\cos\phi}{5}[3\sin\theta + 4\cos\theta].$$

2. Let f be as in Example 4 of Sect. 2.1.3. We showed that f is differentiable at 0 and $df|_0 = 0$. By Theorem 2.1.5, the partial derivatives $\frac{\partial f}{\partial x_i}(0)$ exist and are equal to 0 for all $i = 1, \ldots, k$:

$$\frac{\partial f}{\partial x_i}(0) = df|_0 e^i = 0.$$

We show this directly from the definition of the partial derivative. For $h \neq 0$ real, we have

$$h^{-1}[f(0 + he^i) - f(0)] = h^{-1}[\frac{\sqrt{1+h^2} - 1}{h^2} - \frac{1}{2}].$$

Multiplying the numerator and denominator of the above fraction by $\sqrt{1+h^2}+1$, the numerator becomes $(1+h^2)-1=h^2$, and therefore the above expression is equal to

$$h^{-1}[\frac{1}{\sqrt{1+h^2}+1}-\frac{1}{2}]=(2h)^{-1}\frac{1-\sqrt{1+h^2}}{1+\sqrt{1+h^2}}$$

$$=(2h)^{-1}\frac{1-(1+h^2)}{(1+\sqrt{1+h^2})^2}=-\frac{h}{2(1+\sqrt{1+h^2})^2}\to 0$$

as $h\to 0$. This shows that $\frac{\partial f}{\partial x_i}(0)$ exists and is equal to 0.

It is clear from the properties of limits that linear combinations of two $o(||h||)$ (real valued) functions are $o(||h||)$, and that the product of two $O(||h||)$ functions is $o(||h||)$. This observation implies easily the following proposition, whose proof is left to the reader.

2.1.6 Proposition *Let f, g be real functions differentiable at a point $x\in\mathbb{R}^k$. Then*

(a) Linear combinations $af+bg$ are differentiable at x, and

$$d(af+bg)|_x=a\,df|_x+b\,dg|_x \qquad (a,\,b\in\mathbb{R}).$$

(b) The product fg is differentiable at x, and

$$d(fg)|_x=f(x)dg|_x+g(x)df|_x.$$

(c) If in addition $g(x)\neq 0$, then f/g is differentiable at x and

$$d(\frac{f}{g})|_x=\frac{g(x)df|_x-f(x)dg|_x}{g(x)^2}.$$

The functions $f_i(x)=x_i$, $i=1,\dots,k$, are linear, hence differentiable at every point $x\in\mathbb{R}^k$ (cf. Sect. 2.1.3, Example 2). Since constant functions are differentiable at every point (ibid), it follows from Proposition 2.1.6 that rational functions on $\mathbb{R}^k$ are differentiable at every point of their domain of definition.

2.1.7 Examples

Example 1. Let

$$f(x,y)=\frac{x^3+y^4}{x^2+y^2}$$

for $(x,y)\neq(0,0)$, and $f(0,0)=0$.

The graph of f over the set $E:=\{(x,y)\in\mathbb{R}^2;\ 0<x^2+y^2\leq 1$ is sketched in Fig. 2.1.

We have for $h\neq 0$ real

$$h^{-1}[f(h,0)-f(0,0)]=1, \qquad h^{-1}[f(0,h)-f(0,0)]=h.$$

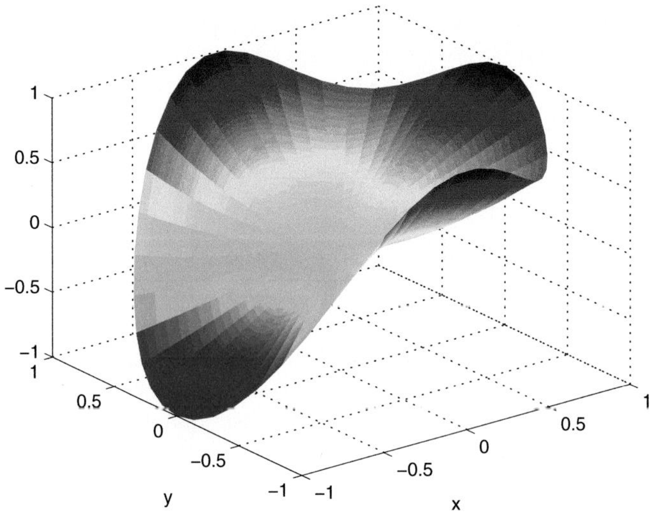

Fig. 2.1 $f(x, y) = \frac{x^3+y^4}{x^2+y^2}$, $(x, y) \in E$

Letting $h \to 0$, we conclude that $\nabla f|_{(0,0)} = (1, 0)$.

If f were differentiable at $(0, 0)$, we would have

$$df|_{(0,0)}(h, k) = (h, k) \cdot \nabla f|_{(0,0)} = (h, k) \cdot (1, 0) = h.$$

Therefore

$$\phi(h, k) = (h^2 + k^2)^{-1}(h^3 + k^4) - h.$$

However $\phi(h, k) \neq o(||(h, k)||)$. Indeed, on the line $k = h$, we have

$$\frac{\phi(h, h)}{||(h, h)||} = \frac{h^2 - h}{2\sqrt{2}\,|h|}.$$

As $h \to 0+$ and $h \to 0-$, the above expression has the two different limits $-1/(2\sqrt{2})$ and $1/(2\sqrt{2})$ respectively. We conclude that f is not differentiable at $(0, 0)$.

Example 2. Let $f(x, y) = x \sin(y^2/x)$ when $x \neq 0$ and $f(0, y) = 0$ for all y. We saw that f is continuous on $\mathbb{R}^2$. Its partial derivatives exist and are continuous on $E = \{(x, y) \in \mathbb{R}^2;\ x \neq 0\}$, and therefore f is differentiable at every point of E, by Theorem 2.1.4. Let $y \neq 0$. For $h \neq 0$, we have

$$h^{-1}[f(h, y) - f(0, y)] = \sin \frac{y^2}{h}.$$

The last expression has no limit as $h \to 0$, hence $f_x(0, y)$ does not exist. By Theorem 2.1.5, f is not differentiable at the points $(0, y)$ with $y \neq 0$.

Finally, consider the point $(0, 0)$. Since $f(h, 0) = f(0, h) = 0$, the differential ratios $h^{-1}[f(h, 0) - f(0, 0)]$ and $h^{-1}[f(0, h) - f(0, 0)]$ vanish for all $h \neq 0$, and therefore $\nabla f(0, 0) = 0$. Hence the zero linear functional is the candidate for $df|_{(0,0)}$. The corresponding error function $\phi := \phi_{(0,0)}$ is

$$\phi(h, k) = f(h, k) - f(0, 0) - df|_{(0,0)}(h, k) = f(h, k).$$

For $(h, k) \in E$,

$$\frac{|\phi(h, k)|}{||(h, k)||} = \frac{|f(h, k)|}{||(h, k)||}$$

$$= \left|\frac{\sin(k^2/h)}{(k^2/h)}\right| \frac{k^2}{||(h, k)||} \leq \frac{k^2}{||(h, k)||} \leq ||(h, k)|| \to 0$$

as $||(h, k)|| \to 0$. We used above the inequalities $|\sin t/t| \leq 1$ and $|k| \leq ||(h, k)||$. Since $\phi(0, k) = f(0, k) = 0 = o(||(h, k)||)$ trivially, we conclude that $\phi(h, k) = o(||(h, k)||)$ as $(h, k) \to (0, 0)$, that is, f is differentiable at $(0, 0)$, and $df|_{(0,0)}$ is the zero linear functional.

Example 3. Let $f : \mathbb{R}^3 \to \mathbb{R}$ be defined by

$$f(x) = \frac{\sin(x_1x_2) + \sin(x_1x_3) + \sin(x_2x_3)}{||x||}$$

for all $x = (x_1, x_2, x_3) \in E := \mathbb{R}^3 \setminus \{0\}$ and $f(0) = 0$. By Theorem 2.1.4 and Proposition 2.1.6, f is differentiable at every point of E.

We discuss below the behavior of f at the point $x = 0$.

Since $|\sin t| \leq |t|$ for all $t \in \mathbb{R}$ and $|x_i| \leq ||x||$, we have

$$|f(x) - f(0)| = |f(x)| \leq \frac{|x_1x_2| + |x_1x_3| + |x_2x_3|}{||x||} \leq 3||x|| \to 0$$

as $x \to 0$. Thus f is continuous at 0, so that f is continuous on $\mathbb{R}^3$, by the preceding observation.

We have $f(h, 0, 0) = f(0, h, 0) = f(0, 0, h) = 0$, and therefore, for real $h \neq 0$,

$$h^{-1}[f(0 + he^j) - f(0)] = h^{-1}f(he^j) = 0 \to 0$$

as $h \to 0$. This shows that the partial derivatives of f exist and are equal to zero at 0. Thus ∇f exists on all of $\mathbb{R}^3$, and $\nabla f|_0 = 0$. We prove now that f is *not differentiable* at 0. Assume f is differentiable at 0. Then by Theorem 2.1.5,

$$df|_0 h = h \cdot \nabla f|_0 = h \cdot 0 = 0$$

for all $h \in \mathbb{R}^3$.

Consider now the error function ϕ_0. We have for all $h \in E$

$$\frac{\phi_0(h)}{||h||} = \frac{f(h) - f(0) - df|_0 h}{||h||} = \frac{\sin(h_1h_2) + \sin(h_1h_3) + \sin(h_2h_3)}{||h||^2}.$$

For $h = t(1, 1, 1)$ with $0 \neq t \in \mathbb{R}$,

$$\frac{\phi_0(h)}{||h||} = \frac{3\sin(t^2)}{3t^2} \to 1$$

as $||h|| = |t|\sqrt{3} \to 0$. Thus $\phi_0(h) \neq o(||h||)$ as $||h|| \to 0$, contradicting the differentiability at 0 hypothesis on f.

This example shows that the requirement in Theorem 2.1.4 that ∇f be continuous at the point $x (= 0$ in our case) cannot be omitted.

2.1.8 Maximal directional derivative Under the hypothesis of Theorem 2.1.5, it follows from (2.8) to (2.10) in the statement of the theorem that for all unit vectors u,

$$\frac{\partial f}{\partial u}(x) = u \cdot \nabla f(x),$$

and consequently, by the Cauchy-Schwarz inequality,

$$\left|\frac{\partial f}{\partial u}(x)\right| \leq ||\nabla f(x)||. \tag{2.11}$$

If the gradient vector $\nabla f(x)$ is not zero, we may choose $u = u^*$, the unit vector in the gradient direction,

$$u^* := ||\nabla f(x)||^{-1} \nabla f(x).$$

We then have

$$\frac{\partial f}{\partial u^*}(x) = u^* \cdot \nabla f(x) = ||\nabla f(x)||.$$

By (2.11), this means that the directional derivatives of f at the given point x attain their maximum absolute value in the direction of the gradient of f at x:

$$\max_u \left|\frac{\partial f}{\partial u}(x)\right| = \frac{\partial f}{\partial u^*}(x). \tag{2.12}$$

The Chain Rule

2.1.9 Vector Valued Functions of One Variable

We consider here an $\mathbb{R}^k$-valued function $x(\cdot)$ of a *single* real variable t defined in a neighborhood $|t - t_0| < \delta$. The *derivative at* t_0, denoted $x'(t_0)$ or $\frac{dx}{dt}(t_0)$, is defined in the usual manner

$$x'(t_0) := \lim_{t \to t_0} \frac{x(t) - x(t_0)}{t - t_0} \tag{2.13}$$

whenever the limit exists. The existence of the limit (2.13) is equivalent to the existence of the corresponding limits for the component functions $x_i(\cdot)$ for all i, that is, to the existence of all the derivatives $x_i'(t_0)$, $i = 1, \ldots, k$, and we then have

$$x'(t_0) = (x_1'(t_0), \ldots, x_k'(t_0)). \tag{2.14}$$

We refer to (2.14) as *component-wise derivation.*

Denote $x^0 = x(t_0)$, and suppose that f is a real function defined in a neighborhood $B(x^0, r)$ of x^0 in $\mathbb{R}^k$. If $x(\cdot)$ has range contained in $B(x^0, r)$ for $|t - t_0| < \delta$, the (real-valued) function $f \circ x := f(x(\cdot))$ is well defined in $|t - t_0| < \delta$. The following theorem formalizes the *chain rule* for the derivative of this composite function.

2.1.10 Theorem *Let $x(\cdot) : (t_0 - \delta, t_0 + \delta) \to \mathbb{R}^k$, and denote $x^0 = x(t_0)$. Suppose*

(a) $x'(t_0)$ exists.
(b) the range of $x(\cdot)$ is contained in some ball $B(x^0, r)$, and $f : B(x^0, r) \to \mathbb{R}$ is differentiable at x^0.

Then the derivative of $f \circ x$ exists at t_0 and is equal to

$$df|_{x^0} x'(t_0). \tag{2.15}$$

By (2.10), we may write Formula (2.15) in the form

$$x'(t_0) \cdot \nabla f|_{x^0}, \tag{2.16}$$

that is,

$$\sum_{i=1}^{k} \frac{\partial f}{\partial x_i}(x(t_0)) \frac{dx_i}{dt}(t_0). \tag{2.17}$$

Formula (2.17) clearly generalizes the usual chain rule for functions of one variable.

Proof Denote $L = df|_{x^0}$. We have

$$f(x^0 + h) - f(x^0) = Lh + \phi(h) \tag{2.18}$$

for all $h \in \mathbb{R}^k$ with $||h|| < \eta$, where $\phi(h) = o(||h||)$. Let $\epsilon > 0$ be given, and set $\epsilon_1 = (||x'(t_0)|| + 1)^{-1}\epsilon$. There exists $0 < \eta_1 \leq \eta$ such that $|\phi(h)| \leq \epsilon_1 ||h||$ whenever $||h|| < \eta_1$. Since $x'(t_0)$ exists, $x(\cdot)$ is continuous at t_0, and therefore $h_t := x(t) - x(t_0)$ is an $\mathbb{R}^k$-vector such that $||h_t|| < \eta_1$ for $|t - t_0| < \delta_1 \leq \delta$. Hence for $t \neq t_0$ in this interval,

$$\left|\frac{\phi(h_t)}{t - t_0}\right| \leq \epsilon_1 \left|\left|\frac{h_t}{t - t_0}\right|\right| = \epsilon_1 \left|\left|\frac{x(t) - x(t_0)}{t - t_0}\right|\right| \to \epsilon_1 ||x'(t_0)||$$

as $t \to t_0$, since the derivative $x'(t_0)$ exists and the norm is continuous. Therefore there exists $0 < \delta_2 \leq \delta_1$ such that $|\phi(h_t)|/|t - t_0| < \epsilon_1(||x'(t_0)|| + 1) = \epsilon$ for $0 < |t - t_0| < \delta_2$. This proves that

$$\lim_{t \to t_0} \frac{\phi(h_t)}{t - t_0} = 0. \tag{2.19}$$

For $0 < |t - t_0| < \delta_1$, we may apply (2.18) to $h = h_t$, because $||h_t|| < \eta_1 \leq \eta$ for such t. Observe that $x^0 + h_t = x(t)$ and use the linearity of L to conclude that

$$\frac{f(x(t)) - f(x(t_0))}{t - t_0} = L\Big(\frac{x(t) - x(t_0)}{t - t_0}\Big) + \frac{\phi(h_t)}{t - t_0}.$$

By the continuity of the linear functional L, the existence of $x'(t_0)$, and (2.19), it follows that the limit of the left hand side exists as $t \to t_0$, that is, the derivative at t_0 of $f(x(\cdot))$ exists, and is equal to $Lx'(t_0)$. □

If $x(\cdot)$ is a function of several variables, we may apply the chain rule by fixing all variables except for one. Ordinary derivatives are then replaced by partial derivatives. We obtain thereby the general chain rule formulated in the next theorem.

2.1.11 Theorem *Suppose $x(\cdot) : \mathbb{R}^l \to \mathbb{R}^k$ is defined in a neighborhood of $u^0 \in \mathbb{R}^l$ and the partial derivatives $\frac{\partial x}{\partial u_j}(u^0)$ exist for $j = 1, \ldots, l$. Let f be a real valued function defined in a ball $B(x^0, r)$ in $\mathbb{R}^k$ containing the range of $x(\cdot)$ (where $x^0 := x(u^0)$), and differentiable at x^0. Then the partial derivatives of $F := f \circ x$ exist at u^0, and*

$$\frac{\partial F}{\partial u_j}\Big|_{u^0} = df|_{x^0}\Big(\frac{\partial x}{\partial u_j}\Big|_{u^0}\Big) \tag{2.20}$$

$$= \sum_{i=1}^{k} \frac{\partial f}{\partial x_i}\Big|_{x^0} \frac{\partial x_i}{\partial u_j}\Big|_{u^0}. \tag{2.21}$$

We may use matrix notation in (2.21). The *Jacobian matrix* of x with respect to u (or *of the map* $u \to x(u)$) is the matrix

$$\Big(\frac{\partial x}{\partial u}\Big) := \Big(\frac{\partial(x_1, \ldots x_k)}{\partial(u_1, \ldots, u_l)}\Big)$$

whose ij-entry is $\frac{\partial x_i}{\partial u_j}$. In case $k = l$, its determinant is called the *Jacobian* of the map $u \to x(u)$, and is denoted $\frac{\partial x}{\partial u}$ or

$$\frac{\partial(x_1, \ldots, x_k)}{\partial(u_1, \ldots, u_k)}.$$

We shall use the notation ∇_u and ∇_x to stress that the independent variable is u and x, respectively. With these notations, we can write (2.21) in the form

$$\nabla_u F\Big|_{u^0} = \nabla_x f\Big|_{x^0} \left(\frac{\partial x}{\partial u}\right)\Big|_{u^0}. \tag{2.22}$$

On the right hand side of (2.22), the $(1 \times k)$-matrix $\nabla_x f$ (at x^0) is multiplied by the $k \times l$ Jacobian matrix (at u^0). The result is the $(1 \times l)$-matrix $\nabla_u F|_{u^0}$.

2.1.12 Example Let f be defined for $(x, y) \neq (0, 0)$ in $\mathbb{R}^2$ by

$$f(x, y) = \log(x^2 + y^2).$$

Here log denotes the *natural* logarithm.

The map

$$(r, \phi) \in I := (0, \infty) \times [0, 2\pi) \to (x, y) = (r \cos \phi, r \sin \phi)$$

which transforms polar coordinates into Cartesian coordinates in the plane, gives rise to the composite function on I

$$F(r, \phi) := f(r \cos \phi, r \sin \phi) = 2 \log r.$$

Its gradient with respect to $(r, \phi) \in I$ is $(2/r, 0)$.

On the other hand, if we use Formula (2.21) to calculate the partial derivatives of the composite function F, we obtain

$$\frac{\partial F}{\partial r} = \frac{2x}{x^2 + y^2} \cos \phi + \frac{2y}{x^2 + y^2} \sin \phi = \frac{2r \cos^2 \phi + 2r \sin^2 \phi}{r^2} = \frac{2}{r},$$

and

$$\frac{\partial F}{\partial \phi} = \frac{2x}{x^2 + y^2}(-r \sin \phi) + \frac{2y}{x^2 + y^2}(r \cos \phi) = 0,$$

in accordance with the previous direct calculation.

The Differential of a Vector Valued Function

We shall discuss in this section the differential of a *vector valued* function

$$f : \mathbb{R}^k \to \mathbb{R}^l,$$

defined in a neighborhood of a point $x^0 \in \mathbb{R}^k$.

Such a function is also called a *vector field*.

We recall that f is continuous at x^0 iff all its component functions f_i are continuous at x^0.

2.1.13 Definition The function $f : R^k \to \mathbb{R}^l$ defined in a neighborhood of x^0 is *differentiable* at x^0 if there exists a *linear map*

$$L : \mathbb{R}^k \to \mathbb{R}^l$$

such that

$$||f(x^0 + h) - f(x^0) - Lh|| = o(||h||) \tag{2.23}$$

for all $h \in \mathbb{R}^k$ with $||h|| < r$ (for a suitable $r > 0$).

The norm on the left hand side is the $\mathbb{R}^l$ Euclidean norm and the norm on the right hand side is the $\mathbb{R}^k$ norm.

2.1.14 Theorem *An $\mathbb{R}^l$-valued function f is differentiable at a point $x^0 \in \mathbb{R}^k$ iff all its component functions f_i $(i = 1, \dots, l)$ are differentiable at that point. In this case, the linear map L in (2.23) is uniquely determined, and is called the differential of f at x^0, denoted $df|_{x^0}$. For any $h \in \mathbb{R}^k$, the vector $df|_{x^0}h$ has the components*

$$(df|_{x^0}h)_i = df_i|_{x^0}h \qquad (i = 1, \dots, l).$$

Proof If $L : \mathbb{R}^k \to \mathbb{R}^l$ is a linear map, define $L_i : \mathbb{R}^k \to \mathbb{R}$, $i = 1, \dots, l$, by

$$L_i h = (Lh)_i, \tag{2.24}$$

where the right hand side of (2.24) denotes the i-th component of the vector Lh. Clearly L_i are linear functionals on $\mathbb{R}^k$.

If $f : \mathbb{R}^k \to \mathbb{R}^l$ is differentiable at x^0 and L is as in (2.23), it follows from the properties of limits that for all $i = 1, \dots, l$,

$$|f_i(x^0 + h) - f_i(x^0) - L_i h| = o(||h||), \tag{2.25}$$

that is, f_i is differentiable at x^0, and its differential at x^0 is uniquely determined as

$$df_i|_{x^0} = L_i. \tag{2.26}$$

Conversely, if f_i is differentiable at x^0 for all $i = 1, \dots, l$, and $L_i := df_i|_{x^0}$, we *define* the linear map $L : \mathbb{R}^k \to \mathbb{R}^l$ by (2.24), that is, the right hand side of (2.24) is defined by the left hand side. Clearly L is a linear map of $\mathbb{R}^k$ into $\mathbb{R}^l$, and (2.23) follows from (2.25). Thus f is differentiable at x^0.

Since the linear functionals L_i are uniquely determined, as we have seen before for real valued functions, the same is true for L, and $(Lh)_i = L_i h = df_i|_{x^0}h$ for all $h \in \mathbb{R}^k$. □

Observe that

$$
\begin{aligned}
(df|_{x^0}h)_i &= df_i|_{x^0}h = h \cdot \nabla f_i|_{x^0} \\
&= \sum_{j=1}^{k} h_j \frac{\partial f_i}{\partial x_j}\Big|_{x^0}
\end{aligned}
\tag{2.27}
$$

for all $i = 1, \ldots, l$. In matrix notation, we can write (2.27) in the form

$$
(df|_{x^0}h)^t = \Big(\frac{\partial f}{\partial x}\Big)\Big|_{x^0} h^t. \tag{2.28}
$$

The superscript t denotes the transpose operation on matrices: the column matrix on the left hand side is the product of the $l \times k$ Jacobian matrix of f with respect to x

$$
(\frac{\partial f}{\partial x}) := \Big(\frac{\partial(f_1, \ldots, f_l)}{\partial(x_1, \ldots, x_k)}\Big)
$$

evaluated at x_0, by the $k \times 1$ (column) matrix h^t.

Note also the vector relation

$$
(df|_{x^0})h = \sum_{j=1}^{k} h_j \frac{\partial f}{\partial x_j}\Big|_{x^0}, \tag{2.29}
$$

for all $h \in \mathbb{R}^k$.

Example. Let $I = (0, \infty) \times (0, 2\pi) \times (0, \pi)$ and let $f : I \to \mathbb{R}^3$ be defined by

$$
f : (r, \phi, \theta) \in I \to (r\cos\phi\sin\theta,\ r\sin\phi\sin\theta,\ r\cos\theta).
$$

The Jacobian matrix has the rows ∇f_i, that is,

$$
(\cos\phi\sin\theta,\ -r\sin\phi\sin\theta,\ r\cos\phi\cos\theta),
$$

$$
(\sin\phi\sin\theta,\ r\cos\phi\sin\theta,\ r\sin\phi\cos\theta),
$$

and

$$
(\cos\theta,\ 0,\ -r\sin\theta).
$$

We use (2.28) to calculate the differential $df|_p$ at any point $p \in I$. For any $h \in \mathbb{R}^3$, $df|_p h$ is the column with the i-th component $h \cdot \nabla f_i|_p$.

2.1.15 Exercises

1. Prove Proposition 2.1.6.
2. Let $f(x, y) = \frac{x^2y^2}{x^4+y^4}$ for $(x, y) \neq (0, 0)$ and $f(0, 0) = 0$.

(a) Show that for any unit vector $u = (u_1, u_2)$,

$$\frac{\partial f}{\partial u}(0, 0) = 0.$$

(b) Show that f is not continuous (hence not differentiable!) at $(0, 0)$.

3. Let g be a real function of one real variable, differentiable at some point $a \neq 0$. Let $x_0 \in \mathbb{R}^k$ be such that $||x_0|| = a$, and define $f(x) = g(||x||)$ for x in a neighborhood of x_0. Prove:

   (a) $\frac{\partial f}{\partial u}(x_0) = [g'(a)/a]\, u \cdot x_0$ for any unit vector u in $\mathbb{R}^k$.
   (b) $\max_u |\frac{\partial f}{\partial u}(x_0)| = |g'(a)|$, and the maximum is attained when $u = x_0/a$.

4. Let $u_j = (\cos j\pi/2, \sin j\pi/2) \in \mathbb{R}^2$, $j = 0, \ldots, 3$. Let $f : \mathbb{R}^2 \to \mathbb{R}$ be differentiable at some point. Calculate the directional derivatives $\frac{\partial f}{\partial u_j}$ at that point (in terms of the partial derivatives of f).
5. Let $f : \mathbb{R}^k \to \mathbb{R}^l$ be differentiable at x, and let $g : \mathbb{R}^l \to \mathbb{R}^m$ be differentiable at $f(x)$. Prove that $g \circ f : \mathbb{R}^k \to \mathbb{R}^m$ is differentiable at x and

$$d(g \circ f)\Big|_x = dg\Big|_{f(x)} \circ df\Big|_x. \tag{2.30}$$

6. Let $f : \mathbb{R}^k \to \mathbb{R}^m$ be defined by

$$f(x) = (\sum_i x_i, \sum_i x_i^2, \ldots, \sum_i x_i^m) \qquad (x \in \mathbb{R}^k).$$

   Given $x \in \mathbb{R}^k$, what is the linear map $df(x) : \mathbb{R}^k \to \mathbb{R}^m$?
7. Let $f : \mathbb{R}^3 \to \mathbb{R}$ be defined by

$$f(x, y, z) = \log(x^2 + y^2 + 4z^2) \qquad (x, y, z) \neq (0, 0, 0).$$

   Let $g : \mathbb{R}^3 \to \mathbb{R}^3$ be defined by

$$g(r, \phi, z) = (r \cos\phi,\ r \sin\phi,\ z) \qquad (r > 0,\ 0 \leq \phi \leq 2\pi, z \in \mathbb{R}).$$

   (a) Find the partial derivatives of $f \circ g$ with respect to r, ϕ, z for $r^2 + z^2 > 0$ and $0 \leq \phi \leq 2\pi$, using the chain rule.
   (b) Express $f \circ g$ explicitly as a function of r, ϕ, z, and verify the results of Part (a) by direct partial differentiation.

8. Let $f : \mathbb{R}^2 \to \mathbb{R}$ be differentiable at the point $(x, y) \neq (0, 0)$. Define $g(r, \theta) = f(r \cos\theta,\ r \sin\theta)$. Prove the formula

$$||\nabla f|_{(x,y)}||^2 = g_r^2 + r^{-2} g_\theta^2, \tag{2.31}$$

where $x = r\cos\theta,\ y = r\sin\theta\ (r > 0, 0 \le \theta \le 2\pi)$.

9. Prove or disprove the differentiability at (0, 0) of the following functions $f : \mathbb{R}^2 \to \mathbb{R}$:

   (a) $f(x, y) = \frac{x^3 - y^2}{||(x,y)||}$ for $(x, y) \neq (0, 0)$ and $f(0, 0) = 0$.

   (b) $f(x, y) = (xy)^{2/3}$.

10. Suppose the real valued function f is defined in a neighborhood of (0, 0) in $\mathbb{R}^2$ and satisfies the following conditions:

    (i) f is differentiable at (0, 0);

    (ii) $\lim_{x\to 0} \frac{f(x,x) - f(x,-x)}{x} = 1$.

Find $f_y(0, 0)$.

2.2 Higher Derivatives

Mixed Derivatives

Let D be an open subset of $\mathbb{R}^k$. If the function $f : D \to \mathbb{R}$ has partial derivatives with respect to all the variables x_i in D, these derivatives are themselves well-defined functions on D. If *their* derivatives exist in D, the latter functions are called partial derivatives of f of second order, denoted

$$\frac{\partial^2 f}{\partial x_j \partial x_i} := \frac{\partial}{\partial x_j}\Big(\frac{\partial}{\partial x_i} f\Big). \tag{2.32}$$

This is the composition of the operators $\frac{\partial}{\partial x_j}$ and $\frac{\partial}{\partial x_i}$ operating on f. For $i = j$, we denote the second order derivative (2.32) by $\frac{\partial^2 f}{\partial x_i^2}$. This is the result of the action of the operator $(\frac{\partial}{\partial x_i})^2$ on f. Multiplication of operators is defined as their composition. Linear combinations of (linear) operators L_j on a vector space are defined on their common domain V by

$$(\sum c_j L_j) f = \sum c_j L_j f \qquad c_j \in \mathbb{R},\ f \in V.$$

2.2.1 Examples

1. The *Laplace operator* or *Laplacian* Δ is defined by

$$\Delta := \nabla \cdot \nabla = \sum_i \Big(\frac{\partial}{\partial x_i}\Big)^2$$

   on the domain V consisting of all real functions f for which the second order derivatives $\frac{\partial^2 f}{\partial x_i^2}$ $(i = 1, \ldots, k)$ exist on some open set in $\mathbb{R}^k$.

2. Let $u : \mathbb{R}^k \to \mathbb{R}$ be defined by

$$u(x) = ||x||^2 := \sum_i x_i^2.$$

Clearly

$$\frac{\partial u}{\partial x_i} = 2x_i \qquad i = 1, \ldots, k.$$

For any positive number p, the function $f := u^{-p}$ on $\Omega = \mathbb{R}^k \setminus \{0\}$ has the derivatives

$$\frac{\partial f}{\partial x_i} = -pu^{-p-1}2x_i \qquad i = 1, \ldots, k,$$

and therefore

$$\frac{\partial^2 f}{\partial x_i^2} = -p(-p-1)u^{-p-2}4x_i^2 - 2pu^{-p-1}.$$

Summing over all i, we obtain

$$\Delta f = 4p(p+1)u^{-p-1} - 2pk\,u^{-p-1} = 2p(2p+2-k)u^{-p-1}.$$

In particular, for $2p = k - 2$, the function

$$f(x) = u^{-p} = \left(\sum x_i^2\right)^{-p}$$

is a solution of the *Laplace equation*

$$\Delta f = 0$$

in Ω. In case $k = 2$, this gives the trivial solution $f(x) = 1$ identically. For all $k > 2$, this is a non-trivial solution. A similar calculation shows that $f = \log u$ is a solution of the Laplace equation in Ω in case $k = 2$.

Inductively, one defines the *mixed derivatives of order* n of f by

$$\frac{\partial^n}{\partial x_{i_1} \ldots \partial x_{i_n}} f := \frac{\partial}{\partial x_{i_1}} \cdots \frac{\partial}{\partial x_{i_n}} f,$$

where the product of the operators is their composition, and $1 \le i_1, \ldots, i_n \le k$ (provided the relevant derivatives exist). The derivative of order 0 of f is f itself, by definition. Given a non-negative integer n, *we say that f is of class C^n in D (or is a C^n-function in D) if it has continuous mixed derivatives of all orders less than or equal to n.*

Under suitable conditions on f, the operators $\frac{\partial}{\partial x_i}$ and $\frac{\partial}{\partial x_j}$ $(i \neq j)$ *commute* when acting on f.

2.2.2 Theorem *Let f be a real valued function of class C^1 in a neighborhood D of the point x^0 in $\mathbb{R}^k$, $k \geq 2$. Suppose that for a certain pair of distinct indices $i, j \in \{1, \ldots, k\}$, the mixed derivative $\frac{\partial^2 f}{\partial x_j \, \partial x_i}$ exists in D and is continuous at x^0. Then $\frac{\partial^2 f}{\partial x_i \, \partial x_j}\Big|_{x^0}$ exists and coincides with $\frac{\partial^2 f}{\partial x_j \, \partial x_i}\Big|_{x^0}$.*

Proof Since only two distinct indices are involved in the statement of the theorem, it suffices to prove the theorem for a function f of two variables, which will be denoted x, y for simplicity. Accordingly, the point x^0 is now (x_0, y_0). We shall also use the simpler notation f_x and f_y for the partial derivatives of f with respect to x and y (respectively). Observe that in accordance with this notation,

$$f_{xy} := (f_x)_y = \frac{\partial^2 f}{\partial y \, \partial x}.$$

The (open) neighborhood D of (x_0, y_0) contains an open square centered at (x_0, y_0), and we may assume without loss of generality that

$$D = \{(x, y) \in \mathbb{R}^2;\ |x - x_0| < a,\ |y - y_0| < a\}.$$

The points $(x, y_0 + t)$ are in D for $|t| < a,\ |x - x_0| < a$. Therefore the functions

$$g_t(x) := t^{-1}[f(x, y_0 + t) - f(x, y_0)] \qquad (0 < |t| < a)$$

are well defined C^1-functions of x in the interval $|x - x_0| < a$, and

$$g_t'(x) = t^{-1}[f_x(x, y_0 + t) - f_x(x, y_0)]$$

in the above ranges of x and t. We apply the MVT for functions of one real variable to the function $g_t(\cdot)$ in the interval $[x_0, x_0 + s]$ (or $[x_0 + s, x_0]$, $0 < |s| < a$). There exists $\theta \in (0, 1)$ (depending on s and t) such that

$$s^{-1}[g_t(x_0 + s) - g_t(x_0)] = g_t'(x_0 + \theta s)$$

$$= t^{-1}[f_x(x_0 + \theta s,\ y_0 + t) - f_x(x_0 + \theta s,\ y_0)]. \tag{2.33}$$

Since the derivatives $f_{xy}(x_0 + \theta s,\ y)$ exist for $y \in [y_0,\ y_0 + t]$ (or $y \in [y_0 + t,\ y_0]$), the MVT for functions of one variable implies the existence of $\theta' \in (0, 1)$ (depending on s, t and θ) such that the expression in (2.33) is equal to

$$f_{xy}(x_0 + \theta s,\ y_0 + \theta' t). \tag{2.34}$$

Since f_{xy} is continuous at (x_0, y_0), it follows from (2.33) to (2.34) that

$$\lim_{(s,t) \to (0,0)} s^{-1}[g_t(x_0 + s) - g_t(x_0)] = f_{xy}(x_0, y_0). \tag{2.35}$$

However the expression on which we applied the lim operation is equal to

$$s^{-1}\Big\{t^{-1}[f(x_0+s,\ y_0+t)-f(x_0+s,\ y_0)]-t^{-1}[f(x_0,\ y_0+t)-f(x_0,\ y_0)]\Big\}.$$

For each fixed s, $0<|s|<a$, the limit as $t\to 0$ of the above expression exists because f_y exists in D, and is equal to

$$s^{-1}\{f_y(x_0+s,\ y_0)-f_y(x_0,\ y_0)\}.$$

Since the limit in (2.35) (as $(s,t)\to(0,0)$) exists, it follows that the (iterated) limit $\lim_{s\to 0}(\lim_{t\to 0})$ of the above expression exists, that is, $f_{yx}(x_0,\ y_0)$ exists, and is equal to the limit as $(s,t)\to(0,0)$ of the same expression, which is $f_{xy}(x_0,\ y_0)$, by (2.35). □

We used at the end of the preceding proof the following statement concerning the equality of the double limit and the iterated limit of a function at a (limit) point $(x_0,\ y_0)$ of its domain of definition:

2.2.3 Proposition *Let* $f:(x,y)\in\mathbb{R}^2\to\mathbb{R}$ *be defined for*

$$0<|x-x_0|,\ |y-y_0|<r.$$

Suppose:

$$\lim_{(x,y)\to(x_0,\ y_0)} f(x,y)=A; \tag{2.36}$$

and

$$\lim_{y\to y_0} f(x,y)=g(x) \tag{2.37}$$

for each x such that $0<|x-x_0|<r$.
Then

$$\lim_{x\to x_0} g(x)=A. \tag{2.38}$$

Thus, under the stated assumptions,

$$\lim_{x\to x_0}\Big(\lim_{y\to y_0} f(x,y)\Big)=\lim_{(x,y)\to(x_0,\ y_0)} f(x,y). \tag{2.39}$$

Proof Given $\epsilon>0$, there exists $\delta\in(0,r)$ such that $|f(x,y)-A|<\epsilon/2$ for all (x,y) such that $|x-x_0|,\ |y-y_0|<\delta$ (by (2.36)). By (2.37), for each x such that $|x-x_0|<\delta$, there exists a real y_x depending on x, such that $|y_x-y_0|<\delta$ and $|f(x,y_x)-g(x)|<\epsilon/2$. Hence

$$|g(x)-A|\le|g(x)-f(x,y_x)|+|f(x,y_x)-A|<\epsilon$$

whenever $|x-x_0|<\delta$. □

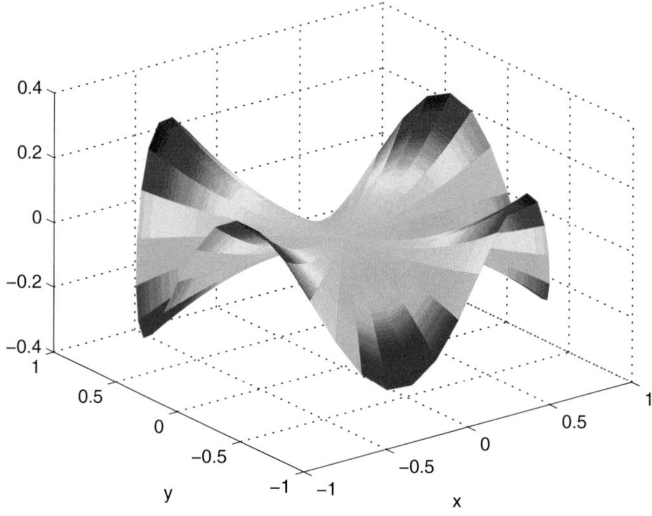

Fig. 2.2 $f(x, y) = xy\,\frac{x^2-y^2}{x^2+y^2}$, $(x, y) \in E$

Example. Let $f : \mathbb{R}^2 \to \mathbb{R}$ vanish at (0, 0) and

$$f(x, y) = xy\frac{x^2 - y^2}{x^2 + y^2} \qquad (x, y) \neq (0, 0).$$

The graph of f over the set $E := \{(x, y);\ 0 < x^2 + y^2 \leq 1\}$ is sketched in Fig. 2.2.

Write

$$f(x, y) = xy - 2\frac{xy^3}{x^2 + y^2} \qquad (x, y) \neq (0, 0).$$

A simple calculation yields

$$f_x(x, y) = y - 2\frac{y^5 - y^3x^2}{(x^2 + y^2)^2} \qquad (x, y) \neq (0, 0).$$

We have for $h \neq 0$ $h^{-1}[f(h, 0) - f(0, 0)] = 0$, hence $f_x(0, 0) = 0$.

Since $f(y, x) = -f(x, y)$ on $\mathbb{R}^2$, changing the roles of x and y yields to the identity

$$f_y(x, y) = -x + 2\frac{x^5 - x^3y^2}{(x^2 + y^2)^2} \qquad (x, y) \neq (0, 0)$$

and $f_y(0, 0) = 0$.

For real $h \neq 0$,

$$\frac{f_x(0,h) - f_x(0,0)}{h} = \frac{h - 2\frac{h^5}{h^4}}{h} = -1 \to -1$$

as $h \to 0$, hence $f_{xy}(0,0) = -1$.

Similarly,

$$\frac{f_y(h,0) - f_y(0,0)}{h} = \frac{-h + 2\frac{h^5}{h^4}}{h} = 1 \to 1$$

as $h \to 0$, hence $f_{yx}(0,0) = 1$.

As a rational function, the function f is certainly of class C^2 in $\mathbb{R}^2 \setminus \{(0,0)\}$ and we saw above that $f_{xy}(0,0)$ and $f_{yx}(0,0)$ both exist but are not equal. This shows that the continuity at x^0 hypothesis of the mixed derivative cannot be omitted in Theorem 2.2.2.

Taylor's Theorem

2.2.4 Definition Given $h \in \mathbb{R}^k$ and $x \in \mathbb{R}^k$, the operator $h \cdot \nabla|_x : f \to \mathbb{R}$ is defined by

$$(h \cdot \nabla)|_x f := h \cdot (\nabla f|_x) = \sum_{i=1}^{k} h_i \frac{\partial f}{\partial x_i}\Big|_x,$$

where the functions f are such that $\nabla f|_x$ exists.

Formally, $h \cdot \nabla|_x$ is the inner product of the vector h and the symbolic vector $\nabla|_x$. Note that if f is differentiable at the point x, then $\nabla f|_x$ exists and

$$(h \cdot \nabla)|_x f = df|_x h \qquad (h \in \mathbb{R}^k). \tag{*}$$

If f is a function of class C^1 in an open set $D \subset \mathbb{R}^k$ containing the line segment

$$\{x(t) := x + th;\ 0 \leq t \leq 1\},$$

then $F := f \circ x$ is well defined on the interval $0 \leq t \leq 1$: $F(t) := f(x + th)$. Since $x'(t) = h$, it follows from Theorems 2.1.4 and 2.1.10 that $F'(t)$ exists for all $t \in [0,1]$ and

$$F'(t) = h \cdot \nabla f|_{x(t)}.$$

Using the operator notation, we have

$$\frac{d}{dt}|_t F = (h \cdot \nabla)|_{x(t)} f \qquad (t \in [0,1]). \tag{2.40}$$

The action of the operator d/dt on F at t coincides with the action of the operator $h \cdot \nabla$ on f at $x(t)$, for each $t \in [0,1]$.

In case f is of class C^n in D, the repeated actions of d/dt on F and on its derivatives at t coincides consequently with the corresponding repeated actions of the operator $h \cdot \nabla$ on f at $x(t) := x + th$, up to n repetitions. Hence

$$(d/dt|_t)^j F = (h \cdot \nabla|_{x(t)})^j f \tag{2.41}$$

for all $j \le n$ and $t \in [0, 1]$.

It follows from Theorem 2.2.2 that

$$\begin{aligned}(h \cdot \nabla)^j &= \Big(h_1 \frac{\partial}{\partial x_1} + \cdots + h_k \frac{\partial}{\partial x_k}\Big)^j \\ &= \sum_{i_1+\ldots+i_k=j} \frac{j!}{i_1!\ldots i_k!} h_1^{i_1} \cdots h_k^{i_k} \frac{\partial^j}{\partial x_1^{i_1} \ldots \partial x_k^{i_k}}\end{aligned} \tag{2.42}$$

as operators on C^n-functions in D, for all $j \le n$. The indices $i_1, \ldots, i_k$ are non-negative integers, varying over all possible values with sum j.

In particular, for $k = 2$, denoting by (x, y) the points of $\mathbb{R}^2$, and letting $h = (s, t) \in \mathbb{R}^2$, we have

$$(h \cdot \nabla)^j = \sum_{i=0}^{j} \binom{j}{i} s^i t^{j-i} \frac{\partial^j}{\partial x^i \partial y^{j-i}}. \tag{2.43}$$

The special case of (2.41) and (2.42) for $j = 2$ will be used in the study of extrema of real valued C^2 functions. We have

$$F''(t) = (h \cdot \nabla)^2 f|_{x(t)} = \sum_{i,j=1}^{k} \frac{\partial^2 f}{\partial x_i \partial x_j}\Big|_{x(t)} h_i h_j. \tag{2.44}$$

The quadratic form on the right hand side of (2.44) can be written in matrix notation as $h\, H\, h^t$, where the column vector h^t is the transpose of the row vector h, and H is the symmetric matrix with ij-entry

$$H_{ij} = \frac{\partial^2 f}{\partial x_i \partial x_j}\Big|_{x(t)}. \tag{2.45}$$

The square matrix with entries equal to the mixed derivatives of second order of f is called *the Hessian matrix of* f. Thus H is the Hessian matrix of f evaluated at the point $x(t) := x + th$.

Using (2.41) and the Taylor formula for the function F of the single real variable t, we obtain Taylor's theorem for functions on $\mathbb{R}^k$.

2.2.5 Theorem *Let f be a real function of class C^{n+1} in the open set $D \subset \mathbb{R}^k$ (where n is a non-negative integer). Suppose D contains the line segment joining the points x and $x+h$ of D. Then there exists $\theta \in (0,1)$ such that*

$$f(x+h) = \sum_{j=0}^{n} \frac{(h \cdot \nabla)^j}{j!} f(x) + \frac{(h \cdot \nabla)^{n+1}}{(n+1)!} f(x+\theta h). \tag{2.46}$$

Formula (2.46) is *Taylor's formula* for functions of several variables. The first sum on the right hand side is a polynomial of degree at most n in the variables $h_1, \ldots, h_k$. The second term is the so-called *remainder in Lagrange form.*

Proof Using the previous notations, F is a C^{n+1}-function of the single variable t in $[0,1]$. By Taylor's theorem for functions of one variables, there exists $\theta \subset (0,1)$ such that

$$F(1) = \sum_{j=0}^{n} \frac{F^{(j)}(0)}{j!} 1^j + \frac{F^{(n+1)}(\theta)}{(n+1)!} 1^{n+1}.$$

Formula (2.46) follows now from (2.41) in Sect. 2.2.4 and the relation $F(t) := f(x+th)$. □

The special case of Taylor's theorem for $n = 0$ is the so-called *Mean Value Theorem for functions on* $\mathbb{R}^k$. We state it separately for later reference:

2.2.6 Theorem *Let f be a real valued function of class C^1 in an open subset D of $\mathbb{R}^k$. Suppose D contains the line segment joining the points x and $x+h$ in D. Then there exists $\theta \in (0,1)$ such that*

$$f(x+h) = f(x) + h \cdot \nabla f(x+\theta h).$$

2.2.7 Other Forms of the Remainder We used above the Lagrange form of the remainder in Taylor's formula for $F : [0,1] \to \mathbb{R}$. Other forms of the remainder (Cauchy, etc.) will produce corresponding forms of the remainder in Taylor's formula for $f : D \subset \mathbb{R}^k \to \mathbb{R}$. See Exercise 1 in Sect. 2.2.14.

We shall now apply Taylor's formula for $n = 1$ to obtain sufficient conditions for local extrema of C^1-functions of several variables.

Local Extrema

2.2.8 Definition Let f be a real valued function defined in a neighborhood D of the point x^0 in $\mathbb{R}^k$. The point x^0 is a *local minimum* (*maximum*) of f if $f(x) \geq f(x^0)$ ($f(x) \leq f(x^0)$, respectively) for all x in a neighborhood of x^0. If x^0 is either a local minimum or a local maximum of f, it is called a *local extremum* of f. The local extremum is *strict* if the above inequalities are strict inequalities for all $x \neq 0$ in a neighborhood of x^0.

Obviously, x^0 is a local maximum of f iff it is a local minimum of $-f$.

We begin with a *necessary condition* for a local extremum.

2.2.9 Theorem *Let f be a real valued function of class C^1 in a neighborhood of $x^0 \in \mathbb{R}^k$. Suppose x^0 is a local extremum for f. Then $\nabla f(x^0) = 0$.*

Proof For each given index $i \in \{1, \ldots, k\}$, fix $x_j = x_j^0$ for all indices $j \neq i$. The function f on these points is a function of the single variable x_i, of class C^1 in an (open) neighborhood of x_i^0, which has an extremum at the point x_i^0. By the known one-dimensional result, the derivative of this function vanishes at x_i^0, that is,

$$\frac{\partial f}{\partial x_i}(x^0) = 0 \tag{2.47}$$

for all $i = 1, \ldots, k$,. □

If ∇f exists in an open set $D \subset \mathbb{R}^k$, the points $x^0 \in D$ such that $\nabla f(x^0) = 0$ are called the *critical points* of f in D. By Theorem 2.2.9, if f is of class C^1 in D, then its local extrema in D are critical points of f. The k scalar equations (2.47) defining critical points x^0 may be used to find the k unknown coordinates x_i^0.

The next theorem gives *sufficient conditions* for a critical point x^0 to be a (strict) local minimum. The condition involves the mixed second derivatives of f at x^0.

We shall need some facts about real quadratic forms on $\mathbb{R}^k$. Given a symmetric $(k \times k)$-matrix $A = (a_{ij})$, the second degree polynomial

$$h \in \mathbb{R}^k \to h\,A\,h^t = \sum_{i,j} a_{ij} h_i h_j \in \mathbb{R}$$

is called the quadratic form associated with the matrix A. The form (or the matrix A) is *positive definite* if $h\,A\,h^t > 0$ for all $h \neq 0$. A sufficient condition for this is that all the *principal minors* M_r of A ($r = 1, \ldots, k$) be positive, where

$$M_r := \det(a_{ij})_{i,j=1}^r.$$

An alternative sufficient condition for positive definiteness of A is that all its *eigenvalues* be positive. These "eigenvalues" are the roots of the characteristic polynomial of A, $p(\lambda) := \det(\lambda I - A)$. We can state now sufficient conditions for a critical point to be a strict local minimum.

2.2.10 Theorem *Let f be of class C^2 in a neighborhood of its critical point x^0 in $\mathbb{R}^k$. Suppose that*

(*) All the principal minors M_j ($j = 1, \ldots, k$) of the Hessian matrix of f at x^0 are positive.

Then x^0 is a strict local minimum of f.

Proof Since f is of class C^2 in a neighborhood D of x^0, all its derivatives of second order are continuous in D. As determinants, the principal minors of the Hessian matrix are algebraic sums of products of their entries, which are continuous in D as we just observed, and therefore these minors are *continuous in D*. By Condition

(*), the latter are positive at x^0, and consequently, by continuity, there exists an x^0-neighborhood $B(x^0, r) \subset D$ such that all the principal minors are positive in it. It follows that the Hessian matrix is *positive definite* in $B(x^0, r)$, that is, if $x \in B(x^0, r)$, then

$$h\, H|_x h^t > 0 \tag{2.48}$$

for all $0 \neq h \in \mathbb{R}^k$. For $0 < ||h|| < r$, the line segment joining x^0 and $x^0 + h$ is contained in $B(x^0, r)$. By Theorem 2.2.5 (for $n = 1$) and (2.44) in Sect. 2.2.4, since $\nabla f(x^0) = 0$, there exists $\theta \in (0, 1)$ such that

$$f(x^0 + h) - f(x^0) = \frac{1}{2} h\, H|_{x^0+\theta h} h^t. \tag{2.49}$$

The point $x^0 + \theta h$ is in $B(x^0, r)$ (since $||h|| < r$), and it follows therefore from (2.48) that the right hand side of (2.49) is positive, that is, $f(x^0 + h) > f(x^0)$ whenever $0 < ||h|| < r$. □

If H is the Hessian matrix of f, then $-f$ has the Hessian matrix $-H$; the latter's j-th principal minor at x^0 is $(-1)^j M_j$ (for M_j as in 2.2.10). We then obtain the following sufficient condition for a strict local maximum of f at x^0:

2.2.11 Corollary *Let f be a real valued function of class C^2 in a neighborhood of its critical point $x^0 \in \mathbb{R}^k$. Let M_j ($j = 1, \dots, k$) be the principal minors of the Hessian matrix of f at x^0. Suppose*
(*) $(-1)^j M_j > 0$ for all $j = 1, \dots, k$.
Then x^0 is a strict local maximum point of f.

2.2.12 Examples 1. Let

$$f(x, y, z) = x^2 + y^2 + 3z^2 - xy + 2xz + yz.$$

The point $x^0 := (x_0, y_0, z_0)$ is a critical point of f iff its components satisfy the three equations
$f_x := 2x - y + 2z = 0$
$f_y := -x + 2y + z = 0$
$f_z := 2x + y + 6z = 0.$
The system's determinant does not vanish (its value is 4). Therefore the above homogeneous linear system has the unique solution $(x_0, y_0, z_0) = (0, 0, 0)$. The Hessian matrix H for f coincides with the (constant) matrix of the above system. Its principal minors are
$M_1 = 2$, $M_2 = 3$, and $M_3 = 4$.
Since they are all positive, we conclude from Theorem 2.2.10 that the point $(0, 0, 0)$ is a strict local minimum point of f.
The point $(0, 0, 0)$ is actually a *global strict minimum point*, that is, $f(x, y, z) > f(0, 0, 0) (= 0)$ for *all* non-zero $(x, y, z) \in \mathbb{R}^3$, because f is a quadratic form with a *positive definite* coefficients matrix H.

2. Let $f(x, y) = x^4 + y^4 - (x + y)^2$. The critical points of f are the solutions of the system of equations

$$f_x(x, y) = 4x^3 - 2(x + y) = 0; \quad f_y(x, y) = 4y^3 - 2(x + y) = 0.$$

Subtracting the equations, we get $4x^3 - 4y^3 = 0$, hence $x^3 = y^3$, and therefore $x = y$. It then follows from anyone of the equations that $x(x^2 - 1) = 0$. The solutions for x are $0, 1, -1$, and we conclude that the critical points of f are $(0, 0)$, $(1, 1)$, and $(-1, -1)$. We have

$$f_{xx}(1, 1) = f_{yy}(1, 1) = 10, \qquad f_{xy}(1, 1) = -2.$$

The principal minors M_1, M_2 of the Hessian at the point $(1, 1)$ have the values 10, 96 respectively; since they are both positive, we conclude that the critical point $(1, 1)$ is a strict local minimum point of f. The same is true for the critical point $(-1, -1)$ (by the same calculation, or from the fact that $f(-x, -y) = f(x, y)$).
At the critical point $(0, 0)$, $M_2 = 0$, and the principal minors test is not applicable. However, for all x such that $0 < |x| < \sqrt{2}$, $f(x, x) - f(0, 0) = f(x, x) = 2x^2(x^2 - 2) < 0$, and for all $x \neq 0$, $f(x, -x) = 2x^4 > 0$. Therefore there is no neighborhood of $(0, 0)$ where $f(x, y) - f(0, 0)$ has a constant sign. This shows that the critical point $(0, 0)$ is *not* a local extremum of f.

2.2.13 The Second Differential Suppose the function $f : D \subset \mathbb{R}^k \to \mathbb{R}$ (D open) is differentiable at every point of D. Then

$$df(\cdot) : D \to (\mathbb{R}^k)', \tag{2.50}$$

where $(\mathbb{R}^k)'$ denotes the *dual space* of $\mathbb{R}^k$, that is, the vector space (over the field $\mathbb{R}$) of all linear functionals on $\mathbb{R}^k$. The latter is isomorphic to $\mathbb{R}^k$ through the isomorphism

$$L \in (\mathbb{R}^k)' \to a \in \mathbb{R}^k,$$

where $a_i = Le^i$, i.e., $Lh = h \cdot a$ for all $h \in \mathbb{R}^k$. It is convenient to *identify* $(\mathbb{R}^k)'$ with $\mathbb{R}^k$. Through this identification, df is then viewed as an $\mathbb{R}^k$-valued vector function on D. If this function is differentiable at some point $x \in D$, its differential $d(df(\cdot))|_x$ is called *the second differential of* f *at* x, and is denoted $d^2 f|_x$:

$$d^2 f|_x := d(df(\cdot))|_x. \tag{2.51}$$

The above isomorphism associates to $df(\cdot)$ the $\mathbb{R}^k$-valued function $\nabla f : D \to \mathbb{R}^k$. By Theorem 2.1.14, the latter is differentiable at $x \in D$ iff its components f_{x_i} are differentiable at x for all $i = 1, \ldots, k$. By Theorem 2.1.4, a sufficient condition for this is that all the partial derivatives $(f_{x_i})_{x_j} := f_{x_i x_j}$ exist in a neighborhood of x and are continuous at x (for $j = 1, \ldots, k$). Therefore, if f is of class C^2 in D, the second differential $d^2 f|_x$ exists for all $x \in D$. Applying (6) in Sect. 2.1.14, we obtain in this case (for all $h \in \mathbb{R}^k$)

$$\begin{aligned} d^2 f|_x h &= d(\nabla f)|_x h \\ &= h\left(\frac{\partial(\nabla f)}{\partial x}\right)\Big|_x = h\,H|_x, \end{aligned} \tag{2.52}$$

where

$$\begin{aligned} H &:= \left(\frac{\partial(\nabla f)}{\partial x}\right) \\ &:= \left(\frac{\partial(f_{x_1},\ldots,f_{x_k})}{\partial(x_1,\ldots,x_k)}\right) = \left(f_{x_i x_j}\right) \end{aligned} \tag{2.53}$$

is precisely the Hessian matrix of f. The action of the linear functional $d^2 f|_x h$ on the vector $u \in \mathbb{R}^k$ is given by

$$(d^2 f|_x h)u = (h\,H|_x)\cdot u = h\,H|_x u^t, \tag{2.54}$$

where the last expression is the matrix product of the row h, the $k \times k$-matrix $H|_x$, and the column u^t. The order can be reversed, since $H|_x$ is a symmetric matrix. The *bilinear form* (2.54) and the corresponding *quadratic form* $h\,H|_x h^t$ are usually associated to the second differential of f at x.

2.2.14 Convex Functions We recall that a differentiable function $f : (a, b) \to \mathbb{R}$ is convex in (a, b) if at each point $x \in (a, b)$, the graph of f is above the tangent line at x. The latter is the linear approximation at x of (the graph of) $f(x+h)$ as a function of h. This interpretation makes sense when the interval (a, b) is replaced by any open set $\Omega \subset \mathbb{R}^k$. If $f : \Omega \to \mathbb{R}$ is differentiable in Ω, then for each $x \in \Omega$, the linear approximation of $f(x+h)$ (as a function of h) is given locally by the "hyperplane" $f(x) + df|_x h = f(x) + h \cdot \nabla f|_x$. Accordingly, the differentiable function f is (strictly) convex in Ω if for all $x \in \Omega$ we have

$$f(x+h) > f(x) + h\cdot\nabla f|_x \qquad (h \neq 0,\ x+h \in \Omega). \tag{2.55}$$

If (2.55) is valid *in some neighborhood* of each point $x \in \Omega$, we say that f is locally convex in Ω.

When f is of class C^2 in Ω, we can use Taylor's theorem with $n = 1$ to obtain sufficient conditions for f to be locally strictly convex in Ω.

2.2.15 Theorem *Let f be a real valued function of class C^2 in the open convex set $\Omega \subset \mathbb{R}^k$. Suppose that all the principal minors M_j of the Hessian matrix H of f are positive in Ω. Then f is strictly convex in Ω.*

Proof Let $x,\ x+h \in \Omega$. Since Ω is convex, the segment joining x and $x+h$ lies in Ω. Since f is of class C^2 in Ω, we can apply Taylor's theorem with $n = 1$. We obtain

$$f(x+h) - [f(x) + h \cdot \nabla f|_x] = \frac{1}{2} h\, H|_{x+\theta h} h^t$$

for some $\theta \in (0, 1)$. Since the principal minors M_j of the Hessian H (at the point $x+\theta h \in \Omega$) are positive, the quadratic form on the right hand side is positive definite, hence

$$f(x+h) > f(x) + h \cdot \nabla f|_x \qquad (x,\ x+h \in \Omega,\ h \neq 0). \quad \square$$

For an arbitrary (not necessarily convex) open set $\Omega \subset \mathbb{R}^k$, we can apply the above theorem in a ball $B(x, r) \subset \Omega$ (for each given $x \in \Omega$), since balls are convex. We then get

2.2.16 Corollary *Let f be a real valued function of class C^2 on the open set $\Omega \subset \mathbb{R}^k$. Suppose its Hessian matrix H is positive definite at all points of Ω (a sufficient condition for this is that all the principal minors of H are positive in Ω). Then f is locally strictly convex in Ω.*

2.2.17 Exercises

1. Let $E = \{(x, y) \in \mathbb{R}^2;\ x \neq 0\}$ and let f be defined on E by $f(x, y) = \arctan(y/x)$. Prove that f satisfies the Laplace equation $\Delta f = 0$ in E. Recall that

$$\Delta = \nabla \cdot \nabla.$$

2. Prove that the function $g(x, y) = \log(x^2 + y^2)$ satisfies the Laplace equation $\Delta g = 0$ in $\mathbb{R}^2 \setminus \{(0, 0)\}$.
3. In Theorem 2.2.5, show that the remainder term can be given the more general form

$$\frac{(1-\theta)^{n+1-p}}{n!p} (h \cdot \nabla)^{n+1} f(x+\theta h), \tag{2.56}$$

for $p > 0$ given and suitable $\theta \in (0, 1)$.
The case $p = n + 1$ is the *Lagrange form* given in Theorem 2.2.5. The case $p = 1$ is *Cauchy's form* of the remainder

$$\frac{(1-\theta)^n}{n!} (h \cdot \nabla)^{n+1} f(x+\theta h). \tag{2.57}$$

The general form (2.56) is the *Schlömilch-Roche form* of the remainder.
Hint: with notation as in the proof of Theorem 2.2.5,

$$F(1) = \sum_{j=1}^{n} \frac{F^{(j)}(0)}{j!} 1^j + \frac{(1-\theta)^{n+1-p}}{n!p} F^{(n+1)}(\theta)\, 1^{n+1},$$

according to the Schlömilch-Roche form of the remainder in the Taylor formula for F.

4. Let $D \subset \mathbb{R}^k$ be an open "starlike" set (that is, there exists a point $p \in D$ such that the line segment
$$\overline{px} := \{p + tx;\ t \in [0, 1]\}$$
is contained in D for all $x \in D$). Suppose $f : D \to \mathbb{R}$ is of class C^1 and $||\nabla f|| \le M$ in D. Prove that
$$|f(x) - f(p)| \le M\,||x - p|| \qquad (x \in D).$$
5. Calculate the Taylor polynomial of degree 2 at the point $(0, 0)$ for the function $f(x, y) = \sin(x^2 + y^2)$. (Do not calculate the remainder.)
6. Show that Condition (*) in Theorem 2.2.10 (and in Corollary 2.2.11) may be replaced by the condition
(**) all the eigenvalues of the Hessian matrix at x^0 are positive (negative, respectively), with the same respective conclusion.
7. Let
$$f(x, y, z) = x^4 + y^4 + z^4 - x^2 - y^2 - z^2 - 2xy - 2xz - 2yz.$$
Find all the critical points of f and determine whether they are local maxima, minima, or neither.
8. Same problem (as No. 7) for the following functions:
   (a) $f(x, y) = (x + y)\,e^{-(x^2+y^2)}$.
   (b) $f(x, y, z) = 7(x^2 + y^2 + z^2) + 10xy + 12xz + 4yz$.
   (c) $f(x, y) = x^2y^3(2 - x - y)$ in the domain $D := \{(x, y);\ x, y > 0\}$.
   (d) $f(x, y) = (x^2 + y^2)\,e^{-(x^2+y^2)}$ in the unit disc $D := \{(x, y);\ x^2 + y^2 < 1\}$.
   (e) $f(x, y) = xy\,(1 - x^2 - y^2)^{1/2}$.
   (f) $f(x, y, z) = x^4 + y^4 + z^4 - 2(x^2 + y^2 + z^2)$.
9. Let $f(x, y) = (y - x^2)(y - 3x^2)$.
   (a) Prove that for any line $g(t) = (a, b)t, t \in \mathbb{R}, (a, b) \in \mathbb{R}^2 \setminus \{(0, 0)\}$ constant, the composite function $f \circ g$ has a local minimum at $t = 0$ (i.e., the function f has a local minimum at $(0, 0)$ along any line through the origin).
   (b) Prove that $(0, 0)$ is a critical point of f but is *not* a local minimum of f.
10. Among all (rectangular) boxes whose edges have a given total length, find the box with maximal volume.
11. Among all boxes with given volume, find the box with minimal surface area.
12. Among all box-shaped uncovered water reservoirs with given capacity, find the reservoir with minimal surface area.
13. Among all boxes inscribed in the ellipsoid with equation
$$x^2 + 2y^2 + 3z^2 = 1,$$
find the coordinates of the box with maximal volume and check the maximality assertion.

14. Among all polygons with $n \geq 3$ edges inscribed in a circle whose centre O is *inside* the polygon, find the polygon with maximal area. Hint: partition the polygons into triangles with common vertex at O (and the other vertices on the circle). Choose as your variables x_i the angles of the triangles at O. In order to prove that the point x is a maximum, show that if A is a $j \times j$ matrix with all entries equal to 1, and I denotes the $j \times j$ unit matrix, then $\det(I + A) = j + 1$.
15. Let $(x_i, y_i) \in \mathbb{R}^2$, $i = 1, \ldots, n$, be given ($n \geq 2$, and $x_i \neq x_j$ for at least one pair (i, j)). Consider the so-called "sum of square deviations"

$$f(s,t) := \sum_i [y_i - (sx_i + t)]^2 \qquad s, t \in \mathbb{R}.$$

Find the point (s^*, t^*) where f attains its absolute minimum (prove your assertion!). The line with equation $y = s^*x + t^*$ is the so-called *least squares line* through the given points. Hint: verify and use the elementary identity

$$n \sum_{i=1}^{n} x_i^2 - \left(\sum_{i=1}^{n} x_i\right)^2 = \frac{1}{2} \sum_{i,j=1}^{n} (x_i - x_j)^2.$$

Chapter 3
Implicit Functions

In this section, we shall prove the *Implicit Function Theorem*, which gives sufficient conditions for the existence and uniqueness of a local solution $y = f(x)$ for a single real equation

$$F(x, y) = 0 \qquad (x \in \mathbb{R}^k,\ y \in \mathbb{R}) \tag{3.1}$$

in a neighborhood of a point (x^0, y^0) of $\mathbb{R}^k \times \mathbb{R}$ where (3.1) is satisfied.

The result is then generalized to the case of a vector equation

$$F(x, y) = 0 \qquad (x \in \mathbb{R}^k,\ y \in \mathbb{R}^l), \tag{3.1$'$}$$

where F is an $\mathbb{R}^l$-valued function, or equivalently, to the case of a system of real equations.

Applications are then given to the local inversion of maps, to extrema with constraints, etc.

To fix the ideas, $\mathbb{R}^k$ is normed with the Euclidean norm, however the results of this section do not depend on the choice of p-norm.

The proofs are based on the *Banach Fixed Point Theorem* for a *contraction* in a complete metric space.

3.1 Fixed Points

3.1.1 Contractions Let X be a metric space with a metric d. A *contraction* on X is a function $T : X \to X$, for which there exists a constant $0 < q < 1$ such that

$$d(Tx, Ty) \le q\, d(x, y)$$

for all $x, y \in X$.

The original version of this chapter was revised. An erratum to this chapter can be found at DOI 10.1007/978-3-319-27956-5_5

S. Kantorovitz, *Several Real Variables*, Springer Undergraduate Mathematics Series, DOI 10.1007/978-3-319-27956-5_3

We used the notation Tx instead of $T(x)$ in order to have fewer parenthesis in the following formulas.

The iterates, or powers, of T are defined inductively by

$$T^1 = T; \qquad T^n = T \circ T^{n-1} \quad (n = 2, 3, \ldots).$$

A contraction T on X is uniformly continuous on X, because

$$d(Tx, Ty) \le q\, d(x, y) \le d(x, y),$$

so that the uniform continuity ϵ, δ-condition is satisfied if we choose $\delta \le \epsilon$.

A *fixed point* for T is a point $x \in X$ such that $Tx = x$. If it exists, it is unique, because if both x, $y \in X$ are fixed points for T, then

$$d(x, y) = d(Tx, Ty) \le q\, d(x, y),$$

that is, $0 \le (1-q)d(x, y) \le 0$, hence $(1-q)d(x, y) = 0$, and therefore $d(x, y) = 0$ since $1-q \ne 0$. Thus $x = y$. The *existence* of a fixed point for T when X is *complete* is the content of the following theorem.

The Banach Fixed Point Theorem

3.1.2 Theorem *Let T be a contraction on the complete metric space X. Then T has a unique fixed point in X.*

Proof Given an arbitrary point $x_0 \in X$, we consider its *orbit* by T, that is, the sequence

$$x_n := T^n x_0 \qquad (n \in \mathbb{N}). \tag{3.2}$$

We claim that

$$d(x_n, x_{n+1}) \le q^n d(x_0, x_1) \qquad (n = 0, 1, 2, \ldots). \tag{3.3}$$

Assuming (3.3) for some n, we get (3.3) for $n + 1$ by applying the contraction condition:

$$d(x_{n+1}, x_{n+2}) = d(Tx_n, Tx_{n+1}) \le q\, d(x_n, x_{n+1})$$

$$\le q\, q^n d(x_0, x_1) = q^{n+1} d(x_0, x_1).$$

Since (3.3) is trivial for $n = 0$, (3.3) follows for all $n = 0, 1, 2, \ldots$ by induction.

Next, for any positive integers $m > n$, we claim that

$$d(x_n, x_m) \le (q^n + \cdots + q^{m-1})d(x_0, x_1). \tag{3.4}$$

Assuming (3.4) for some $m > n$, we get (3.4) for $m + 1$ by using the triangle inequality and (3.3) for m:

$$d(x_n, x_{m+1}) \leq d(x_n, x_m) + d(x_m, x_{m+1})$$

$$\leq (q^n + \cdots + q^{m-1})d(x_0, x_1) + q^m d(x_0, x_1) = (q^n + \cdots + q^m)d(x_0, x_1).$$

Since (3.4) reduces to (3.3) for $m = n + 1$, (3.4) follows for all $m > n$ by induction on m starting with $m = n + 1$.

By (3.4), for all $m > n$,

$$d(x_n, x_m) \leq (q^n + q^{n+1} + \cdots)d(x_0, x_1)$$
$$= \frac{q^n}{1-q} d(x_0, x_1) \to 0$$

as $n \to \infty$, since $q < 1$. Hence $\{x_n\}$ is a Cauchy sequence. Since X is complete, $x := \lim x_n$ exists in X. By continuity of T (cf. 3.1.1), we have

$$Tx = \lim Tx_n = \lim x_{n+1} = x,$$

that is, x is a fixed point for T. □

The Space $C(X)$

In order to use Banach's Fixed Point Theorem to prove the Implicit Function Theorem, we shall first define the complete metric space on which Banach's theorem is applied.

3.1.3 Definition Let X be a compact metric space. The set of all real valued continuous functions on X is denoted by $C(X)$. With the operations of addition and multiplication by real scalars defined pointwise, $C(X)$ is a vector space over the field $\mathbb{R}$.

For each $f \in C(X)$, the function $|f|$, defined pointwise, is continuous (cf. 1.4.4), hence bounded, by the compactness of X (cf. Corollary 1.4.7). Therefore

$$||f|| := \sup |f|(X) < \infty. \tag{3.5}$$

Actually, the above supremum is a *maximum*, assumed by the continuous function $|f|$ on X (cf. 1.4.8).

We verify easily that $|| \cdot ||$ defined by (3.5) is a norm on $C(X)$.

Thus $C(X)$ *is a normed space.*

Convergence of a sequence $\{f_n\} \subset C(X)$ to $f \in C(X)$ in this normed space is equivalent to its *uniform convergence* on X, that is, for every $\epsilon > 0$, there exists $n(\epsilon) \in \mathbb{N}$ such that

$$|f_n(x) - f(x)| < \epsilon \tag{3.6}$$

for all $n > n(\epsilon)$ and for all $x \in X$.

The uniformity refers to the *independence* of $n(\epsilon)$ on x.

Indeed, if $f_n \to f$ in the metric space $C(X)$, then given $\epsilon > 0$, there exists $n(\epsilon) \in \mathbb{N}$ such that $||f_n - f|| < \epsilon$ for all integers $n > n(\epsilon)$. Then for these n and for all $x \in X$,

$$|f_n(x) - f(x)| = |f_n - f|(x) \leq ||f_n - f|| < \epsilon.$$

On the other hand, if $n(\epsilon)$ is as in (3.6), then taking the supremum of the left hand side of (3.6) with $\epsilon/2$ over all $x \in X$, we obtain that for all $n > n^*(\epsilon) := n(\epsilon/2)$,

$$||f_n - f|| = \sup_{x \in X} |f_n(x) - f(x)| \leq \frac{\epsilon}{2} < \epsilon,$$

that is, $f_n \to f$ in the normed space $C(X)$.

3.1.4 Theorem *$C(X)$ is a Banach space.*

Proof We must prove that the normed space $C(X)$ is *complete*. Let $\{f_n\}$ be a Cauchy sequence in $C(X)$. Thus, given $\epsilon > 0$, there exists $n(\epsilon) \in \mathbb{N}$ such that

$$||f_n - f_m|| < \epsilon \tag{3.7}$$

for all $n, m > n(\epsilon)$.

Therefore, for all such integers n, m,

$$|f_n(x) - f_m(x)| = |f_n - f_m|(x) \leq ||f_n - f_m|| < \epsilon$$

for each $x \in X$. This means that for each $x \in X$, the real sequence $\{f_n(x)\}$ is Cauchy, and since $\mathbb{R}$ is complete (cf. 1.3.6 for $k = 1$), the sequence converges in $\mathbb{R}$. We may then *define* $f(x)$ as the limit

$$f(x) := \lim f_n(x) \qquad (x \in X).$$

We shall prove that the function $f : X \to \mathbb{R}$ is *continuous on* X by using a so-called "$\epsilon/3$ argument". Given $\epsilon > 0$, we have by (3.7) (with $\epsilon/3$):

$$|f_n(x) - f_m(x)| < \frac{\epsilon}{3} \tag{3.8}$$

for all $n, m > n(\epsilon/3)$ and for all $x \in X$.

Letting $m \to \infty$, we obtain

$$|f_n(x) - f(x)| \leq \frac{\epsilon}{3} \tag{3.9}$$

for all $n > n(\epsilon/3)$ and for all $x \in X$.

Therefore, whenever $n > n(\epsilon/3)$, we have for all $x, y \in X$

$$|f(x) - f(y)| \le |f(x) - f_n(x)| + |f_n(x) - f_n(y)| + |f_n(y) - f(y)|$$
$$\le \frac{2\epsilon}{3} + |f_n(x) - f_n(y)|. \tag{3.10}$$

Fixing $n > n(\epsilon/3)$, the continuity of the function f_n implies its uniform continuity on the *compact* metric space X (cf. 1.4.11). Therefore there exists $\delta > 0$ such that

$$|f_n(x) - f_n(y)| < \frac{\epsilon}{3} \tag{3.11}$$

for all $x, y \in X$ such that $d(x, y) < \delta$. By (3.10) and (3.11), we have $|f(x) - f(y)| < \epsilon$ whenever $d(x, y) < \delta$. This proves that $f \in C(X)$.

Taking in (3.9) the supremum over all $x \in X$, we obtain

$$||f_n - f|| \le \frac{\epsilon}{3} < \epsilon$$

for all $n > n^*(\epsilon) := n(\epsilon/3)$, that is, $f_n \to f$ in the normed space X. □

3.1.5 The Space $C(X, Y)$

For the application of the Banach Fixed Point theorem to a *system of simultaneous implicit equations*, we need to consider $\mathbb{R}^q$-valued continuous functions f for $q > 1$. More generally, for any *compact* metric space X and for any *Banach space* Y, let $C(X, Y)$ be the set of all continuous functions $f : X \to Y$; it is a normed space over $\mathbb{R}$ for the operations of addition and multiplication by real scalars defined pointwise, and for the norm

$$||f|| := \sup_{x \in X} ||f(x)||. \tag{3.12}$$

The norm is continuous on Y, so that $||f(\cdot)||$ is continuous on the *compact* metric space X, hence *bounded* on X. Therefore the supremum in (3.12) is finite, and we verify easily that it satisfies the norm properties.

The proof of Theorem 3.1.4 applies word for word to the present case, if we replace absolute values by norms and if we use the completeness of Y instead of the completeness of $\mathbb{R}$. Then we have

3.1.6 Theorem *Let X be a compact metric space, and let Y be a Banach space. Then $C(X, Y)$ is a Banach space.*

By 1.3.3, we have the following consequence, which will allow us to apply the Banach Fixed Point Theorem in the proof of the Implicit Function Theorem.

3.1.7 Corollary *Let X be a compact metric space, Y a Banach space, and $b > 0$. Let $C_b(X, Y)$ denote the closed ball with center 0 and radius b in the space $C(X, Y)$, that is*

$$C_b(X, Y) := \{f \in C(X, Y);\ ||f|| \le b\}.$$

Then $C_b(X, Y)$ *is a complete metric space. In particular,* $C_b(X)$ *is a complete metric space.*

3.2 The Implicit Function Theorem

In the following, we shall view $\mathbb{R}^{k+1}$ as the Cartesian product $\mathbb{R}^k \times \mathbb{R}$, and denote its elements as ordered pairs (x, y), where $x \in \mathbb{R}^k$ and $y \in \mathbb{R}$.

The key for using Banach's Fixed Point Theorem to prove the Implicit Function Theorem (IFT) is Lipschitz' condition.

Lipschitz' Condition

3.2.1 Definition A real function f defined on a subset D of $\mathbb{R}^{k+1}$ is *Lipschitz with respect to* y in D (uniformly with respect to x) if there exists a constant $q > 0$ such that

$$|f(x, y) - f(x, y')| \le q\,|y - y'|$$

for all points $(x, y),\ (x, y') \in D$.

The constant q is called a Lipschitz constant for f in D.

The main argument in the proof of the IFT is contained in the following lemma, which states that the equation $f(x, y) = y$ has locally a unique solution $y = \phi(x)$ in a neighborhood of the solution $(0, 0)$, if f is continuous and Lipschitz with Lipschitz constant $q < 1$ in a neighborhood of $(0, 0)$. The proof of this lemma uses Banach's Fixed Point Theorem.

3.2.2 Lemma *Let*

$$D := \{(x, y) \in \mathbb{R}^k \times \mathbb{R};\ |x_i| < a,\ |y| \le b \quad (i = 1, \ldots, k)\},$$

where $a, b > 0$. *Suppose*

(a) $f : D \to \mathbb{R}$ *is continuous in* D *and* $f(0, 0) = 0$;
(b) f *is Lipschitz with respect to* y *in* D *with a Lipschitz constant* $q < 1$.

Then there exists $a' \in (0, a)$, *such that for each* x *in the cell*

$$I := \{x \in \mathbb{R}^k;\ |x_i| \le a' \quad (i = 1, \ldots, k)\},$$

the equation

$$f(x, y) = y$$

has a unique solution y *with* $|y| \le b$. *This unique solution, which is necessarily a real valued function* $\phi : x \in I \to y \in [-b, b]$ *is continuous and* $\phi(0) = 0$.

Proof Let

$$D_0 := \{x \in \mathbb{R}^k;\ |x_i| < a \quad (i = 1, \dots, k)\}.$$

The function $f_0 : D_0 \to \mathbb{R}$ defined by $f_0(x) = f(x, 0)$ is continuous on D_0 and $f_0(0) = f(0, 0) = 0$ by hypothesis. Thus, given $\epsilon := (1 - q)b$, there exists $\delta > 0$ such that $|f_0(x)|\ (= |f_0(x) - f_0(0)|) < \epsilon$ for all $x \in D_0$ such that $||x||\ (= ||x - 0||) < \delta$. Choose $a' > 0$ such that

$$a' < \min\,(a, \frac{\delta}{\sqrt{k}}),$$

and let I be as in the statement of the lemma. Then for all $x \in I$, we have $x \in D_0$ (because $a' < a$) and $||x|| \le \sqrt{k}a' < \delta$, and therefore

$$|f_0(x)| < (1 - q)b \qquad (x \in I). \tag{3.13}$$

The cell I is a compact metric space (cf. Theorem 1.2.8); we consider the complete metric space $C_b(I)$ (cf. Corollary 3.1.7). Recall that its metric is induced by the norm $||\cdot||$ on the normed space $C(I)$, which is the supremum norm. Thus, if $\psi \in C_b(I)$ and $x \in I$, then

$$|\psi(x)| \le \sup_{z \in I} |\psi(z)| = ||\psi|| \le b,$$

hence $(x,\ \psi(x)) \in D$, and therefore $f(x,\ \psi(x))$ is well defined for all $x \in I$. Clearly, this function is the composition $f \circ \tilde{\psi}$, where $\tilde{\psi} : I \to D$ is defined by $\tilde{\psi}(x) := (x,\ \psi(x))$. By the continuity of $\tilde{\psi}$ on I and of f on D, $f \circ \tilde{\psi}$ is *continuous* on I (cf. Theorem 1.4.4). Since f is Lipschitz with respect to y, we have for all $x \in I$,

$$|f(x,\ \psi(x))| \le |f(x,\ \psi(x)) - f(x, 0)| + |f(x, 0)|$$

$$\le q\,|\psi(x) - 0| + |f_0(x)| < qb + (1 - q)b = b,$$

where we used (3.13) in the last estimate.

Hence $||f \circ \tilde{\psi}|| \le b$, and therefore

$$f \circ \tilde{\psi} \in C_b(I).$$

We then define the function

$$T : C_b(I) \to C_b(I)$$

by

$$T\psi = f \circ \tilde{\psi},$$

that is,

$$(T\psi)(x) = f(x,\ \psi(x)) \qquad (x \in I). \tag{3.14}$$

We verify that T is a contraction on $C_b(I)$. Let $\psi,\ \chi \in C_b(I)$. Then for all $x \in I$, the points $(x,\ \psi(x))$ and $(x,\ \chi(x))$ are in D. Since f is Lipschitz with respect to y on D, we obtain for all $x \in I$

$$|(T\psi)(x) - (T\chi)(x)| = |f(x,\ \psi(x)) - f(x,\ \chi(x))| \le q\,|\psi(x) - \chi(x)|.$$

Taking the supremum over all $x \in I$, we obtain

$$||T\psi - T\chi|| \le q\,||\psi - \chi||.$$

Since $q < 1$ by hypothesis, T is indeed a contraction of the complete metric space $C_b(I)$. By the Banach Fixed Point theorem, there exists a unique $\phi \in C_b(I)$ such that $T\phi = \phi$. This means that $f(x, \phi(x)) = \phi(x)$, ϕ is continuous, and $|\phi(x)| \le b$ for all $x \in I$. Thus $y = \phi(x)$ is a continuous solution of the equation $f(x, y) = y$ for $x \in I$, and $|y| \le b$.

If y is *any* solution of the latter equation (with $x \in I$ and $|y| \le b$), then since the points $(x,\ \phi(x))$ and (x, y) are in D for all $x \in I$, we have by the Lipschitz condition (for all $x \in I$)

$$|y - \phi(x)| = |f(x, y) - f(x, \phi(x))| \le q\,|y - \phi(x)|,$$

hence, since $1 - q > 0$,

$$0 \le (1 - q)|y - \phi(x)| \le 0,$$

that is, the last expression vanishes, and so $y = \phi(x)$ for all $x \in I$. This proves the uniqueness of the solution and its continuity on I, since $\phi \in C_b(I)$. By Condition (a), $y = 0$ is a solution for $x = 0$. Hence $\phi(0) = 0$ by uniqueness of the solution.□

In the next lemma, we give a sufficient condition for f to be Lipschitz, and an estimate for an appropriate Lipschitz constant.

3.2.3 Lemma *Let $\Omega \subset \mathbb{R}^{k+1}$ be an open set with the property that whenever $(x, y),\ (x, y') \in \Omega$, the line segment joining these points lies in Ω. Let $f : \Omega \to \mathbb{R}$ be such that f_y exists and is bounded by $M > 0$ in Ω. Then f is Lipschitz with respect to y in Ω, with Lipschitz constant M.*

Proof Let $(x, y),\ (x, y') \in \Omega$, $y < y'$, and consider $f(x, \cdot)$ as a function of a single real variable on the interval $[y, y']$. All the points (x, t) with $t \in [y, y']$ belong to Ω by hypothesis, so the restriction of f to these points is well defined. Its derivative is precisely $f_y(x, \cdot)$; it exists in the said interval and is bounded there by M. By the MVT for real functions of one real variable,

$$f(x, y') - f(x, y) = f_y(x, \tilde{y})(y' - y),$$

where $y < \tilde{y} < y'$. Hence

$$|f(x, y) - f(x, y')| \leq M |y - y'|. \qquad \square$$

We shall apply now the last two lemmas to prove the *Implicit Function Theorem* for a single real equation $F(x, y) = 0$.

The Implicit Function Theorem

3.2.4 Theorem *Let F be a real valued function on $\Omega \subset \mathbb{R}^k \times \mathbb{R}$, where Ω is the open cell centred at the point (x^0, y^0)*

$$\Omega := \{(x, y) \in \mathbb{R}^k \times \mathbb{R};\ |x_i - x_i^0| < a,\ |y - y^0| < b,\ i = 1, \dots, k\}.$$

Suppose

(i) F is continuous in Ω;
(ii) F_y exists and is continuous in Ω;
(iii) $F(x^0, y^0) = 0$ and $F_y(x^0, y^0) \neq 0$.

Then there exist a', b', $0 < a' < a$, $0 < b' < b$, such that for each x in the cell

$$I := \{x \in \mathbb{R}^k;\ |x_i - x_i^0| \leq a',\ i = 1, \dots, k\},$$

the equation

$$F(x, y) = 0 \tag{3.15}$$

has a unique solution $y = \phi(x)$ with $|y - y^0| \leq b'$. The function ϕ is continuous on I and $\phi(x^0) = y^0$.

Proof Shifting the coordinates ($x \to x' = x - x^0$, $y \to y' = y - y^0$), we may assume without loss of generality that $(x^0, y^0) = (0, 0)$. By (iii),

$$c := F_y(0, 0) \neq 0.$$

We may then define

$$f(x, y) := y - \frac{1}{c} F(x, y) \qquad ((x, y) \in \Omega). \tag{3.16}$$

Equation (3.15) is then *equivalent to the equation*

$$f(x, y) = y, \tag{3.17}$$

and it remains to deal with (3.17) by means of Lemma 3.2.2. We verify that the hypothesis of the lemma are satisfied. The hypothesis on F translate to the following properties of f:

(a) f is continuous in Ω, and $f(0,0)=0$;
(b) f_y exists and is *continuous* in Ω, and since $f_y = 1-(1/c)F_y$, we have $f_y(0,0) = 0$.

Fix an arbitrary number q, $0<q<1$. It follows from (b) that there exist positive numbers $\alpha<a$ and $\beta<b$ such that $|f_y(x,y)|\le q$ for all (x,y) such that $|x_i|<\alpha$ and $|y|<\beta$. By Lemma 3.2.3, f is Lipschitz with respect to y in this open cell, with Lipschitz constant q. By Lemma 3.2.2, there exist positive numbers $a'<\alpha(<a)$ and $b'<\beta(<b)$, such that for each $x\in I:=\{x\in\mathbb{R}^k;\ |x_i|\le a'\ (i=1,\dots,k)\}$, there exists a unique solution $y=\phi(x)$ of (3.17) with $|y|\le b'$; ϕ is *continuous* on I. By (a), $y=0$ is a solution of (3.17) for $x=0$; by the uniqueness of the solution, we have necessarily $\phi(0)=0$. □

We assumed in the above theorem that F_y exists and is continuous in Ω. If we assume in addition that F_{x_i} exist and are continuous in Ω for all $i=1,\dots,k$, we prove below that the solution function ϕ is of class C^1 in a neighborhood of x^0, and we obtain a formula for its partial derivatives in terms of the partial derivatives of F.

3.2.5 Theorem *Suppose the hypothesis are as in Theorem 3.2.4, and assume in addition that F_{x_i} exist and are continuous in Ω for all $i=1,\dots,k$ (that is, F of class C^1 in Ω). Then the unique solution ϕ is of class C^1 in a neighborhood U of x^0, and*

$$\phi_{x_i} = -\frac{F_{x_i}(x,\ \phi(x))}{F_y(x,\ \phi(x))}$$

for all $x\in U$ and $i=1,\dots,k$.

Proof As in the preceding proof, we assume without loss of generality that $(x^0,y^0)=(0,0)$. Since $F_y(0,0)\neq 0$ and F_y is continuous in Ω, there exists a neighborhood W of $(0,0)$ contained in Ω such that $F_y(x,y)\neq 0$ for all $(x,y)\in W$. Let a', b' be as in the statement of Theorem 3.2.4, and let

$$W_1 := \{(x,y)\in W;\ |x_i|<a',\ |y|<b'\quad (i=1,\dots,k)\}.$$

Since the unique solution ϕ (cf. Theorem 3.2.4) is continuous in I and $\phi(0)=0$, the function $\tilde\phi := (\cdot,\ \phi(\cdot))$ (cf. proof of Lemma 3.2.2) is continuous in I and $\tilde\phi(0)=(0,0)$. Therefore there exists an $\mathbb{R}^k$-neighborhood U of 0 contained in I such that $(x,\ \phi(x))\in W_1$ for all $x\in U$.

Fix $x\in U$ and $i\in\{1,\dots,k\}$.

Since $\tilde\phi$ is continuous in I, there exists $\delta>0$ such that $\tilde\phi(x+he^i)=(x+he^i,\phi(x+he^i))\in W_1$ for all real h with $|h|<\delta$. Given h, denote

$$u_h := \tilde\phi(x+he^i)-\tilde\phi(x)\quad (\in\mathbb{R}^k\times\mathbb{R}), \tag{3.18}$$

that is,

$$\tilde\phi(x+he^i)=\tilde\phi(x)+u_h.$$

The neighborhood W_1 ($\subset \Omega$) clearly contains the line segment joining any two points in it. Therefore by the Mean Value theorem (Theorem 2.2.6), there exists a real number θ, $0 < \theta < 1$ (depending on h) such that

$$\begin{aligned} F(x + he^i, \phi(x + he^i)) &= F(\tilde{\phi}(x + he^i)) = F(\tilde{\phi}(x) + u_h) \\ &= F(\tilde{\phi}(x)) + u_h \cdot \nabla F(\tilde{\phi}(x) + \theta u_h). \end{aligned} \tag{3.19}$$

By the definition of ϕ, the (extreme) left hand side and the first summand on the (extreme) right hand side vanish. Since

$$u_h = (x + he^i, \phi(x + he^i)) - (x, \phi(x)) = (he^i, \phi(x + he^i) - \phi(x)),$$

it follows from (3.19) that

$$0 = he^i \cdot \nabla_x F(\tilde{\phi}(x) + \theta u_h) + [\phi(x + he^i) - \phi(x)]F_y(\tilde{\phi}(x) + \theta u_h). \tag{3.20}$$

The first summand on the right hand side of (3.20) is equal to

$$hF_{x_i}(\tilde{\phi}(x) + \theta u_h).$$

The argument of F_y is in W_1, and therefore $F_y \neq 0$ at that point. Dividing, we then get (for $0 < |h| < \delta$)

$$h^{-1}[\phi(x + he^i) - \phi(x)] = -\frac{F_{x_i}}{F_y}\bigg|_{\tilde{\phi}(x)+\theta u_h}. \tag{3.21}$$

We have by (3.18)

$$||\theta u_h|| = \theta ||u_h|| \leq ||u_h|| = ||\tilde{\phi}(x + he^i) - \tilde{\phi}(x)|| \to 0$$

as $h \to 0$, by the continuity of $\tilde{\phi}$ in W_1. The points $\tilde{\phi}(x) + \theta u_h$ are in Ω hence F_{x_i} and F_y are *continuous* at these points by the C^1 assumption, and therefore, as $h \to 0$, the numerator and denominator in (3.21) converge to $F_{x_i}(\tilde{\phi}(x)) = F_{x_i}(x, \phi(x))$ and $F_y(x, \phi(x)) \neq 0$ respectively. Consequently $\phi_{x_i}(x)$ exists and is equal to $-\frac{F_{x_i}(x,\phi(x))}{F_y(x,\phi(x))}$. Thus

$$\phi_{x_i} = -\frac{F_{x_i} \circ \tilde{\phi}}{F_y \circ \tilde{\phi}}$$

on U. Since $\tilde{\phi}$ is continuous on U, with range contained in W_1, F_{x_i} and F_y are continuous on W_1, and $F_y \neq 0$ on W_1, we conclude that ϕ_{x_i} is continuous on U. □

Example. Consider the equation

$$z^3 + y - xz = 0.$$

The function $F : \mathbb{R}^3 \to \mathbb{R}$ defined by the left hand side of the equation is a polynomial, so that the smoothness conditions of Theorem 3.2.5 (and 3.2.4, a fortiori) are trivially satisfied in $\mathbb{R}^3$.

Let $p = (3, -2, 2)$. We have $F(p) = 0$ and $F_z(p) = (3z^2 - x)|_p = 9 \neq 0$. By Theorem 3.2.4, the equation $F = 0$ defines locally a function $z = \phi(x, y)$ in a neighborhood of the point $(3, -2)$. By Theorem 3.2.5, the function ϕ is of class C^1 in a neighborhood V of $(3, -2)$ in $\mathbb{R}^2$, $\phi(3, -2) = 2$, and for some $\epsilon > 0$ and all $(x, y, x) \in V \times (-\epsilon, \epsilon)$,

$$\nabla\phi(x, y) = -\frac{1}{3z^2 - x}(F_x, F_y) = -\frac{1}{3z^2 - x}(-z, 1),$$

with $z = \phi(x, y)$. In particular

$$\nabla\phi(3, -2) = -\frac{1}{9}(-2, 1) = (\frac{2}{9}, \frac{-1}{9}).$$

We may use the preceding formula for $\nabla\phi$ in V to calculate higher derivatives of ϕ in V. For example, for all $(x, y) \in V$,

$$\begin{aligned}\phi_{yy}(x, y) &= \frac{\partial}{\partial y}[x - 3\phi(x, y)^2]^{-1} \\ &= -[x - 3\phi(x, y)^2]^{-2}[-6\phi(x, y)\,\phi_y(x, y)].\end{aligned}$$

In particular, we get $\phi_{yy}(3, -2) = -4/243$.

3.2.6 Exercises

1. Show that the equation

$$x^5 + y^5 + z^5 = 2 + xyz$$

determines in a neighborhood of the point $(1, 1, 1)$ a unique function $z = z(x, y)$ of class C^1, and calculate its partial derivatives with respect to x and y at the point $(1, 1)$.

2. (a) Show that the equation $x^4 = 2xy + 3y^2 z$ defines in a neighborhood of the point $(x, y, z) = (1, 1, -1/3)$ a real valued function $\phi : (y, z) \to x$ from a neighborhood of $(1, -1/3)$ to a neighborhood of 1, such that ϕ is of class C^1 and $\phi(1, -1/3) = 1$.
   (b) Calculate $\nabla\phi(1, -1/3)$.
3. (a) Show that the equation

$$e^{z-x-y^2} + z^3 + 2z = \cos(x^2 + 4xy - z)$$

defines a unique real-valued function $z = \phi(x, y)$ of class C^1 in a neighborhood of the point $(0, 0)$ in $\mathbb{R}^2$, such that $\phi(0, 0) = 0$.
   (b) Calculate $\nabla\phi(0, 0)$.

3.3 System of Equations

In this section, we generalize the Implicit Function Theorem for one equation $F(x, y) = 0$ with one "unknown" y to the case of s simultaneous equations with s unknowns $y_1, \dots, y_s$:

$$F_r(x_1, \dots, x_k; y_1, \dots, y_s) = 0 \qquad (r = 1, \dots, s). \tag{3.22}$$

It will be convenient to use vector notation. The space $\mathbb{R}^{k+s}$ will be viewed as the Cartesian product $\mathbb{R}^k \times \mathbb{R}^s$, and the argument of F_r in (3.22) is then denoted (x, y), points of $\mathbb{R}^k \times \mathbb{R}^s$. The system of scalar equations (3.22) is equivalent to the single vector equation $F(x, y) = 0$, where

$$F := (F_1, \dots, F_s) : D \subset \mathbb{R}^k \times \mathbb{R}^s \to \mathbb{R}^s$$

is a given vector valued function.

We formulate first the Lipschitz Condition for vector valued functions, and generalize Lemma 3.2.3 to the present situation.

Definition Let

$$f : D \subset \mathbb{R}^k \times \mathbb{R}^s \to \mathbb{R}^s.$$

We say that f satisfies the Lipschitz Condition with respect to y (with Lipschitz constant $q > 0$) if uniformly in x

$$||f(x, y) - f(x, y')|| \leq q\, ||y - y'|| \tag{3.23}$$

whenever $(x, y),\ (x, y') \in D$.

The norms in (3.23) are $\mathbb{R}^s$ (Euclidean) norms; for simplicity of notation, we use the same notation for all norms, unless confusion may not be avoided by the context.

The desired generalization of Lemma 3.2.3 follows.

3.3.1 Lemma *Let $D \subset \mathbb{R}^k \times \mathbb{R}^s$ be open, with the property: if $(x, y),\ (x, y') \in D$, then the line segment joining these points lies in D. Let $f : D \to \mathbb{R}^s$ be such that the partial derivatives f_{y_j} exist and are continuous in D for all $j = 1, \dots, s$. Suppose that for all $r = 1, \dots, s$, the component functions f_r of f satisfy the condition*

$$||\nabla_y f_r|| \leq \omega_r \tag{3.24}$$

on D, where ω_r are positive constants. Denote $\omega := (\omega_1, \dots, \omega_s)$. Then f is Lipschitz with respect to y on D, with Lipschitz constant $q = ||\omega||$.

Recall that ∇_y denotes the ∇ operator with respect to the variables y_j, $j = 1, \dots, s$.

Proof For any two points (x, y), $(x, y') \in D$, the hypothesis allows us to use the Mean Value theorem (Theorem 2.2.6) on each component f_r, $r = 1, \ldots, s$, of the vector function f. Write $(x, y) = (x, y') + (0, y - y')$; there exist real numbers θ_r, $0 < \theta_r < 1$, such that for all $r = 1, \ldots, s$

$$f_r(x, y) - f_r(x, y') = (0, y - y') \cdot \nabla f_r((x, y') + \theta_r(0, y - y'))$$

$$= (y - y') \cdot \nabla_y f_r((x, y' + \theta_r(y - y')).$$

By the Cauchy-Schwarz inequality

$$|f_r(x, y) - f_r(x, y')| \le ||y - y'|| \, ||\nabla_y f_r((x, y' + \theta_r(y - y'))|| \le \omega_r ||y - y'||$$

by (3.24), since the points $(x, y' + \theta_r(y - y'))$ are on the segment joining (x, y) and (x, y'), and belong therefore to D. Hence

$$||f(x, y) - f(x, y')|| \le ||\omega|| \, ||y - y'||. \qquad \square$$

We generalize next Lemma 3.2.2 to the vector case. Absolute values are replaced by the appropriate norms, and the space $C_b(I)$ by the complete metric space $C_b(I, \mathbb{R}^s)$ (cf. Corollary 3.1.7). The proof of the following generalized lemma is then identical to the proof of Lemma 3.2.2.

3.3.2 Lemma *Let $f : \mathbb{R}^k \times \mathbb{R}^s \to \mathbb{R}^s$ be defined and continuous on the set*

$$D := \{(x, y) \in \mathbb{R}^k \times \mathbb{R}^s;\ |x_i| < a,\ ||y|| \le b,\ (i = 1, \ldots, k)\},$$

and $f(0, 0) = 0$. Suppose f is Lipschitz with respect to y in D (uniformly in x), with Lipschitz constant $q < 1$. Then there exists a positive number $a' < a$ such that for each x in the cell

$$I := \{x \in \mathbb{R}^k;\ |x_i| \le a' \quad (i = 1, \ldots, k)\}, \tag{3.25}$$

the equation

$$f(x, y) = y \tag{3.26}$$

has a unique solution $y \in \mathbb{R}^s$ with $||y|| \le b$. The unique solution $y = \phi(x)$ is continuous in I and $\phi(0) = 0$.

The Implicit Function Theorem for Systems

We are now ready to prove the Implicit Function Theorem for a system of s equations (with s "unknown" $y_1, \ldots, y_s$).

3.3.3 Theorem *Let $F : \mathbb{R}^k \times \mathbb{R}^s \to \mathbb{R}^s$ satisfy the following conditions:*

(i) F is defined and continuous in an open cell Ω centred at the point $(x^0.y^0) \in \mathbb{R}^k \times \mathbb{R}^s$, given by

$$\Omega := \{(x, y) \in \mathbb{R}^k \times \mathbb{R}^s;\ |x_i - x_i^0| < a,\ |y_j - y_j^0| < b,\ (i = 1, \dots, k;\ j = 1, \dots, s)\}; \tag{3.27}$$

(ii) the partial derivatives F_{y_j} exist and are continuous in Ω for all $j = 1, \dots, s$;
(iii) one has

$$F(x^0, y^0) = 0; \qquad \frac{\partial(F_1, \dots, F_s)}{\partial(y_1, \dots, y_s)}(x^0, y^0) \neq 0. \tag{3.28}$$

Then there exist positive numbers $a' < a$ and $b' < b$ such that for each x in the $\mathbb{R}^k$-cell

$$I = \{x \in \mathbb{R}^k;\ |x_i - x_i^0| \leq a',\ i = 1, \dots, k\},$$

the equation

$$F(x, y) = 0 \tag{3.29}$$

has a unique solution y in the $\mathbb{R}^s$-cell

$$J := \{y \in \mathbb{R}^s;\ |y_j - y_j^0| \leq b' \quad (j = 1, \dots, s)\}.$$

This unique solution defines a function $y = \phi(x) : I \to J$, which is continuous on I and satisfies $\phi(x^0) = y^0$.

Proof As before, we assume without loss of generality that $(x^0, y^0) = (0, 0)$. Then

$$\Omega = \{(x, y) \in \mathbb{R}^k \times \mathbb{R}^s;\ |x_i| < a,\ |y_j| < b\ (i = 1, \dots, k,\ \ j = 1, \dots, s)\}, \tag{3.27'}$$

and the hypothesis (3.28) is

$$F(0, 0) = 0; \qquad \det A \neq 0, \tag{3.28'}$$

where $A := (a_{rj})$ denotes the $s \times s$ Jacobian matrix of F with respect to y at $(0, 0)$, that is

$$a_{rj} = \frac{\partial F_r}{\partial y_j}(0, 0) \qquad (r, j = 1, \dots, s). \tag{3.30}$$

For convenience in using matrix notation, we shall view vectors as *columns*. Since A is non-singular (by (3.28′)), we may define $f : \Omega \to \mathbb{R}^s$ by

$$f := y - A^{-1} F. \tag{3.31}$$

The right hand side makes sense when F is considered as a column. Since A is non-singular, Eq. (3.29) is equivalent to the equation

$$f(x, y) = y. \tag{3.32}$$

We verify that f satisfies the hypothesis of Lemma 3.3.2.

By (3.28′), $f(0, 0) = 0$. For any fixed $\beta < b$, f is continuous in the set

$$D := \{(x, y) \in \mathbb{R}^k \times \mathbb{R}^s;\ |x_i| < a,\ ||y|| \le \beta\ (i = 1, \dots, k)\}.$$

Since A is a constant matrix, the hypothesis (ii) implies that the partial derivatives f_{y_j} exist in Ω,

$$f_{y_j} = e^j \quad A^{-1} F_{y_j}, \tag{3.33}$$

hence are continuous in Ω (also by (ii)), for all $j = 1, \dots, s$.

We can rewrite (3.33) in the form

$$\Big(\frac{\partial f_r}{\partial y_j}\Big) = Id - A^{-1}\Big(\frac{\partial F_r}{\partial y_j}\Big),$$

where Id denotes the $s \times s$ identity matrix. Evaluating at (0, 0), we obtain

$$\Big(\frac{\partial f_r}{\partial y_j}\Big)\Big|_{(0,0)} = 0,$$

where 0 on the right hand side denotes the $s \times s$ zero matrix.

Fix an arbitrary positive number $q < 1$. Since $\frac{\partial f_r}{\partial y_j}$ are continuous in Ω and vanish at (0, 0), there exist positive numbers $\alpha < a$ and $\beta' < \beta$ such that $||\nabla_y f_r|| \le q/\sqrt{s}$ (for all $r = 1, \dots, s$) whenever $|x_i| < \alpha$ and $|y_j| < \beta'$ (for all i, j). Hence by Lemma 3.3.1, f is Lipschitz with respect to y in the open cell

$$\{(x, y);\ |x_i| < \alpha,\ |y_j| < \beta' \quad (i = 1, \dots, k;\ j = 1, \dots, s)\},$$

with Lipschitz constant

$$\frac{q}{\sqrt{s}}\,||(1, \dots, 1)|| = q < 1.$$

Thus f satisfies indeed the hypothesis of Lemma 3.3.2 for all (x, y) such that $|x_i| < \alpha$ and $|y_j| \le b'$, with b' any positive number smaller than β' (hence $b' < b$). It follows from Lemma 3.3.2 that there exists a positive number $a' < \alpha$ (hence $a' < a$) such that for each $x \in I := \{x \in \mathbb{R}^k;\ |x_i| \le a'\ (\forall i = 1, \dots k)\}$, the Eq. (3.32) (hence the equivalent equation (3.29)) has a unique solution y with $|y_j| \le b'$ $(\forall j = 1, \dots, s)$. The solution $y = \phi(x)$ is continuous on the cell I, and $\phi(0) = 0$. □

Next we generalize Theorem 3.2.5 to the case of a system of equations.

3.3.4 Theorem *If the hypothesis (i)–(ii) of Theorem 3.3.3 are strengthened by requiring that F be of class C^1 with respect to all its variables in a neighborhood of the point (x^0, y^0) in $\mathbb{R}^k \times \mathbb{R}^s$, then the unique solution ϕ is of class C^1 in a neighborhood of x^0 in $\mathbb{R}^k$, and the partial derivatives of the column ϕ in that neighborhood are given by the formula*

$$\phi_{x_i}\Big|_x = -\Big(\frac{\partial F}{\partial y}\Big)^{-1} F_{x_i}\Big|_{(x,\,\phi(x))}. \tag{3.34}$$

On the right hand side of (3.34), the Jacobian matrix of F with respect to y and the column F_{x_i} are evaluated at the point $(x, \phi(x))$.

Proof As in the proof of Theorem 3.2.5, we apply the Mean Value theorem to the real components F_r of F, $r = 1, \ldots, s$. For x (*fixed*) in a suitable neighborhood V of (x^0, y^0), h small enough (see proof of 3.2.5), $i = 1, \ldots, k$, and $r = 1, \ldots, s$, there exist real numbers θ_{ri} (depending on x and h), $0 < \theta_{ri} < 1$, such that

$$F_r(x + he^i,\ \phi(x + he^i)) = F_r(x,\ \phi(x))$$

$$+(he^i,\ \phi(x+he^i)-\phi(x))\cdot\nabla F_r((x,\ \phi(x))+\theta_{ri}(he^i,\ \phi(x+he^i)-\phi(x))). \tag{3.35}$$

The left hand side and the first summand on the right hand side vanish identically in V by the definition of ϕ. Denote the argument of ∇F_r by z_{ri}, that is

$$z_{ri} = z_{ri}(h) := (x,\ \phi(x)) + \theta_{ri}(he^i,\ \phi(x + he^i) - \phi(x)). \tag{3.36}$$

Equation (3.35) reduces to (cf. proof of 3.2.5)

$$0 = h\frac{\partial F_r}{\partial x_i}(z_{ri}) + \sum_{j=1}^{s}[\phi_j(x + he^i) - \phi_j(x)]\frac{\partial F_r}{\partial y_j}(z_{ri}).$$

For $h \neq 0$ with absolute value small enough, we then have the s linear equations (for each given i)

$$\sum_{j=1}^{s}\frac{\partial F_r}{\partial y_j}(z_{ri})\Big[\frac{\phi_j(x + he^i) - \phi_j(x)}{h}\Big] = -\frac{\partial F_r}{\partial x_i}(z_{ri}), \tag{3.37}$$

$r = 1, \ldots, s$, for the s "unknowns" in square brackets. Using matrix notation with ϕ and F written as column vectors, the above system of linear equations is equivalent to the vector equation

$$\Big(\frac{\partial F_r}{\partial y_j}(z_{ri})\Big)\, h^{-1}[\phi(x + he^i) - \phi(x)] = -\Big(\frac{\partial F_r}{\partial x_i}(z_{ri})\Big). \tag{3.38}$$

The continuity of ϕ in a neighborhood of x^0 (cf. Theorem 3.3.3) implies that $z_{ri} \to (x,\ \phi(x))$ for x in that neighborhood and for all r, i, when $h \to 0$. Since determinants are algebraic sums of products of their entries, it follows from the C^1 hypothesis on F that the determinant of the above linear system of equations converges to

$$\frac{\partial(F_1, \ldots, F_s)}{\partial(y_1, \ldots, y_s)}(x,\ \phi(x))$$

when $h \to 0$. At the point $(x^0,\ \phi(x^0)) = (x^0,\ y^0)$, the above Jacobian does not vanish, by hypothesis. Therefore it does not vanish in a neighborhood of $(x^0,\ y^0)$. Hence, for x in a neighborhood of x^0 and h small enough, the matrix of the system is invertible, and we have for each fixed $i = 1, \ldots, k$

$$h^{-1}[\phi(x + he^i) - \phi(x)] = -\Big(\frac{\partial F_r}{\partial y_j}(z_{ri})\Big)^{-1}\Big(\frac{\partial F_r}{\partial x_i}\Big|_{z_{ri}}\Big). \tag{3.39}$$

As $h \to 0$, the entries of the inverse matrix in (3.39) converge to the corresponding entries of $\Big(\partial F/\partial y\Big)^{-1}\Big|_{(x,\phi(x))}$, by the C^1 hypothesis and the expression of the entries of the inverse matrix as ratios of determinants, which are themselves algebraic sums of products of their entries. The entries of the column vector on the right hand side converge to the corresponding entries of the column $\partial F/\partial x_i\Big|_{(x,\phi(x))}$. We conclude that the partial derivative $\frac{\partial \phi}{\partial x_i}(x)$ exists and is given by the formula in the statement of Theorem 3.3.4. From this formula and the continuity with respect to x of the entries of both matrices on its right hand side, we deduce that the derivatives ϕ_{x_i} are continuous in a neighborhood of x^0, that is, the $\mathbb{R}^s$-valued function ϕ is of class C^1 in an $\mathbb{R}^k$-neighborhood of x^0. □

Example. We consider the system of equations

$$2x - y^2 - z^2 = 0; \quad x^2 - 4y - 2z^2 = 0.$$

The functions F_j defined by the left hand sides of the equations are polynomials, so that the smoothness conditions of Theorem 3.3.4 (and 3.3.3) are trivially satisfied in $\mathbb{R}^3$. Let $p = (4, 2, 2)$. Clearly $F(p) = 0$, and the Jacobian matrix $A := (\frac{\partial(F_1,F_2)}{\partial(x,y)})$ has the rows $(2. -2y)$ and $(2x, -4)$. The Jacobian's value at the point p is 24. Therefore, by Theorem 3.3.3, there exists a neighborhood U of $z = 2$ in $\mathbb{R}$ and a unique function $z \in U \to \phi(z) = (x, y) \in V \subset \mathbb{R}^2$, such that (x, y, z) is a solution of the system of equations in $V \times U$. Moreover $\phi(2) = (4, 2)$, ϕ is of class C^1 in U, and for all $z \in U$,

$$\phi'(z) = -A^{-1}F_z\Big|_{(\phi(z),z)},$$

where F is written as a column. In particular, $\phi'(2) = (2/3)(1, -1)$.

An immediate corollary of the last two theorems is the local inverse map theorem.

The Local Inverse Map Theorem

3.3.5 Theorem *Let*

$$f : \mathbb{R}^k \to \mathbb{R}^k$$

be defined and of class C^1 in a neighborhood W of x^0, and suppose

$$\frac{\partial(f_1, \ldots, f_k)}{\partial(x_1, \ldots, x_k)}(x^0) \neq 0. \tag{3.40}$$

Then there exist $\mathbb{R}^k$-cells U, V centred at x^0 and $y^0 := f(x^0)$ respectively, such that the inverse map g of f is well defined and of class C^1 on V into U.

Proof Consider the function

$$F(x, y) := f(x) - y : \tilde{W} := W \times \mathbb{R}^k \to \mathbb{R}^k, \tag{3.41}$$

and denote $y^0 := f(x^0)$. The function F is of class C^1 in $\tilde{W}$,

$$F(x^0, y^0) = f(x^0) - y^0 = 0,$$

and

$$\left.\frac{\partial(F_1, \ldots, F_k)}{\partial(x_1, \ldots, x_k)}\right|_{(x^0, y^0)} = \left.\frac{\partial(f_1, \ldots, f_k)}{\partial(x_1, \ldots, x_k)}\right|_{x^0} \neq 0.$$

Thus F satisfies the hypothesis of Theorem 3.3.4, and a fortiori those of Theorem 3.3.3. Note that the unknown of the system is x. Hence there exists a cell U centred at x^0 and a cell V centred at y^0 such that for each $y \in V$, the equation $F(x, y) = 0$ (that is, $y = f(x)$) has a unique solution $x = g(y)$ lying in U, and the function g is of class C^1 in V. □

3.3.6 The Jacobian of a Composed Map Let

$$f : U \subset \mathbb{R}^k \to \mathbb{R}^k$$

be of class C^1 in the open subset U of $\mathbb{R}^k$, with values in the open subset $V \subset \mathbb{R}^k$, and suppose

$$g : V \to \mathbb{R}^k$$

is of class C^1. By the Chain Rule (Theorem 2.1.10), the composed map

$$h := g \circ f : U \to \mathbb{R}^k$$

is of class C^1 in U, and denoting

$$y = f(x); \qquad z = g(y),$$

the partial derivatives $\frac{\partial h_i}{\partial x_j}$ (i.e., $\frac{\partial z_i}{\partial x_j}$) are given by the formula

$$\frac{\partial h_i}{\partial x_j} = \sum_{r=1}^{k} \frac{\partial h_i}{\partial y_r} \frac{\partial y_r}{\partial x_j}, \tag{3.42}$$

for $i, j = 1, \ldots, k$. Formula (3.42) means that the Jacobian matrix of h with respect to x is the *product* of the Jacobian matrix of h with respect to y and the Jacobian matrix of $y = f(x)$ with respect to x, in this order. Since the determinant of the product of two square matrices is the product of their determinants, we conclude that

$$\frac{\partial(z_1, \ldots, z_k)}{\partial(x_1, \ldots, x_k)} = \frac{\partial(z_1, \ldots, z_k)}{\partial(y_1, \ldots, y_k)} \frac{\partial(y_1, \ldots, y_k)}{\partial(x_1, \ldots, x_k)}. \tag{3.43}$$

In particular, if $y = f(x)$ satisfies the hypothesis of the Local Inverse Map theorem (Theorem 3.3.5), then for $U,\ V$ as in that theorem, let $g : V \to U$ be the inverse map $x = g(y)$. The composed map $h = g \circ f$ is the identity map on U, i.e., $h(x) = x$ on U. The Jacobian matrix of h with respect to x is the identity matrix; hence the left hand side of (3.43) is equal to 1. Therefore the non-vanishing of the Jacobian $\frac{\partial f}{\partial x}$ in U is *necessary* for the existence of the inverse map g (of class C^1) in V, and we conclude from (3.43) that for all $x \in U$

$$\frac{\partial(x_1, \ldots, x_k)}{\partial(y_1, \ldots, y_k)} = \left[\frac{\partial(y_1, \ldots, y_k)}{\partial(x_1, \ldots, x_k)}\right]^{-1}. \tag{3.44}$$

If we use our usual vector notation $x = (x_1, \ldots, x_k)$, etc., formula (3.44) takes on the very suggestive form

$$\frac{\partial x}{\partial y} = \left[\frac{\partial y}{\partial x}\right]^{-1}.$$

Example. Consider the maps

$$g : (x, y, z) \in (0, \infty)^3 \to (u, v, w) = (xz \log(xy),\ (x+y+z)^{-1},\ \cos(x+y+z))$$

and

$$f : (u, v, w) \in \mathbb{R}^3 \to \left(\frac{\sin u \sin v}{1 + \sin^2 w}, \frac{\cos u \cos v}{1 + \cos^2 w}, \frac{uv^5}{1 + e^{w^2}}\right).$$

The map g is of class C^1 in its domain of definition in $\mathbb{R}^3$. Since v and w are symmetric expressions in x, y, z, $\nabla v = v_x(1, 1, 1)$ and $\nabla w = w_x(1, 1, 1)$ (no calculation needed!). Factoring out v_x and w_x from the second and third rows of the Jacobian $\frac{\partial(u,v,w)}{\partial(x,y,z)}$ (respectively), we see that the latter is equal to $v_x w_x$ multiplied by a determinant with second and third rows both equal to $(1, 1, 1)$. The Jacobian vanishes therefore identically in $(0, \infty)^3$. Since the non-vanishing of the Jacobian is a necessary condition for the existence of a local inverse map (of class C^1), as we observed above, we conclude that g does not have a C^1 local inverse in the neighborhood of any point of its domain.

Let $\psi := f \circ g$ (with domain $\Omega = (0, \infty)^3$). Since g and f are of class C^1 in Ω and $\mathbb{R}^3$ respectively, we may use the product formula to calculate the Jacobian of ψ with respect to (x, y, z):

$$\frac{\partial \psi}{\partial(x, y, z)} = \frac{\partial f}{\partial(u, v, w)} \frac{\partial(u, v, w)}{\partial(x, y, z)} = 0$$

at any point $(x, y, z) \in \Omega$, because the second factor vanishes identically in Ω. Again, no calculation needed!

3.3.7 Exercises

1. In the example of Sect. 3.3.4, calculate $\phi''(2)$.
2. Consider the system of equations

$$2(x^2 + y^2) - z^2 = 0; \qquad x + y + z - 2 = 0.$$

   (a) Prove that the system defines a unique function $\phi : z \to (x(z), y(z))$ from a neighborhood U of $z = 2$ to a neighborhood V of $(1, -1)$, and ϕ is of class C^1 in U.
   (b) Calculate $\phi'(2)$.
   (c) Show that ϕ is of class C^2 in U and calculate $\phi''(2)$.

3.4 Extrema with Constraints

We shall apply the Implicit Function Theorem to find a necessary condition for local extrema under constraints. The method is used to find such possible extrema. Sufficient conditions will not be discussed.

3.4.1 Theorem *Let*

$$F : W \subset \mathbb{R}^k \times \mathbb{R}^s \to \mathbb{R}$$

be of class C^1 in the neighborhood W of (x^0, y^0). Suppose (x^0, y^0) is a local extremum of F under the constraint

$$G(x, y) = 0, \tag{3.45}$$

where $G : W \to \mathbb{R}^s$ is of class C^1, and

$$\left. \frac{\partial(G_1, \ldots, G_s)}{\partial(y_1, \ldots, y_s)} \right|_{(x^0, y^0)} \neq 0. \tag{3.46}$$

Then there exists a constant vector $c \in \mathbb{R}^s$ such that

$$\nabla(F - c \cdot G)\Big|_{(x^0, y^0)} = 0. \tag{3.47}$$

Note that the vector equation (3.47) is equivalent to $k + s$ scalar equations for its components. Together with the vector equation (3.45), which is valid at (x^0, y^0), and is equivalent to s scalar equations, we have $k + 2s$ scalar equations for the $k + 2s$ unknown $x_1^0, \dots, x_k^0, y_1^0, \dots, y_s^0$, and $c_1, \dots, c_s$, where c_j are the components of the vector c.

The parameters c_j are usually called *Lagrange multipliers*, because the method is due to Lagrange, and for each j, c_j multiplies the constraint function G_j in (3.47) when the inner product is expressed in terms of the components, that is, $c \cdot G = \sum c_j G_j$.

Proof The function G satisfies the hypothesis of Theorem 3.3.4, and a fortiori, those of Theorem 3.3.3. Therefore (3.45) has a unique solution $y = \phi(x)$ in a neighborhood $W_1 \subset W$ of (x^0, y^0), ϕ is of class C^1 in a neighborhood U of x^0, and

$$\phi_{x_i}\Big|_x = A^{-1} G_{x_i}\Big|_{(x,\phi(x))} \tag{3.48}$$

for $x \in U$. Here A denotes the Jacobian matrix of G with respect to y, and G is written as a column vector on the right hand side of (3.48).

By our assumption, x^0 is a local extremum of the C^1 function $F \circ \tilde{\phi}$ in the usual sense, where we used the notation $\tilde{\phi}(x) := (x, \phi(x))$ as before. Therefore

$$\nabla F \circ \tilde{\phi}(x^0) = 0. \tag{3.49}$$

Since F and $\tilde{\phi}$ are of class C^1 in the appropriate neighborhoods, we may apply the chain rule to the components of the left hand side of (3.49) (cf. Theorems 2.1.10 and 2.1.5). Since $\phi(x^0) = y^0$, we obtain the k equations

$$F_{x_i}(x^0, y^0) + \sum_{j=1}^{s} F_{y_j}(x^0, y^0)(\phi_j)_{x_i}(x^0) = 0, \tag{3.50}$$

$i = 1, \dots, k$.

Rewrite (3.50) using matrix notation with $\nabla_y F$ a *row* and ϕ_{x_i} a *column* as before:

$$F_{x_i}\Big|_{(x^0, y^0)} + \nabla_y F\Big|_{(x^0, y^0)} \phi_{x_i}(x^0) = 0. \tag{3.51}$$

Substituting (3.48) in (3.51) and using the associative law of matrix multiplication, we obtain

$$\left(F_{x_i} - [(\nabla_y F) A^{-1}] G_{x_i}\right)\Big|_{(x^0, y^0)} = 0, \tag{3.52}$$

for $i = 1, \dots, k$.

Denote by c the constant s-dimensional row vector

$$c := [(\nabla_y F)\, A^{-1}]\Big|_{(x^0, y^0)},$$

and let

$$H := F - c \cdot G.$$

Then (3.52) becomes

$$H_{x_i}(x^0, y^0) = 0 \qquad (i = 1, \ldots, k). \tag{3.53}$$

Since $F \circ \tilde{\phi}$ and $G \circ \tilde{\phi}$ are functions of x only, we have trivially

$$H_{y_j}(x^0, y^0) = 0 \qquad (j = 1, \ldots, s).$$

Together with (3.53), this gives

$$(\nabla H)(x^0, y^0) = 0,$$

as desired. □

3.4.2 Examples

1. Among all boxes with given area, find the box of maximal volume.
Denote the given area by $2B$, and let x, y, z denote the lengths of the edges. We wish to maximize the volume $F(x, y, z) = xyz$ under the constraint $G(x, y, z) := xy + xz + yz - B = 0$. The functions F, G are polynomials, so they trivially satisfy the C^1 requirements in the open set $W = (0, \infty)^3$. We consider $H := F - cG$ (c a real Lagrange multiplier). The necessary condition of Theorem 3.4.1 gives the equations

$$H_x = yz - c(y + z) = 0$$

$$H_y = xz - c(x + z) = 0$$

$$H_z = xy - c(x + y) = 0.$$

The point (x^*, y^*, z^*) for which F is maximal under the given constraint is found among the solutions of the above system of equations. Multiplying these equations by x, y, z respectively, we get

$$xyz - c(xy + xz) = 0$$

$$xyz - c(xy + yz) = 0$$

$$xyz - c(xz + yz) = 0.$$

Since $xyz > 0$ in W, any of the above equations shows that $c \neq 0$. Subtracting the first two equations and dividing by c, we get $(x-y)z = 0$, hence $x = y$ (since $z > 0$ in W). Similarly, the last two equations imply that $y = z$. Therefore $x = y = z$. The constraint equation becomes $3x^2 = B$, and we conclude that there is a *unique* possible local extremum of F under the said constraint, namely

$$(x^*, y^*, z^*) = \sqrt{\frac{B}{3}}(1, 1, 1).$$

This *unique* solution, namely a *cube*, is actually the *global* maximum of F in W under the given constraint. We show this by the following argument.

By the constraint, $xy < xy + xz + yz = B$, hence $y < B/x$, and similarly $z < B/x$ in W. Therefore $F(x, y, z) = xyz < B^2/x$ in W. Since we can interchange the variables, we may replace x by either y or z in the above estimate. If $(x, y, z) \in W \setminus (0, R)^3$ with $R > 3\sqrt{3B}$, at least one of the variables, say x, is $\geq R$, hence $F(x, y, z) < B^2/x \leq B^2/R < (B/3)^{3/2} = F(x^*, y^*, z^*)$.

The set

$$E := \{(x, y, z) \in [0, R]^3;\ xy + xz + yz = B\},$$

is closed and bounded, hence compact. Therefore the continuous function F attains its maximum on E at a point $(x_0, y_0, z_0) \in E$. This point is necessarily an interior point of $[0, R]^3$, because the values of F on the faces of this cube are either 0 or where shown to be smaller than the value at the point (x^*, y^*, x^*), which belongs to E (since $x^* = y^* = z^* = \sqrt{B/3} < R/9$). It follows that (x_0, y_0, z_0) is a critical point of $F|_E$. Since (x^*, y^*, z^*) is the unique critical point of the latter function, it follows that for all $(x, y, z) \in (0, R)^3$, under the given constraint, $F(x, y, z) \leq F(x_0, y_0, z_0) = F(x^*, y^*, z^*)$. Together with the estimate obtained above for points in $W \setminus (0, R)^3$, we conclude that $F(x, y, z) \leq F(x^*, y^*, z^*)$ (under the given constraint) for all points $(x, y, z) \in W$. This shows that the point $(x^*, y^*, z^*) = \sqrt{B/3}(1, 1, 1)$ is the global maximum point of F in W under the given constraint.

2. Let

$$E = \{x \in \mathbb{R}^k;\ x_i \geq 0,\ \sum_i x_i = a\},$$

where $a > 0$ is a given constant. We wish to maximize the function $f(x) = \prod_i x_i$ on E.

If $x_i = 0$ for *some* i, $f(x) = 0$. Since there are points $x \in E$ for which $f(x) > 0$, the value 0 of f is not its maximum value on E. Therefore if x is a maximum point of f in E, we have $x_i > 0$ for all i. By Theorem 3.4.1, we have the necessary condition

$$\nabla[\prod_i x_i - c(\sum_i x_i - a)] = 0,$$

for some real constant c. This vector equation is equivalent to the k scalar equations

$$\prod_{i;i\neq j} x_i = c, \qquad j = 1, \ldots, k.$$

Since $x_j \neq 0$ for all j, we get equivalent equations by multiplying the j-th equation by x_j:

$$\prod_i x_i = cx_j, \qquad j = 1, \ldots, k.$$

The left hand side is positive. Hence $c > 0$, and therefore $x_j = f(x)/c$ for all j. In particular, there is a unique solution x to the above system of equations; all its components are equal, and since their sum is equal to a for $x \in E$, we have $kx_j = a$, hence $x_j = a/k$ for all $j = 1, \ldots, k$.

We wish to show that the point $x^* := (a/k, \ldots, a/k)$ is the global maximum of f on E.

If $x \in E$, $x_j \leq \sum_i x_i = a$ for all j, hence $||x|| \leq a\sqrt{k}$. This shows that the set E is bounded.

The function $g(x) = \sum_i x_i$ is linear, hence continuous on $\mathbb{R}^k$. The set $\mathbb{R} \setminus \{a\}$ is open in $\mathbb{R}$. Therefore $g^{-1}(\mathbb{R} \setminus \{a\})$ is open, that is, the set $\{x;\ g(x) \neq a\}$ is open, and its complement $\{x;\ g(x) = a\}$ is then closed. Since the set $\{x;\ x \geq 0\}$ is closed, the intersection of the last two sets, which is precisely E, is closed. We conclude that E is a closed bounded set in $\mathbb{R}^k$, hence a compact set. Since f is continuous (it is a polynomial!), it assumes its maximum on the compact set E. However f vanishes on ∂E. Therefore any maximum point of f on E is necessarily a solution of the above system of equations. Since the solution x^* is unique, it must be the maximum point of f in E. This shows that $f(x) \leq f(x^*) = (a/k)^k$ for all $x \in E$. Noting that $a = \sum_i x_i$ for all $x \in E$, and taking the non-negative k-th root, we obtain the inequality

$$\Big(\prod_{i=1}^{k} x_i\Big)^{1/k} \leq \frac{\sum_{i=1}^{k} x_i}{k}$$

for all $x \in \mathbb{R}^k$ with non-negative components. This is the so-called *geometric-arithmetic mean* inequality.

3.4.3 Exercises

1. Solve Exercise 10 in Sect. 2.2.14 by using the method of Lagrange multipliers.
2. Let Γ be the intersection curve of the sphere with equation

$$x^2 + y^2 + z^2 = 9$$

and the hyperboloid with equation

$$x^2 + y^2 - z^2 = 1.$$

Find the distance $d(P, \Gamma)$.
Hint: recall that

$$d(P, \Gamma) := \min\{d(P, X);\ X := (x, y, z) \in \Gamma\}.$$

Find the minimum of $d(P, X)^2$.

3. Find the distance from $(0, 0)$ to the hyperbola $7x^2 + 8xy + y^2 = 45$.
4. Find the rectangle with given area that has the smallest perimeter.
5. Among all boxes with edges summing up to a given constant, find the box with maximal volume.
6. Given a point $P := (\alpha, \beta, \gamma)$ and a plane π with the equation

$$ax + by + cz + d = 0,$$

find a formula for the distance $d(P.\pi)$. See the hint in Exercise 2.

7. Find the local extrema of the function $f(x, y, z) = x^2 + y^2 + z^2$ under each one of the constraints given below:
   (a) $(x/a)^2 + (y/b)^2 + (z/c)^2 = 1$, where $a > b > c > 0$.
   (b) $x^4 + y^4 + z^4 = 1$.
8. Find a rectangle with maximal area inscribed in the ellipse $x^2 + 2y^2 = 1$.
9. Find the global maximum and minimum of the function $f(x, y, z) = xy + yz$ under the constraints $x^2 + y^2 = 1$ and $y^2 + z^2 = 4$.

3.5 Applications in $\mathbb{R}^3$

In this section, we shall apply the general tools of the last two chapters to some geometric concepts in three real dimensions. The content of the section is more elementary and somewhat intuitive, but relies on some basic facts from Linear Algebra. We shall proceed at a rather slow pace, and the presentation does not necessarily use the shortest way to its conclusions.

For simplicity of notation, vectors in $\mathbb{R}^3$ will be denoted (x, y, z). The unit vectors in the direction of the coordinate axis will be denoted $i := (1, 0, 0)$, $j := (0, 1, 0)$, and $k := (0, 0, 1)$. The norm is the Euclidean norm.

3.5.1 Surfaces Let

$$F : W \subset \mathbb{R}^3 \to \mathbb{R}$$

be defined and of class C^1 in the neighborhood W of the point (x_0, y_0, z_0), and

$$F(x_0, y_0, z_0) = 0 : \qquad F_z(x_0, y_0, z_0) \neq 0. \tag{3.54}$$

We are interested in the *surface* (or *manifold*) M consisting of all the solutions $(x, y, z) \in W$ of the equation

$$F(x, y, z) = 0, \tag{3.55}$$

that is,

$$M := \{(x, y, z) \in W;\ F(x, y, z) = 0\}. \tag{3.56}$$

The equation (3.55) is called *the equation of the surface M*.

By the Implicit Function Theorem (cf. Theorems 3.2.4 and 3.2.5), the hypothesis on F imply that (3.55) is locally equivalent to the explicit equation

$$z = \phi(x, y), \tag{3.57}$$

for (x, y, z) in an appropriate neighborhood W_1 of (x_0, y_0, z_0) in $\mathbb{R}^3$, and M is then locally the *graph* of ϕ over a neighborhood U of (x_0, y_0) in $\mathbb{R}^2$, that is, the set

$$\{(x, y, \phi(x, y));\ (x, y) \in U\}. \tag{3.58}$$

Locally, the function ϕ is uniquely determined and of class C^1, and

$$\phi_x = -\frac{F_x}{F_z}; \qquad \phi_y = -\frac{F_y}{F_z}. \tag{3.59}$$

The left hand side in the above relations is evaluated at $(x, y) \in U$, while the right hand side is evaluated at the corresponding point $(x, y, \phi(x, y))$ on the graph M.

Since the variables can be interchanged, similar conclusions are valid if the hypothesis $F_z \neq 0$ at (x_0, y_0, z_0) is replaced by the non-vanishing of anyone of the partial derivatives of F at that point, i.e., by the non-vanishing of ∇F at (x_0, y_0, z_0).

Examples. 1. The equation of the sphere of radius R centred at $(0, 0, 0)$ is

$$M: \quad x^2 + y^2 + z^2 - R^2 = 0. \tag{3.60}$$

Since $\nabla F = 2(x, y, z)$ vanishes only at $(0, 0, 0)$, which is not a point on the sphere, the surface M can be represented *locally* as the graph of a unique function expressing one variable in term of the other two. The word "locally" cannot be omitted in the above assertion, because (3.60) does not have "globally" a *unique* solution for anyone of its variables in terms of the other two. There are in fact *two* C^1 solutions, $z = (R^2 - x^2 - y^2)^{1/2}$ and $z = -(R^2 - x^2 - y^2)^{1/2}$ for (x, y) in the open disc: $x^2 + y^2 < R^2$.

2. The same analysis can be made more generally for the *ellipsoid*, whose "canonical equation" is

$$\frac{x^2}{a^2} + \frac{y^2}{b^2} + \frac{z^2}{c^2} = 1 \qquad (a, b, c > 0),$$

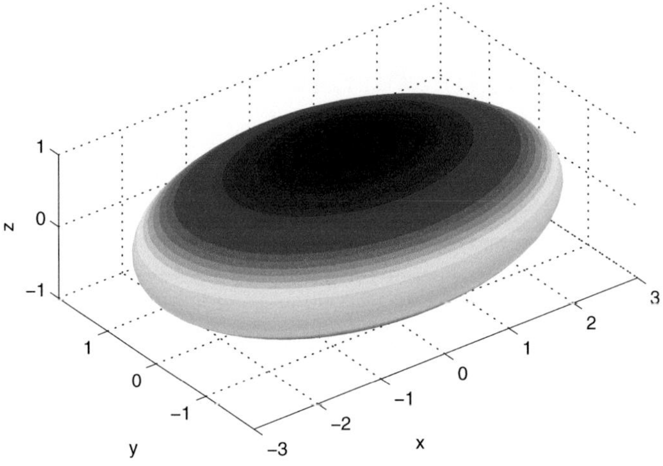

Fig. 3.1 $\frac{x^2}{9} + \frac{y^2}{4} + z^2 = 1$

or for the *one-sheeted* and the *two-sheeted hyperboloids*, whose canonical equations are respectively

$$\frac{x^2}{a^2} + \frac{y^2}{b^2} - \frac{z^2}{c^2} = 1 \quad and \quad \frac{x^2}{a^2} - \frac{y^2}{b^2} - \frac{z^2}{c^2} = 1.$$

These surfaces are sketched in Figs. 3.1, 3.2 and 3.3.

On the other hand, the *elliptic* and the *hyperbolic paraboloids*, whose canonical equations are respectively

$$\frac{x^2}{a^2} + \frac{y^2}{b^2} = 2pz \quad and \quad \frac{x^2}{a^2} - \frac{y^2}{b^2} = 2pz \qquad (a, b, p > 0),$$

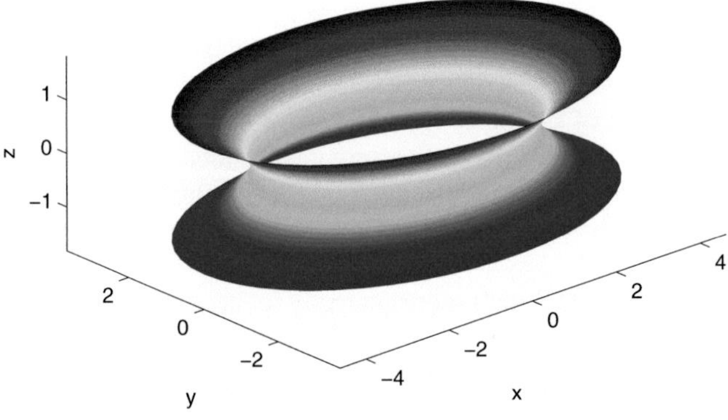

Fig. 3.2 $\frac{x^2}{9} + \frac{y^2}{4} - z^2 = 1$

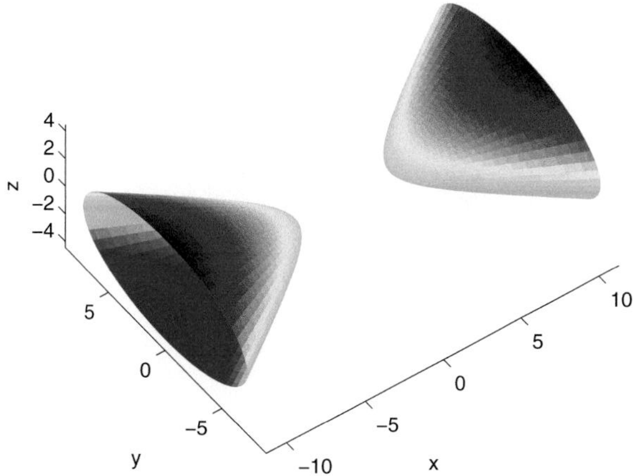

Fig. 3.3 $\frac{x^2}{9} - \frac{y^2}{4} - z^2 = 1$

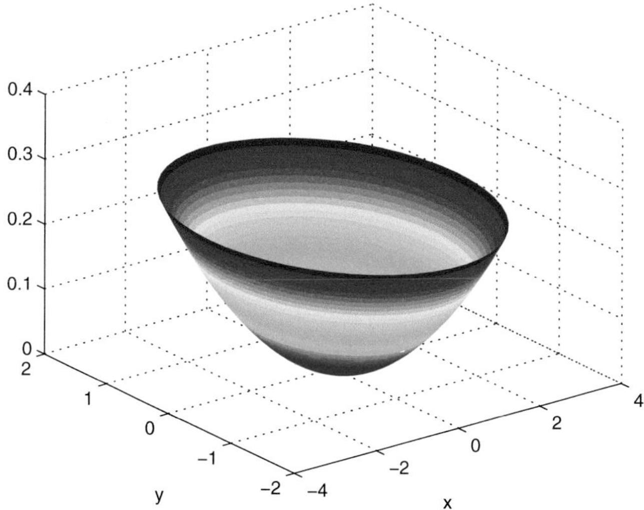

Fig. 3.4 $\frac{x^2}{9} + \frac{y^2}{4} = 4z$

are globally the graphs of functions $z = \phi(x, y)$. These graphs are sketched in Figs. 3.4 and 3.5.

The representation (3.58) of M as the graph of a function $\phi : U \subset \mathbb{R}^2 \to \mathbb{R}$ is a special case of a *parametric representation* (or *parametrization*) of the surface M, in which M is represented as the range of an $\mathbb{R}^3$-valued C^1 function defined in an open set $V \subset \mathbb{R}^2$, that is, a vector valued function of *two parameters*, say s and t:

$$f : (s, t) \in V \to f(s, t) \in \mathbb{R}^3. \tag{3.61}$$

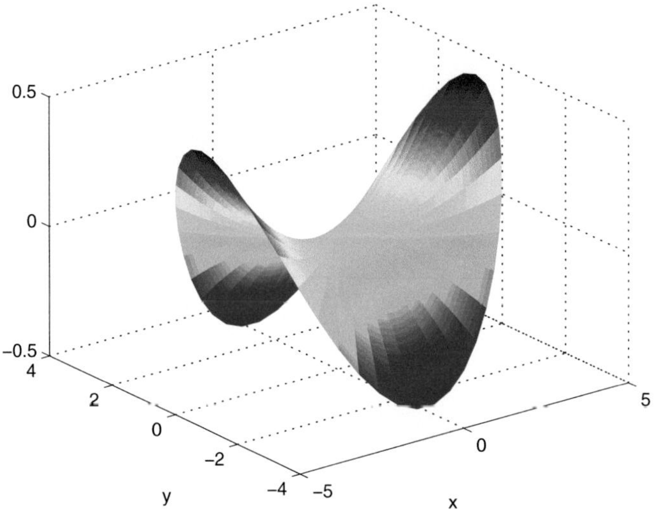

Fig. 3.5 $\frac{x^2}{9} - \frac{y^2}{4} = 4z$

In the special parametric representation (3.58), the parameters are x and y, (x, y) varies in U, and

$$f(x, y) = (x, y, \phi(x, y)) \in \mathbb{R}^3.$$

Examples. 1. *Parametric representation of the sphere.* The sphere M considered above can be parametrized by using the latitude θ and the longitude ϕ of a point on it. We take the latitude to range from 0 at the North Pole to π at the South Pole, while the longitude ranges from 0 to 2π. The orthogonal projection of the point $(x, y, z) \in M$ onto the xy-plane is at a distance $R \sin\theta$ from the origin, so that its plane *polar* coordinates are $[R \sin\theta,\ \phi]$. The corresponding Cartesian coordinates in the xy-plane are then $(R \sin\theta \cos\phi,\ R \sin\theta \sin\phi)$. The orthogonal projection z of the given point onto the z axis is $R \cos\theta$. Thus we have the parametrization of M by means of the function

$$f(\theta, \phi) = (R \sin\theta \cos\phi,\ R \sin\theta \sin\phi,\ R \cos\theta), \tag{3.62}$$

where the two parameters θ, ϕ vary such that (θ, ϕ) are in the cell $[0, \pi] \times [0, 2\pi]$. This is the so-called *polar* or *spherical* parametrization of the sphere M.

2. *Parametric representation of a plane M.* Let $p^i = (x_i, y_i, z_i)$ $(i = 1, 2, 3)$ be three non-colinear given points in M. The vectors $u := p^2 - p^1$ and $v := p^3 - p^1$ are linearly independent (since the given points are not co-linear) and belong to the plane M. Consequently, a point $X := (x, y, z) \in \mathbb{R}^3$ belongs to M iff $X - p^1$ is a linear combination of u and v:

$$M = \{X \in \mathbb{R}^3;\ X = p^1 + su + tv, \quad (s, t) \in \mathbb{R}^2\}. \tag{3.63}$$

The parametric representation (3.63) of the plane M as the range of an $\mathbb{R}^3$ valued function $f : \mathbb{R}^2 \to \mathbb{R}^3$ is achieved with the function $f(s,t) = p^1 + su + tv$, ($(s,t) \in \mathbb{R}^2$), where p^1, u, v are given as above.

The vector equation in (3.63) is equivalent to three scalar equations for the components. Consider these equations as a system of three homogeneous linear equations for the three unknowns s, t, and -1. Since they have solutions $(s,t,-1) \neq (0,0,0)$, the system's determinant must vanish. The columns of this determinant are $X - p^1, u$, and v, written as columns. Since the determinant's value is unchanged by transposing the corresponding matrix, it will be more convenient to write the necessary condition we obtained on X for being a point of M, as the vanishing of the determinant with the *rows* $X - p^1$, u, and v. Expanding the latter determinant with respect to the first row, we get the equation

$$A \cdot (X - p^1) = 0, \tag{3.64}$$

where A denotes the vector *whose components are the algebraic cofactors of the elements of the determinant's first row*. Symbolically, we can write A as the determinant with rows $A_1 = (i, j, k)$, $A_2 = u$, and $A_3 = v$. This is a "symbolic" determinant, because the components of the row A_1 are *vectors*, rather than scalars. The vector A is called the *cross product* or *vector product* of u and v in this order, and is denoted $u \times v$.

Conversely, suppose $X - p^1$ satisfies (3.64). Equation (3.64) is a system of one homogeneous linear equation for the three unknown coordinates of $X - p^1$, and the matrix of the system has rank 1 ($A \neq 0$, because u and v are linearly independent). The solution space of the system has then the algebraic dimension $3 - 1 = 2$. Two obvious solutions are $X - p^1 = u$ and $X - p^1 = v$, because the determinant expressed by the left hand side of (3.64) has then two equal rows, hence vanishes. Since these solutions are linearly independent, they are a basis for the 2-dimensional solution space of the system. Consequently, the latter's elements $X - p^1$ can be represented as arbitrary linear combinations of u and v, that is, $X - p^1 = su + tv$ with $s, t \in \mathbb{R}$, and (3.63) is satisfied. Thus the parametric representation (3.63) of the plane is equivalent to the representation of the plane by means of the implicit equation (3.64), where $A \neq 0$ is a constant vector, and p^1 is a given point on the plane. We call (3.64) the equation of the plane through p^1 with *normal vector* A. In this context, "normal" means orthogonal to the plane, i.e., to all the vectors $X - p^1$ lying in the plane; the orthogonality is the geometric interpretation of the vanishing of the inner product in (3.64). In our particular construction, $A = u \times v$ with u, v two linearly independent vectors lying in the plane; however *any* normal vector will do, since the meaning of (3.64) is precisely the orthogonality of the vector A and any vector $X - p^1$ lying in the plane.

3.5.2 The Cross Product In the preceding section, we defined the cross product $u \times v$ of two vectors u, v in $\mathbb{R}^3$ as the symbolic determinant with rows (i, j, k), u, and v, in this order. The properties of the determinant as a function of its rows imply the following properties of the cross product.

(a) The cross product is a bilinear function of if its factors.
(b) $v \times u = -u \times v$ (*anti-commutativity*).
(c) For any vector a, the so-called *triple product* $a \cdot (u \times v)$ is equal to the determinant with rows a, u, v, in this order.
In particular, taking $a = u$ or $a = v$ in (c), we get a determinant with two equal rows, hence the triple product vanishes. Consequently:
(d) The cross product is orthogonal to its factors, hence to the plane they determine, if they are not co-linear. In other words, $u \times v$ is a normal vector to that plane.
(e) The cross product vanishes if its factors are linearly dependent. In particular, $u \times u = 0$ for any vector u.

A trivial calculation gives

$$i \times j = k; \quad j \times k = i; \quad k \times i = j. \tag{3.65}$$

As suggested by (3.65), the standard convention is to consider $u \times v$ as a "positively oriented" normal vector to the plane determined by the factors (when u, v are linearly independent).

Geometric model.

We recall informally the standard geometric model for vectors as (cosets of) directed segments $\overline{AB}$, where A is the initial point and B is the final point of the segment, both points in $\mathbb{R}^3$. The coseting is done relative to the equivalence relation between directed segments, whereby two such segments $\overline{AB}$ and $\overline{CD}$ $(A, B, C, D \in \mathbb{R}^3)$ are equivalent if there is a translation of the space which maps one segment onto the other, i.e., if $ABDC$ is a parallelogram. The components u_i of the vector u are the orthogonal projections of u onto the coordinate axis. Componentwise addition of the vectors u, v corresponds to the elementary triangle law (or parallelogram law) of addition of forces: if $\overline{AB}$ and $\overline{BC}$ are representatives of (the cosets) u and v respectively, in which B is both the final point of the first segment and the initial point of the second segment, then $u + v$ is represented by the directed segment $\overline{AC}$.

If u is the vector represented by the directed segment from the point $X \in \mathbb{R}^3$ to the point $Y \in \mathbb{R}^3$, then $u_i = Y_i - X_i$. In particular, the ordered triple (X_1, X_2, X_3) represents *both* the point X and the vector represented by the directed segment from the point $0 := (0, 0, 0)$ to the point X; the latter is called the *position vector* of the point X. Any point in $\mathbb{R}^3$ represents the zero vector 0.

The Euclidean norm of u defined in terms of the components coincides with the length of any directed segment representing u. Componentwise multiplication of the vector u by the scalar -1 produces the vector $-u$ such that $u + (-u) = 0$; geometrically, if u is represented by the directed segment $\overline{XY}$, then $-u$ is represented by the directed segment $\overline{YX}$.

For $t > 0$, componentwise multiplication of the vector u by the scalar t corresponds to a directed segment on the same line, with the same orientation, and with length multiplied by t. For $t < 0$, $tu = (-t)(-u)$; and of course $tu = 0$ for

$t = 0$. The vectors u, v are linearly independent in the algebraic sense iff they are represented by non-colinear non-zero segments.

For simplicity of language, we identify from now on the vector u with any directed segment representing it.

Denote by θ the angle from u to v, where we assume that the vectors are not colinear. The vector $w = v - u$ gives the third edge of the triangle determined by the vectors u and v. By the cosine theorem of trigonometry,

$$||w||^2 = ||u||^2 + ||v||^2 - 2||u|| \, ||v|| \cos\theta.$$

On the other hand, by the bilinearity of the dot product,

$$||w||^2 = w \cdot w = (v - u) \cdot (v - u) = ||v||^2 + ||u||^2 - 2u \cdot v.$$

Comparing the last two equations, we get

$$u \cdot v = ||u|| \, ||v|| \cos\theta. \tag{3.66}$$

We see directly that (3.66) is also valid when u, v are colinear, i.e., when one vector is a scalar multiple of the other. In retrospect, Eq. (3.66) shows that for non-zero vectors u, v, orthogonality in the sense above, namely the relation $u \cdot v = 0$, is equivalent to the relation $\cos\theta = 0$, which is the usual geometric meaning of orthogonality, namely $\theta = \pi/2$ or $\theta = 3\pi/2$ in the range $(0, 2\pi)$ of θ.

The area S of the parallelogram with sides u, v is $||u|| \, ||v|| \, |\sin\theta|$, hence by (3.66)

$$\begin{aligned} S^2 = ||u||^2||v||^2(1 - \cos^2\theta) &= ||u||^2 \, ||v||^2 - [||u|| \, ||v|| \cos\theta]^2 \\ &= ||u||^2||v||^2 - (u \cdot v)^2. \end{aligned} \tag{3.67}$$

Writing (3.67) in terms of the components u_i and v_i of u and v respectively and doing some straightforward algebraic computation, we find out that the right hand side of (3.67) is equal to $||u \times v||^2$. Hence

$$||u \times v|| = S. \tag{3.68}$$

3.5.3 Tangent Vectors Let $a < b$ be real numbers, and let $f : [a, b] \to \mathbb{R}^3$ be continuous. The range γ of f is called a curve (or path) in $\mathbb{R}^3$; f is a parametric representation (or parametrization) of the curve. If f is of class C^1, the curve is said to be *smooth*.

Examples. 1. Given $A, B \in \mathbb{R}^3$, if $f(t) = A + t(B - A), 0 \le t \le 1$, the corresponding curve γ is the line segment from A to B. The same function with the parameter $t \in \mathbb{R}$ unrestricted, has the line through A and B as its range; the equation

$$X = A + t(B - A) \qquad (t \in \mathbb{R}) \tag{3.69}$$

is then called the parametric representation of that line. More generally, the equation

$$X = A + tu \qquad (t \in \mathbb{R}) \tag{3.70}$$

is the parametric representation of the line through the point A, which contains (a representative of) the given non-zero vector u.
2. Given $\theta_0 \in [0, \pi]$ and $R > 0$, the curve γ corresponding to the function

$$f(\phi) = (R \sin\theta_0 \cos\phi,\; R \sin\theta_0 \sin\phi,\; R \cos\theta_0),$$

($\phi \in [0, 2\pi]$) is the θ_0-latitude circle on the sphere of radius R centred at 0. Similarly, if we fix the longitude $\phi = \phi_0$, the curve corresponding to the function

$$g(\theta) = (R \sin\theta \cos\phi_0,\; R \sin\theta \sin\phi_0,\; R \cos\theta),$$

($\theta \in [0, \pi]$) is the ϕ_0-longitude semi-circle on the sphere. The two curves meet at the unique point on the sphere with the spherical coordinates $[R, \theta_0, \phi_0]$.

Let γ be parametrized by the differentiable function $f : t \in [a, b] \to \mathbb{R}^3$, let $s, t \in [a, b]$, $s \neq t$. The vector $f(t) - f(s)$ is the chord segment from the point $f(s)$ to the point $f(t)$ on γ; therefore the corresponding differential ratio is a vector in the direction of that chord, or the opposite direction, if $t - s < 0$. The limit as $t \to s$ (i.e., $f'(s)$) is then a vector on the line tangent to γ at the point $f(s)$. We *define* $f'(s)$ as *the tangent vector to* γ *at the point* $f(s)$. By (3.70), the tangent line to γ at the point $f(s)$ has the parametric representation

$$X = f(s) + t\, f'(s) \qquad (t \in \mathbb{R}). \tag{3.71}$$

If we replace the parameter t by a parameter τ such that $\tau \to t(\tau)$ is a C^1 increasing function mapping the interval $[\alpha, \beta]$ onto $[a, b]$, we get a *re-parametrization* of the same curve γ as the ordered range of the vector valued function

$$f \circ t : [\alpha, \beta] \to \mathbb{R}^3.$$

By the chain rule

$$(f \circ t)'(\tau_0) = t'(\tau_0) f'(t_0),$$

where $t_0 := t(\tau_0)$, $\tau_0 \in [\alpha, \beta]$. The tangent vector $f'(t_0)$ at the point $p^0 = f(t_0)$ on γ has been multiplied by the positive scalar $t'(\tau_0)$, so that we get the same tangent vector at $p^0 = (f \circ t)(\tau_0)$, up to a change of magnitude. In any case, the tangent line to the curve γ at p^0 is unchanged by the re-parametrization.

3.5.4 Tangent Plane Consider a surface $M \subset \mathbb{R}^3$ parametrized by a C^1 function $f : V \subset \mathbb{R}^2 \to \mathbb{R}^3$ (V open), and let $(s_0, t_0) \in V$. The curves parametrized by $f(\cdot, t_0)$ and $f(s_0, \cdot)$ (with the parameters s and t in adequate intervals centred at s_0 and t_0 respectively) lie on M, and go through the point $f(s_0, t_0)$. The tangent vectors to these curves at this point are $f_s(s_0, t_0)$ and $f_t(s_0, t_0)$ respectively. If they are linearly independent (equivalently, if their cross product does not vanish), they span the *tangent plane of* M *at the point* $f(s_0, t_0)$. By (3.64), the equation of the tangent plane to the surface M at the point $f(s_0, t_0)$ is then

$$(f_s \times f_t)\Big|_{(s_0,t_0)} \cdot (X - f(s_0, t_0)) = 0. \tag{3.72}$$

When M is given as the graph of a C^1 function

$$\phi : U \subset \mathbb{R}^2 \to \mathbb{R}^3,$$

we look at the corresponding parametrization of M by the function

$$f(x, y) := (x,\ y,\ \phi(x, y)) : U \to \mathbb{R}^3.$$

The parameters are x and y. At the point

$$p^0 := (x_0, y_0, z_0) = (x_0, y_0,\ \phi(x_0, y_0))$$

on M, the tangent vectors f_x and f_y are

$$(1, 0, \phi_x(x_0, y_0)); \qquad (0, 1,\ \phi_y(x_0, y_0)).$$

Their cross product is

$$A = (-\phi_x,\ -\phi_y,\ 1)\Big|_{(x_0,y_0)}. \tag{3.73}$$

This is a normal vector to the tangent plane at p^0; it is also called a normal vector to the *surface* M at the point p^0. The Eq. (3.72) of the tangent plane to M at p^0 takes now the form

$$z = z_0 + \phi_x\Big|_{(x_0,y_0)}(x - x_0) + \phi_y\Big|_{(x_0,y_0)}(y - y_0). \tag{3.74}$$

If M is given by the equation

$$F(x, y, z) = 0$$

where F is a C^1 real valued function in a neighborhood of the point $p^0 = (x_0, y_0, z_0) \in M$ and $\nabla F(p^0) \neq 0$, then, assuming $F_z(p^0) \neq 0$ without loss of generality, the surface M is locally the graph of a uniquely determined C^1 function ϕ:

$$M = \{(x, y,\ \phi(x, y));\ \ (x, y) \in U\}$$

(cf. (3.58)). It follows from (3.59) and from (3.74) that the equation of the tangent plane to the surface M at the point $p^0 \in M$ is

$$F_x(p^0)(x - x_0) + F_y(p^0)(y - y_0) + F_z(p^0)(z - z_0) = 0, \tag{3.75}$$

that is, in vector notation,

$$\nabla F(p^0) \cdot (X - p^0) = 0. \tag{3.76}$$

Note that (3.76) means that the gradient vector $\nabla F(p^0)$ is a normal vector to M at the point $p^0 \in M$.

Remark Given $c \in \mathbb{R}$, the *level surface* M_c of the function $F : W \subset \mathbb{R}^3 \to \mathbb{R}$ at level c is defined by

$$M_c := \{(x, y, z) \in W;\ \ F(x, y, z) = c\}. \tag{3.77}$$

Assuming as before that F is of class C^1 in a neighborhood W of p^0, the gradient vector $\nabla F(p^0)$ ($p^0 \in M_c$) is a normal vector at p^0 to the level surface M_c of F. Indeed, if $G := F - c$, the equation of M_c is $G(x, y, z) = 0$ and $\nabla G = \nabla F$, so that the conclusion follows from the preceding observation.

Example. Let $F(x, y, z) = x^2 + y^2 + z^2$. The level surface M_c is empty for $c < 0$, it is the singleton $\{0\}$ for $c = 0$, and it is the sphere centred at the origin with radius $\sqrt{c}$ if $c > 0$. In the later case, the point $p^0 = \sqrt{c/3}(1, 1, 1)$ is on M_c. We have $\nabla F(p^0) = 2\sqrt{c/3}(1, 1, 1)$. This vector, hence also the vector $(1, 1, 1)$, is normal to the sphere M_c. The equation of the tangent plane to M_c at p^0 is then $(1, 1, 1) \cdot (X - p^0) = 0$, that is,

$$(x - \sqrt{\frac{c}{3}}) + (y - \sqrt{\frac{c}{3}}) + (z - \sqrt{\frac{c}{3}}) = 0.$$

3.5.5 Exercises

1. (a) Let M be a quadratic surface in $\mathbb{R}^3$, given by the general equation

$$xAx^t + 2xb^t = c,$$

where $x = (x_1, x_2, x_3) \in \mathbb{R}^3$, x^t is the transpose of x, $b \in \mathbb{R}^3$, $c \in \mathbb{R}$, and A in a non-zero 3×3 real symmetric matrix. Let us say that the point $p \in M$ is *permissible* if $Ap^t + b^t \neq 0$. Show that the equation of the tangent plane to M at a permissible point $p \in M$ is

$$xAp^t + (x + p)b^t = c.$$

Formally, the quadratic form xAx^t and the linear form $2xb^t$ in the equation of the surface are replaced in the equation of the tangent plane at p by the corresponding *bilinear form* xAp^t and linear form $(x+p)b^t$, respectively.

(b) Let M be the circular cylinder with radius r and axis along the x_2-axis. Find the equation of the tangent plane to M at the point $p = (r/2)(1, \sqrt{2}, \sqrt{3})$.

(c) Let M be the circular cone

$$M = \{x \in \mathbb{R}^3;\ x_1^2 + x_2^2 - 2x_3 = 0\}.$$

Find a point $p \in M$ such that the tangent plane to M at p is orthogonal to the vector $(1, 1, -2)$.

2. Let M be the surface parametrized by

$$f(s,t) = ((1+s)\cos t,\ (1-s)\sin t,\ s) : \mathbb{R}^2 \to \mathbb{R}^3.$$

Find the equations of the *two* tangent planes to M at the point $(1, 0, 1) \in M$.

3. The circle with equation $(y-r)^2 + z^2 = a^2$ in the yz-plane $(r > a)$ is rotated about the z-axis. The surface M generated in this manner is a *torus*.

(a) Show that a parametrization of M is given by the function

$$f(s,t) := \Big((r + a\cos s)\cos t,\ (r + a\cos s)\sin t,\ a\sin s\Big) : \mathbb{R}^2 \to \mathbb{R}^3.$$

What are the geometric definitions of the parameters?

(b) Find the unit normal to M when $r = 2a$ and $s = t = \pi/4$.

4. Let $x = f(t) : [a, b] \to \mathbb{R}^3$ be a parametrization of a curve γ in $\mathbb{R}^3$, and suppose $f'(c)$ exists for some $c \in [a, b]$. Find

(a) a parametric representation of the tangent line l to γ at the point $p := f(c)$.

(b) the equation of the plane through the point p, orthogonal to l.

5. Let l be the intersection of the quadratic surfaces with equations $x = y^2$ and $x = 1 - z^2$ in an x, y, z system of coordinates.

(a) Find parametrizations of the curves in l. Hint: if $x = t$, then $t \in [0, 1]$ for points of l, and the corresponding x, y may be obtained as (real) square roots; alternatively, since $t \in [0, 1]$, we may set $t = \cos^2 s$ and get parametrizations not involving square roots.

(b) Find a parametrization of the tangent line to the curve in l through the point $(1/2,\ 1/\sqrt{2},\ -1/\sqrt{2})$.

6. The function $F : \mathbb{R}^k \to \mathbb{R}$ is homogeneous of degree n if

$$F(tx) = t^n F(x) \qquad (x \in \mathbb{R}^k,\ t \in \mathbb{R}).$$

(a) Prove that if $F : \mathbb{R}^k \to \mathbb{R}$ is homogeneous of degree n and of class C^1, then

$$x \cdot \nabla F|_x = nF(x) \qquad (x \in \mathbb{R}^k).$$

(This is Euler's formula.)

(b) If F is homogeneous of degree n and of class C^m $(m \geq 1)$, then

$$(x \cdot \nabla)^r F = n(n-1)\ldots(n-r+1)F$$

for all $r = 1, \ldots, m$.

(c) Let $F : \mathbb{R}^3 \to \mathbb{R}$ be homogeneous of degree n and of class C^1, and let M_c be the level surface of F at level c. If $p \in M_c$ is such that $\nabla F|_p \neq 0$, prove that the equation of the plane tangent to M_c at p is

$$\nabla F|_p \cdot x = nc.$$

7. Let $x, y, z, w \in \mathbb{R}^3$. Prove
   (a) $(x \times y) \cdot (z \times w) = (x \cdot z)(y \cdot w) - (x \cdot w)(y \cdot z)$.
   (b) $(x \times y) \times z = (x \cdot z)\, y - (y \cdot z)\, x$.
   (c) $(||y||\, x + ||x||\, y) \cdot (||y||\, x - ||x||\, y) = 0$.
   (d) $\det(x, y, z) = x \cdot (y \times z)$.
   (e) $\det(x, y, x \times y) = ||x \times y||^2$.

3.6 Application to Ordinary Differential Equations

In this section, we shall use the Banach Fixed Point Theorem to prove one of the versions of the Existence and Uniqueness Theorem for ODE (Ordinary Differential Equations).

3.6.1 The Setting In order to avoid repetitions, we fix the setting as follows. For convenience, the norm we choose on $\mathbb{R}^k$ is $||\cdot|| = ||\cdot||_1$.

Consider the compact set $I \subset \mathbb{R} \times \mathbb{R}^k$ given by

$$I := \{(x, y) \in \mathbb{R} \times \mathbb{R}^k;\ |x - x^0| \leq \alpha,\ ||y - y^0|| \leq \beta\},$$

where $\alpha, \beta > 0$.

We are given a function $f : I \to \mathbb{R}^k$. To avoid trivialities, f is not identically 0. A *local solution* in a neighborhood of (x^0, y^0) of the ODE

$$y' = f(x, y) \tag{3.78}$$

is an $\mathbb{R}^k$-valued function $y = y(x)$ defined and differentiable in a closed neighborhood $U = \{x \in \mathbb{R};\ |x - x^0| \leq a\}$, $0 < a \leq \alpha$, with values in the closed neighborhood $V = \{y \in \mathbb{R}^k;\ ||y - y^0|| \le b\}$, $0 < b \le \beta$, (hence $U \times V \subset I$ and $f(x, y(x))$ is well-defined for all $x \in U$), such that

$$y'(x) = f(x,\ y(x)) \qquad (x \in U). \tag{3.79}$$

If the local solution y satisfies also the *initial condition*

$$y(x^0) = y^0, \tag{3.80}$$

it is called a local solution of the *Cauchy Initial Value Problem*

$$y' = f(x, y) \qquad y(x^0) = y^0 \tag{IVP}$$

in the compact neighborhood $U \times V$ of (x^0, y^0).

Existence and Uniqueness Theorem

3.6.2 Theorem *Let*

$$I = \{(x, y) \in \mathbb{R} \times \mathbb{R}^k;\ |x - x^0| \le \alpha,\ ||y - y^0|| \le \beta\}$$

and suppose $f \in C(I, \mathbb{R}^k)$ is Lipschitz with respect to y in I, uniformly in x. Let

$$0 < a \le \min(\alpha, \frac{\beta}{M}), \qquad b = \beta,$$

where $M \ge ||f||$ (the norm of f in $C(I, \mathbb{R}^k)$). Then the Cauchy Initial Value Problem (IVP) has a unique local solution in J, where

$$J = \{(x, y);\ |x - x^0| \le a,\ ||y - y^0|| \le b\}.$$

Proof By the Fundamental Theorem of the Differential and Integral Calculus, y is a local solution of (IVP) in $U \times V$ (see notation in Sect. 3.6.1) if and only if it is continuous from U to V and satisfies the *integral equation*

$$y(x) = y^0 + \int_{x^0}^{x} f(t,\ y(t))\,dt \qquad (x \in U), \tag{3.81}$$

where integration of a vector function is performed component-wise.

Indeed, (IVP) and (3.81) imply each other, by integration and differentiation, respectively. Moreover, it follows from (3.81) that

$$||y(x) - y^0|| = ||y(x) - y^0||_1 \le \Big|\int_{x^0}^{x} ||f(t,\ y(t))||_1\,dt\Big|$$

$$\le M\,|x - x^0| \le M\,a \le \beta$$

whenever $|x - x^0| \leq a$, that is, y maps U into V, as needed in the definition of a local solution of (IVP) in $U \times V$.

We shall apply the following lemma, which is a simple corollary of the Banach Fixed Point Theorem.

Lemma *Let X be a complete metric space, and let $T : X \to X$ be such that T^N is a contraction for some $N \in \mathbb{N}$. Then there exists a unique fixed point for T is X.*

Proof of Lemma By the Banach Fixed Point Theorem, there exists a unique fixed point p of T^N in X. Since $T^N p = p$, it follows that

$$T^N(Tp) = T^{N+1}p = T(T^N p) = Tp,$$

that is, Tp is also a fixed point of T^N. By uniqueness, we conclude that $Tp = p$, that is, p is a fixed point of T in X. If q is also a fixed point of T in X, then $T^N q = q$, and therefore $q = p$ by the uniqueness of the fixed point of the contraction T^N. □

We shall apply the lemma to the complete metric space

$$X := \{y \in C(U, \mathbb{R}^k);\ ||y - y_0|| \leq b\}. \tag{3.82}$$

The norm above is the norm of the Banach space $C(U,\ \mathbb{R}^k)$, that is,

$$||y - y^0|| := \sup_{x \in U} ||y(x) - y_0||_1,$$

and U is the closed interval $[x_0 - a,\ x_0 + a]$. Thus X is the closed ball with radius b and center at the constant function y^0 in the Banach space $C(U,\ \mathbb{R}^k)$. Hence X is indeed a complete metric space (cf. Proposition in Sect. 1.3.7 and Theorem 3.1.6).

Fix $y \in X$ and $x \in U$. If t is a number between x^0 and x, then $t \in U$ (hence $|t - x^0| \leq \alpha$ because $a \leq \alpha$), and

$$||y(t) - y^0||_1 \leq ||y - y^0|| \leq b (= \beta).$$

This shows that $(t,\ y(t)) \in I$, and therefore $f(t,\ y(t))$ is well defined. Moreover $f(\cdot,\ y(\cdot))$ is continuous for such t. Consequently the following integral exists:

$$T(y)(x) := y^0 + \int_{x^0}^{x} f(t,\ y(t))\, dt. \tag{3.83}$$

By continuity of the integral with respect to its upper limit, the function $T(y) : x \in U \to T(y)(x)$ is continuous, and clearly $T(y)(x^0) = y^0$. Moreover, for all $x \in U$,

$$||T(y)(x) - y^0|| = ||\int_{x^0}^{x} f(t,\ y(t))\, dt|| \leq \left| \int_{x_0}^{x} ||f(t,\ y(t))||\, dt \right|$$

$$\leq M\,|x - x^0| \leq M\,a \leq \beta = b.$$

Taking the supremum over all x in the closed interval U, we get

$$||T(y) - y^0|| \leq b.$$

Hence $T(y) \in X$. Thus the function $T : y \in X \to T(y)$ maps X into itself.

Our next step is to show that there exists $N \in \mathbb{N}$ such that T^N is a contraction on X. Let $y, z \in X$. For each $x \in U$,

$$||T(y)(x) - T(z)(x)|| = ||\int_{x^0}^{x} [f(t, y(t)) - f(t, z(t))]\, dt||$$

$$\leq \left| \int_{x^0}^{x} ||f(t, y(t)) - f(t, z(t))||\, dt \right| \leq K \left| \int_{x^0}^{x} ||y(t) - z(t)||\, dt \right|, \tag{3.84}$$

where K is a Lipschitz constant for f with respect to y in I. We claim that for all $n \in \mathbb{N}$ and $x \in U$

$$||T^n(y)(x) - T^n(z)(x)|| \leq \frac{K^n |x - x^0|^n}{n!} ||y - z||. \tag{3.85}$$

We prove (3.85) by induction on n. For $n = 1$, (3.85) follows from (3.84), because the last integrand is bounded by $||y - z||$. Assuming (3.85) for some n, we get from this hypothesis and (3.84)

$$||T^{n+1}(y)(x) - T^{n+1}(z)(x)|| = ||T(T^n(y))(x) - T(T^n(z))(x)||$$

$$\leq K \left| \int_{x^0}^{x} ||T^n(y)(t) - T^n(z)(t)||\, dt \right|$$

$$\leq K \left| \int_{x^0}^{x} \frac{K^n |t - x^0|^n}{n!} ||y - z||\, dt \right| = \frac{K^{n+1} |x - x^0|^{n+1}}{(n+1)!} ||y - z||,$$

and (3.85) follows by induction. Taking the supremum over all $x \in U$, we conclude that

$$d(T^n(y), T^n(z)) \leq \frac{(Ka)^n}{n!} d(y, z). \tag{3.86}$$

Fix $0 < q < 1$. Since $(Ka)^n/n! \to 0$, there exists $N \in \mathbb{N}$ such that $(Ka)^N/N! < q$, and it follows from (3.86) that T^N is a contraction on X. By the lemma, there exists a unique fixed point y of T in X. By (3.81) and (3.83), y is the unique local solution of (IVP) in J. □

3.6.3 Remark It is clear from the proof that the same result is valid when U is an interval of the form $[x^0, x^0 + a]$ or $[x^0 - a, x^0]$.

If $f \in C(I, \mathbb{R}^k)$ is of class C^1 with respect to the variables y_j, it follows from Lemma 3.3.1 that f is Lipschitz with respect to y, and we then have the following

3.6.4 Corollary *The conclusions of the Existence and Uniqueness theorem for (IVP) are valid under the following conditions (a)–(b):*

(a) $f \in C(I, \mathbb{R}^k)$.
(b) f *is of class* C^1 *with respect to* y *in* I.

3.6.5 The n-th Order ODE Next we consider Cauchy's initial value problem for the n-th order ODE in normal form

$$y^{(n)} = f(x, y, y', \ldots, y^{(n-1)}), \tag{3.87}$$

where

$$f : D \subset \mathbb{R} \times \mathbb{R}^n \to \mathbb{R}$$

is a given function, $n \geq 2$, and D is a domain. The independent variable is x. The unknown *scalar* function is $y = y(x)$. Given a point $(x^0, \eta) \in D$, Cauchy's initial condition is

$$(y, y', \ldots, y^{(n-1)})|_{x^0} = \eta. \tag{3.88}$$

Cauchy's Initial Value Problem (IVP) consists of finding a function y differentiable up to the n-th order in a neighborhood of x^0, such that

$$(x,\ y(x),\ y'(x), \ldots, y^{(n-1)}(x)) \in D$$

and

$$y^{(n)}(x) = f(x,\ y(x),\ y'(x), \ldots, y^{(n-1)}(x)) \tag{3.89}$$

for all x in that neighborhood, and (3.88) is satisfied. Such a function is called a local solution of the Cauchy Initial Value Problem

$$y^{(n)} = f(x, y, y', \ldots, y^{(n-1)}) \qquad (y, y', \ldots, y^{(n-1)})|_{x^0} = \eta. \tag{3.90}$$

Given f as above, define the $\mathbb{R}^n$-valued function

$$F : (x, y) \in D \to F(x, y) \in \mathbb{R}^n$$

by

$$F(x, y) = (y_2, \ldots, y_n,\ f(x, y)), \tag{3.91}$$

where $y = (y_1, \ldots, y_n)$ as usual. Consider the first order IVP

$$y' = F(x, y), \qquad y(x^0) = \eta. \tag{3.92}$$

If y is a local (vector) solution of the IVP (3.92), then its component y_1 is necessarily a local solution of the IVP (3.90). On the other hand, if y is a local (scalar) solution of (3.90) in a neighborhood of (x^0, η), then $(y, y', \ldots, y^{(n-1)})$ is a local solution of (3.92) in that neighborhood. In this sense, the initial value problems (3.90) and (3.92) are *equivalent*. The following result is then obtained;

3.6.6 Theorem *Let f be a real valued continuous function in the compact neighborhood*

$$I = \{(x, y) \in \mathbb{R} \times \mathbb{R}^n;\ |x - x^0| \leq \alpha,\ ||y - \eta|| \leq \beta\},$$

of the given point $(x^0, \eta) \in \mathbb{R} \times \mathbb{R}^n$. Suppose that f is Lipschitz with respect to y in I, uniformly in x. Let $M = \max_I |f| + ||\eta|| + \beta$ and $a = \min[\alpha, \beta/M]$. Then there exists a unique solution y to the IVP (3.90) in the interval $[x^0 - a,\ x^0 + a]$.

Proof The vector function F defined by (3.91) is continuous in I, since each of its components $y_2, \ldots, y_n$ is trivially continuous on $\mathbb{R} \times \mathbb{R}^n$ and f is continuous on I by hypothesis. We have

$$||F|| := \max_I ||F(x, y)||_1 = \max_I [|y_2| + \cdots + |y_n| + |f(x, y)|]$$

$$\leq \max_I [||y||_1 + |f(x, y)|] \leq ||\eta||_1 + \beta + \max_I |f(x, y)| = M.$$

Finally, F is Lipschitz with respect to y in I, uniformly in x. Indeed, if (x, y) and (x, z) are in I, then uniformly in x

$$||F(x, y) - F(x, z)|| = \sum_{i=2}^{n} |y_i - z_i| + |f(x, y) - f(x, z)|$$

$$\leq ||y - z|| + K\, ||y - z|| = (1 + K)\, ||y - z||,$$

where K is a Lipschitz constant for f.

Thus F satisfies the conditions of Theorem 3.6.2, and the result follows. □

3.6.7 Linear ODE Consider the vector equation $y' = f(x, y)$ with $f : D \subset \mathbb{R} \times \mathbb{R}^k \to \mathbb{R}^k$, for the unknown $\mathbb{R}^k$-valued function y of the independent variable x. The equation is said to be *linear* if

$$f_i(x, y) = \sum_{j=1}^{k} a_{ij}(x) y_j + b_i(x) \qquad (i = 1, \ldots, k), \tag{3.93}$$

where the coefficients a_{ij} and b_i are continuous functions of x in an interval U : $|x - x^0| \leq \alpha$. For an arbitrary $\beta > 0$, f is continuous in the set

$$I = \{(x, y) \in \mathbb{R} \times \mathbb{R}^k;\ |x - x^0| \leq \alpha,\ ||y - y^0|| \leq \beta\},$$

for any given $y^0 \in \mathbb{R}^k$. In addition, f is trivially of class C^1 with respect to y in I. By Corollary 3.6.4, the conclusion of the Existence and Uniqueness theorem are valid in the present case. Since β is arbitrary, we may take $a = \alpha$. We then have the following result:

3.6.8 Theorem *Suppose the coefficients a_{ij} and b_i of the linear equation $y' = f(x, y)$ (with f given by (3.93)) are continuous in the interval $U : |x - x^0| \leq \alpha$. Then for any $y^0 \in \mathbb{R}^k$, there exists a unique solution $y : x \in U \to y(x) \in \mathbb{R}^k$ of the IVP*

$$y' = f(x, y), \qquad y(x^0) = y^0. \tag{3.94}$$

3.6.9 Linear Equation of Order n The linear ODE of order n (in normal form) is the equation

$$y^{(n)} = \sum_{i=0}^{n-1} a_i y^{(i)} + b, \tag{3.95}$$

where the coefficients a_i, b are given functions of the independent variable x in some interval.

The vector ODE of first order which is equivalent to the linear scalar ODE of order n (3.95) (cf. Sect. 3.6.5) is clearly linear, and its coefficients are continuous functions of x in any interval where the coefficients a_i, b of (3.95) are continuous. Therefore, by Theorem 3.6.8, we have the following

3.6.10 Theorem *Let a_i, b $(i = 0, \ldots, n-1)$ be continuous real valued functions of x in the interval $U : |x - x^0| \leq \alpha$. Then for any $\eta \in \mathbb{R}^n$, there exists a unique solution $y : x \in U \to \mathbb{R}$ to the IVP*

$$y^{(n)} = \sum_{i=0}^{n-1} a_i y^{(i)} + b \qquad (y, y', \ldots, y^{(n-1)})|_{x^0} = \eta.$$

3.6.11 Homogeneous Linear Equation The vector ODE of first order (cf. Sect. 3.6.7) can be written in matrix notation as

$$y' = Ay + b, \tag{3.96}$$

where A is the $k \times k$ matrix (a_{ij}), b is the *column* vector with components b_i, and y is the *column* with components y_i. The use of columns is convenient because of the definition of matrix multiplication. The *homogeneous equation* associated with (3.96) is the equation

$$y' = Ay. \tag{H}$$

If y, z are solutions of (H) and α, β are constant scalars, then $w := \alpha y + \beta z$ is also a solution of (H), since

$$w' = \alpha y' + \beta z' = \alpha Ay + \beta Az = Aw.$$

This shows that the set of all solutions of (H) (in the interval U) is a vector space over the real field $\mathbb{R}$ (or the complex field $\mathbb{C}$, if we consider solutions with complex components). The vector space operations are defined pointwise. We call this vector space *the solution space* of (H), and denote it by $\mathcal{S} := \mathcal{S}(A)$. We shall prove that its dimension (over $\mathbb{R}$) is k.

3.6.12 Theorem *Let A be a $k \times k$ matrix whose entries are continuous functions of x in the closed interval $U : |x - x^0| \le \alpha$. Then the dimension of the solution space $\mathcal{S}(A)$ for the equation (H) is equal to k.*

Proof Let $\{e^j;\ j = 1, \dots, k\}$ be the standard basis of $\mathbb{R}^k$, written as columns. For each $j \in \{1, \dots, k\}$, the IVP

$$y' = Ay \qquad y(x^0) = e^j$$

has a unique solution $y = y^j$ in U, by Theorem 3.6.8. Suppose c is a constant column such that $\sum_j c_j y^j = 0$, the zero function on U. Evaluating at x^0, we get $c = \sum c_j e^j = 0$. This proves that the elements y^j of $\mathcal{S}$ are linearly independent. Next, let $y \in \mathcal{S}$, denote $\eta := y(x^0)$, and define

$$z := \sum_j \eta_j y^j.$$

Then $z \in \mathcal{S}$ and $z(x^0) = \sum_j \eta_j e^j = \eta$, that is, both y and z are solutions in U of the IVP

$$y' = Ay, \qquad y(x^0) = \eta.$$

By uniqueness of the solution (cf. Theorem 3.6.8), $y = z = \sum_j \eta_j y^j$, that is, the solutions y^j, $j = 1, \dots, k$, span the solution space $\mathcal{S}$. We conclude that y^j $(j = 1, \dots, k)$ is a basis for $\mathcal{S}$. In particular, the dimension of $\mathcal{S}$ is k. □

3.6.13 Remark If $y^j \in \mathcal{S}$ $(j = 1, \dots, s)$ are such that $y^j(t)$ are linearly dependent $\mathbb{R}^k$ vectors for *some* $t \in U$, then y^j are linearly dependent *as elements of* $\mathcal{S}$.

Indeed, our assumption means that there exist real constants c_j, not all zero, such that $\sum c_j y^j(t) = 0$. For *these* c_j, define $z := \sum c_j y^j$. Then $z \in \mathcal{S}$ and $z(t) = 0$. Suppose $x^0 \le t \le x^0 + \alpha$, to fix the ideas. Since both z and the zero function satisfy the above conditions, it follows from the uniqueness of the solution of Cauchy's IVP that $z = 0$ identically in the interval $[t, x^0 + \alpha]$, hence throughout U (cf. Remark 3.6.3). This means that y^j $(j = 1, \dots, s)$ are linearly dependent elements of $\mathcal{S}$.

3.6.14 Fundamental Matrix A streamline presentation of the structure of the solution space $\mathcal{S}(A)$ of the homogeneous equation (H) uses the concept of the *fundamental matrix*.

Definition A fundamental matrix in the interval $U : |x - x^0| \le \alpha$ for the matrix A as in Theorem 3.6.12 is a $k \times k$ matrix Y whose columns are a basis for the solution space $\mathcal{S}(A)$ of the homogeneous equation (H) in U.

We need some simple observations on the calculus of matrix-valued functions. Let $\mathbb{M}_k$ denote the vector space of all $k \times k$ matrices over the field $\mathbb{R}$. A function $Y : U \to \mathbb{M}_k$ is said to be differentiable in U if each entry of Y is differentiable in U. When this is the case, the derivative Y' is defined entry-wise. A simple calculation shows that for any differentiable functions $Y,\ Z : U \to \mathbb{M}_k$ and $a, b \in \mathbb{R}$, the functions $aY + bZ$ and YZ are differentiable in U, and

$$(aY + bZ)' = aY' + bZ', \qquad (YZ)' = Y'Z + YZ'.$$

Note that in the rule for the derivative of a product of matrix functions, the factors are not necessarily square matrices, provided that their product is well-defined.

Also $Y : U \to \mathbb{M}_k$ is a constant matrix iff $Y' = 0$ in U. If $C : U \to \mathbb{M}_k$ is constant on U, then $(CY)' = CY'$ and $(YC)' = Y'C$ on U for any differentiable function $Y : U \to \mathbb{M}_k$.

The columns of AY are the products Ay^j, where y^j are the columns of Y. If $Y : U \to \mathbb{M}_k$, then y^j are differentiable and $(y^j)' = Ay^j$ in U for all j if and only if Y is differentiable and $Y' = AY$ in U.

Observation. *Let Y be a fundamental matrix for A in U. Then*

(i) *Y is differentiable and $Y' = AY$ in U, and*
(ii) *Y is non-singular at every point of U.*

Proof (i) was observed in the comments preceding the statement of the proposition. To prove (ii), suppose $Y(t)$ is singular at some point $t \in U$. Then $y^j \in \mathcal{S}$ are such that $y^j(t)$ are dependent vectors in $\mathbb{R}^k$, but this implies that y^j are dependent elements of $\mathcal{S}$ by Remark 3.6.13, contradicting the assumption that Y is a fundamental matrix.□

Conversely, if Y is a (differentiable) solution in U of the matrix differential equation

$$Y' = AY \tag{3.97}$$

and is non-singular at every point of U, then its columns y^j are linearly independent elements of $\mathcal{S}$, that is, Y is a fundamental matrix for A in U. We conclude that

3.6.15 Theorem *The function $Y : U \to \mathbb{M}_k$ is a fundamental matrix for A in U if and only if Y is differentiable, non-singular, and $Y' = AY$, at every point of U.*

Let Y be a fundamental matrix for A in U, and let C be a non-singular constant matrix (in U). Then YC is a fundamental matrix for A in U. Indeed, YC is non-singular at every point of U as the product of two non-singular matrices. Moreover $(YC)' = Y'C = (AY)C = A(YC)$, and our claim follows from Theorem 3.6.15. On the other hand, if Z is any fundamental matrix for A in U, consider the matrix function $V := Y^{-1}Z$. It follows from the formula for the elements of Y^{-1} that V is differentiable in U, and since $YV = Z$, we have

$$AZ = Z' = (YV)' = Y'V + YV' = (AY)V + YV' = A(YV) + YV' = AZ + YV'.$$

Hence $YV' = 0$, and therefore $V' = 0$ identically in U, because Y is non-singular at every point of U. This proves that $V = C$, a constant matrix in U; C is non-singular, as the product of the two non-singular matrices Y^{-1} and Z, and $Z = YC$.

In conclusion, we have the following

Proposition *The most general fundamental matrix for A in U is YC, where Y is any particular fundamental matrix for A (in U) and C is an arbitrary constant non-singular matrix.*

3.6.16 Exercises

In the following linear equations, the assumptions and notation are as in the preceding sections.

1. Let Y be a fundamental matrix for $A \in \mathbb{M}_k$ in U. Prove:
   (a) The elements y of the solution space $\mathcal{S}(A)$ are given by the formula $y = Yc$, where c is an arbitrary constant column vector. We refer to a formula representing the solutions of an ODE as the *general solution of the given equation*. Thus $y = Yc$ is the general solution of the equation $y' = Ay$.
   (b) The solution of the IVP
   $$y' = Ay, \qquad y(x^0) = y^0 \tag{3.98}$$
   is $y = Y\,[Y(x^0)]^{-1}y^0$.
2. Consider the non-homogeneous linear equation
   $$y' = Ay + b, \tag{NH}$$
   where the matrix function $A : U \to \mathbb{M}_k$ and the column vector function $b : U \to \mathbb{R}^k$ have continuous entries. Prove
   (a) If z is a particular solution of (NH), then the general solution of (NH) is $y = z + Yc$, where Y is a fundamental matrix for A and c is an arbitrary constant (column) vector.
   (b) A particular solution of (NH) is given by
   $$z(x) = Y(x) \int_{x^0}^{x} [Y(t)]^{-1} b(t)\, dt. \qquad (x \in U)$$
3. Prove that Y is a fundamental matrix for A in U iff $Z := (Y^t)^{-1}$ is well-defined and is a fundamental matrix for $-A^t$ in U.
4. Let $A \in \mathbb{M}_k$. The continuous function $y \to ||Ay||$ of $\mathbb{R}^k$ into $[0, \infty)$ attains its maximum on the closed unit ball of $\mathbb{R}^k$, because that ball is compact. Define
   $$||A|| = \max_{||y|| \le 1} ||Ay||. \tag{3.99}$$

Prove:

(a) $||Ay|| \le ||A||\,||y||$ for all $y \in \mathbb{R}^k$.
(b) $\mathbb{M}_k$ is a normed space for the norm $A \to ||A||$. We fix this norm on $\mathbb{M}_k$.
(c) $||AB|| \le ||A||\,||B||$ for all $A, B \in \mathbb{M}_k$, and $||I|| = 1$ for the identity matrix I.
Properties (b)–(c) are summarized by saying that $\mathbb{M}_k$ is a *normed algebra* over the field $\mathbb{R}$.
(d) In any normed algebra, multiplication is continuous. Hint: prove first the inequality

$$||AB - CD|| \le ||A - C||\,||B|| + ||C||\,||B - D||$$

for any four elements A, B, C, D of the normed algebra.
(e) $A_n \in \mathbb{M}_k$ converge to A in the normed algebra $\mathbb{M}_k$ as $n \to \infty$ if and only if the entries a_{ij}^n converge (in $\mathbb{R}$) to the corresponding entries a_{ij} of A. Hint: the j-th column of A_n is $A_n e^j$, and $||A_n e^j - Ae^j|| \le ||A_n - A||$, where the norms are the $\mathbb{R}^k$-norm on the left hand side and the $\mathbb{M}_k$-norm on the right hand side.
(f) The normed space $\mathbb{M}_k$ is complete. Hint: rely on the completeness of $\mathbb{R}^k$. Do not use Part (e).
A complete normed algebra is called a *Banach algebra*. Thus $\mathbb{M}_k$ is a Banach algebra. This is a special case of a Banach space.

5. (a) Given $A \in \mathbb{M}_k$, prove that the series

$$\sum_{n=0}^{\infty} \frac{x^n}{n!} A^n \tag{3.100}$$

converges absolutely for all $x \in \mathbb{R}$, hence converges in the Banach space $\mathbb{M}_k$ (cf. Sect. 1.3.7, Exercise 4(a)). The sum of the above series is denoted e^{xA}.
(b) Immitating the proof for usual power series, prove that power series with coefficients in $\mathbb{M}_k$ can be differentiated term-by-term in their open interval of absolute convergence.
(c) Prove that $(e^{xA})' = A\,e^{xA} = e^{xA}A$ for all $x \in \mathbb{R}$.
(d) If $A, B \in \mathbb{M}_k$ commute (i.e., $AB = BA$), prove that B commutes with e^{xA} for all $x \in \mathbb{R}$.
(e) Let $F(x) = e^{-xA}e^{x(A+B)}$, with A, B as in (d). Prove that $F'(x) = B\,F(x)$, and conclude that $[e^{-xB}F(x)]' = 0$ for all x. Therefore the matrix function $x \to e^{-xB}F(x)$ is constant, and since its value at $x = 0$ is the identity matrix I, it follows that

$$e^{-xB}e^{-xA}e^{x(A+B)} = I. \tag{3.101}$$

(f) Take $B = 0$ in (3.101) to deduce that $e^{-xA}e^{xA} = I$ for all x.
Hence $Y(x) := e^{xA}$ is invertible, that is, non-singular, and $(e^{xA})^{-1} = e^{-xA}$. By Part (c), Y is a fundamental matrix on $\mathbb{R}$ for the constant matrix A.

(g) If $A \in \mathbb{M}_k$ is constant, the solution on $\mathbb{R}$ of the IVP

$$y' = Ay, \qquad y(x^0) = y^0$$

is

$$y(x) = e^{(x-x^0)A} y^0,$$

and the general solution on $\mathbb{R}$ of the homogeneous equation $y' = Ay$ is $y(x) = e^{xA}c$, where c is an arbitrary constant column in $\mathbb{R}^k$.

(h) If $A \in \mathbb{M}_k$ is constant and $b : U \to \mathbb{R}^k$ is continuous on the closed interval U, then the solution on U of the IVP

$$y' = Ay + b, \qquad y(x^0) = y^0$$

is

$$\tilde{y}(x) = y^0 + \int_{x^0}^{x} e^{(x-t)A} b(t)\, dt, \qquad x \in U,$$

and the general solution in U of the equation $y' = Ay + b$ is $y(x) = \tilde{y}(x) + e^{xA}c$, where c is an arbitrary constant column in $\mathbb{R}^k$.

6. Let $A \in \mathbb{M}_k$ have distinct eigenvalues λ_j, $j = 1, \ldots, k$. It is known from Linear Algebra that if q^j is an eigenvector corresponding to λ_j (for $j = 1, \ldots, k$), then the vectors q^j are linearly independent, and the (non-singular) matrix Q with the columns q^j satisfies the relation

$$Q^{-1} A\, Q = \Lambda,$$

where Λ denotes the diagonal matrix with λ_j in its diagonal,

$$\Lambda := \operatorname{diag}(\lambda_1, \ldots, \lambda_k).$$

Prove

(a) The fundamental matrix $Y(x) := e^{xA} Q$ for A in $\mathbb{R}$ (cf. Exercise 5(f)) is equal to $Q\, e^{x\Lambda} = Q \operatorname{diag}(e^{\lambda_1 x}, \ldots, e^{\lambda_k x})$. Conclude that the columns y^j of the fundamental matrix Y defined above are given by

$$y^j = e^{\lambda_j x} q^j, \qquad (j = 1, \ldots, k).$$

(This formula gives a basis for the solution space of the homogeneous equation $y' = Ay$.)

(b) Find a fundamental matrix and write a basis for the solution space on $\mathbb{R}$ for the system

$$y_1' = 3y_1 + y_2; \qquad y_2' = y_1 + 3y_2.$$

(c) Same as (b) for the system

$$y_1' = 2y_1 - y_2; \qquad y_2' = 3y_1 - 2y_2.$$

(d) Same as (b) for the system

$$y_1' = y_1 - y_2 + 4y_3; \quad y_2' = 3y_1 + 2y_2 - y_3; \quad y_3' = 2y_1 + y_2 - y_3.$$

7. Suppose $A \in \mathbb{M}_k$ has diagonal blocks $A_1, \ldots, A_m$ and zero entries outside the blocks. We use the notation $A = \text{diag}\,(A_1, \ldots, A_m)$ to state this fact.
   (a) Prove that $e^{xA} = \text{diag}\,(e^{xA_1}, \ldots, e^{xA_m})$.
   (b) It is known from Linear Algebra that any matrix $A \in \mathbb{M}_k$ is similar (over the complex field $\mathbb{C}$) to a matrix of the form $J = \text{diag}\,(J_0, \ldots, J_m)$, where some of the blocks listed could be missing. The matrix J_0 is diagonal with all the simple characteristic roots of A in its diagonal. For $1 \le s \le m$, the *Jordan block* J_s is an $h \times h$ matrix (with $h > 1$) of the form $B = \lambda I + N$, where λ is a characteristic root of A, I is the $h \times h$ identity matrix, and N is the $h \times h$ matrix with all entries equal to zero except for the entries of the first super-diagonal, which are equal to 1. The similarity mentioned above means that there exists a non-singular matrix Q (with complex entries in general) such that $Q^{-1} A\, Q = J$. The matrix J is called the Jordan canonical form of A. Prove that the fundamental matrix $Y(x) = e^{xA} Q$ for A on $\mathbb{R}$ is equal to $Q\,\text{diag}\,(e^{xJ_0}, \ldots, e^{xJ_m})$.
   (c) $e^{xB} = e^{\lambda x} e^{xN} = e^{\lambda x} \sum_{r=0}^{h-1} (x^r/r!) N^r$.
   (d) $[N^r]_{st} = \delta_{s+r,t}$, where δ is Kronecker's delta.
   (e) $[e^{xN}]_{st} = x^{t-s}/(t-s)!$ for $t \ge s$, and equals zero otherwise.
   (f) Find a fundamental matrix for the system

$$y_1' = y_2 + y_3; \quad y_2' = y_1 + y_3; \quad y_3' = y_1 + y_2.$$

8. Suppose $A : U \to \mathbb{R}^k$ has continuous entries, and $y \in \mathbb{C}^k$ solves the equation $y' = Ay$. Define $\Re y$ and $\Im y$ componentwise (i.e., $(\Re y)_i = \Re y_i$, and similarly for $\Im y$).
   (a) Prove that $\Re y$ and $\Im y$ are solutions of the equation $z' = Az$.
   (b) Find a basis consisting of vectors with real components for the solution space of the system

$$y_1' = 2y_1 - 5y_2; \quad y_2' = y_1 - 2y_2.$$

9. Find the general solution of the following systems (cf. Exercise 5(h)).
   (a) $y_1' = y_1 + y_2 + e^{-2x}; \quad y_2' = 4y_1 - 2y_2 - 2e^x$.
   (b) $y_1' = 2y_1 - y_2 + e^x; \quad y_2' = 3y_1 - 2y_2 - e^x$.

10. Consider the linear equation $y' = ay + b$ in one dimension (i.e., $k = 1$), with $a, b : U \to \mathbb{R}$ continuous. Denote by z an indefinite integral of a in U (that is, $z' = a$).
    (a) Prove that the general solution of the above linear equation is given by the formula
    $$y = e^z[\int b\,e^{-z}dx + c],$$
    where $\int v\,dx$ denotes an indefinite integral of v in U and c is an arbitrary constant.
    (b) Find the general solution of the following linear equations
    (b1) $y' = x^2y + e^{x^3/3} \quad (x \in \mathbb{R})$.
    (b2) $y' + y = (1 + e^{2x})^{-1} \quad (x \in \mathbb{R})$.
    (b3) $y' = y/x + \log^m x \quad (x > 0)$.
    (b4) $y' + \cos x\; y = \sin 2x \quad (x \in \mathbb{R})$.
    (b5) $y' = y/x^2 + e^{-1/x} \cot x$.
11. Consider the Bernoulli equation
    $$y' = ay + by^p,$$
    where $p > 0$, $p \neq 1$, is a parameter, $a, b : U \to \mathbb{R}$ are given continuous functions, and $y : U \to \mathbb{R}$ is the unknown function.
    (a) Prove that y is a nowhere vanishing solution of the Bernoulli equation with parameter p if and only if $z := y^{-p+1}/(-p + 1)$ is a solution of the linear equation
    $$z' = (-p + 1)a\,z + b.$$
    (b) Find the general nowhere vanishing solution of the equation $y' = xy + y^2$.
12. Let $h : V \to \mathbb{R}$ be continuous in the interval V, and $h(t) \neq t$ in V.
    (a) Prove that y is a solution of the equation $y' = h(y/x)$ such that $y/x \in V$ if and only if the function $z := y/x$ is a solution of the equation $(h(z) - z)^{-1}z' = 1/x$.
    (b) Solve $y' - y/x = e^{y/x} \quad (x > 0)$.
    (c) Solve $y' - y/x = y^r x^{-r} \quad (x > 0)$, where $r > 1$ is a constant.
13. *Notation*. The derivation operator D is defined on functions y differentiable in some interval U by $Dy = y'$. Its power D^n is defined inductively on n times differentiable functions y in U by $D^n y = D(D^{n-1}y)$, $D^0 = I$, where I is the identity operator, $Iy = y$ for all y. If $a : U \to \mathbb{R}$ is continuous, the operator $a D^n$ is defined by $(a D^n)y = a\,(D^n y)$. The sum of operators S, T acting on functions is defined as usual on the intersection of their domain by $(S + T)y = Sy + Ty$.

Thus, the general linear ordinary differential operator of order n with continuous coefficients a_j may be written as

$$L = \sum_{j=0}^{n} a_j D^j.$$

The operator is in normal form if $a_n = 1$ identically in U.
The homogeneous equation associated with the operator L is

$$Ly = 0. \tag{H}$$

The set of all solutions of (H) is a vector space over the field $\mathbb{R}$ (or the complex field $\mathbb{C}$, when we consider complex valued solutions) for the usual pointwise operations. This follows from the fact that L is a linear operator, that is, $L(\alpha y + \beta z) = \alpha Ly + \beta Lz$ for any scalars α, β and any functions y, z in the domain of L, which is the space of all functions differentiable up to the order n in U. We denote by $\mathcal{S} = \mathcal{S}(L)$ the vector space of all solutions of (H) in U.

(a) Let y_j $(j = 1, \ldots, n)$ be the unique solution in $U : |x - x_0| \le \alpha$ of the Cauchy IVP

$$Ly = 0, \qquad y^{(k)}(x_0) = \delta_{k,j-1} \quad (k = 0, \ldots, n-1).$$

Prove that $y_1, \ldots, y_n$ are linearly independent elements of $\mathcal{S}$.

(b) Given $y \in \mathcal{S}$, define

$$z = \sum_{k=0}^{n-1} y^{(k)}(x_0) y_{k+1}.$$

Prove that $w := y - z$ solves the Cauchy IVP

$$Lw = 0, \qquad w^{(i)}(x_0) = 0 \quad (i = 0, \ldots, n-1).$$

Conclude that $y = z$ (in U), and therefore (cf. Part (a)) $y_1, \ldots, y_n$ is a basis for $\mathcal{S}$ and the dimension of $\mathcal{S}$ is n.

(c) Let $y := (y_1, \ldots, y_n) \in \mathcal{S}^n$. The *Wronskian matrix* $(W(y))$ is the matrix whose rows are $y^{(i)}$, $i = 0, \ldots, n-1$. The Wronskian is $W(y) := \det(W(y))$. Prove that $W := W(y)$ satisfies the differential equation

$$W' + a_{n-1} W = 0 \qquad (x \in U).$$

(d) For any given point $x_1 \in U$, prove that

$$W(y)|_x = W(y)|_{x_1} \exp\Big(-\int_{x_1}^{x} a_{n-1}(t)\, dt\Big) \qquad (x \in U).$$

Therefore either W vanishes identically in U or never vanishes there.

(e) Let $y \in \mathcal{S}^n$ be such that $W(y)|_{x_1} \neq 0$ at some $x_1 \in U$. Suppose there exists $0 \neq c \in \mathbb{R}^n$ such that $c \cdot y = 0$ identically in U. Deriving this equation up to the $(n-1)$-th order, deduce from the algebraic linear equations obtained in this manner that $c = 0$. Conclude that $y_1, \ldots, y_n$ is a basis for $\mathcal{S}$.

(f) Suppose $W(y)|_{x_1} = 0$ at some point $x_1 \in U$. Then $W = 0$ identically in U (cf. Part (d)), and in particular $W|_{x_0} = 0$. The system of algebraic linear equations in Part (e) evaluated at x_0 has then a solution $c \neq 0$. With this c, define $z = c \cdot y$. Verify that z is a solution of the Cauchy IVP

$$Lz = 0, \qquad z^{(i)}(x_0) = 0 \quad (i = 0, \ldots, n-1).$$

Conclude that $z = 0$, and therefore $y_1, \ldots, y_n$ are linearly dependent elements of $\mathcal{S}$.

By Parts (e)–(f), *the n solutions $y_1, \ldots, y_n$ of (H) are a basis for the solution space $\mathcal{S}$ if and only if $W(y) \neq 0$ in U.*

(g) Let $b : U \to \mathbb{R}$ be continuous. Let $y_1, \ldots, y_n$ be a basis of $\mathcal{S}$ and $y := (y_1, \ldots, y_n)$. The operator L operates on y componentwise (like the derivatives of order at most n). Thus $Ly = 0$. Let

$$u^t := \int_{x_0}^{x} b(s)(W)^{-1}(s)(e^n)^t ds \qquad x \in U,$$

where $(W) := (W(y))$ and $e^n = (0, \ldots, 0, 1)$. Thus $(W)(u^t)' = b(e^n)^t$. Deduce that

$$D^j(yu^t) = (D^j y)\, u^t \qquad j = 0, \ldots, n-1,$$

and $D^n(yu^t) = (D^n y)\, u^t + b$. Therefore $L(yu^t) = (Ly)\, u^t + b = b$. Conclude that the general solution in U of the non-homogeneous equation

$$Lw = b \tag{NH}$$

is given by the formula

$$w(x) = y(x)\,\Big(\int_{x_0}^{x} b(s)(W)^{-1}(s)(e^n)^t\, ds + c^t\Big),$$

where c is an arbitrary constant vector in $\mathbb{R}^n$. (Multiplications are matrix multiplications.)

(h) Find the general solution of $y'' - y' = e^x$.

14. Let $p(t) = \sum_{j=0}^{n} a_j t^j$ be a monic polynomial (i.e., $a_n = 1$) with real coefficients. We denote by $p(D)$ the differential operator obtained formally by replacing t by D in the definition of $p(t)$. This is a linear differential operator of order n.

(a) Prove the formula

$$p(D)e^{tx} = p(t)\, e^{tx}.$$

Conclude that if t is a root of p (that is, $p(t) = 0$), then e^{tx} is a solution of the homogeneous equation $p(D)y = 0$.

(b) Suppose p has n distinct roots t_j (not necessarily real). Prove that $e^{t_j x}$, $j = 1, \ldots, n$ is a basis for the solution space $\mathcal{S}$ of the above equation. Hint: the Wronskian of these solutions is related to the Vandermonde determinant of the roots.

(c) Prove

$$\frac{\partial^j}{\partial t^j} p(D)e^{tx} = p(D)x^j e^{tx}.$$

Then deduce from (a) and Leibnitz' formula for the j-th derivative of a product that

$$p(D)x^j e^{tx} = \sum_{i=0}^{j} p^{(i)}(t)x^{j-i} e^{tx}.$$

If t_1 is a root of p of multiplicity $m_1 \geq 2$, then $p^{(i)}(t_1) = 0$ for $i = 0, \ldots, m_1 - 1$, and the above formula implies that $p(D)x^j e^{t_1 x} = 0$ for all $j = 0, \ldots, m_1 - 1$. Therefore the m_1 linearly independent solutions $x^j e^{t_1 x}$ of the equation $p(D)y = 0$ are associated with the root t_1 of multiplicity m_1 of the polynomial p. If $t_1, \ldots, t_s$ are all the distinct roots of p, with respective multiplicities $m_1, \ldots, m_s$, then we get the $m_1 + \ldots + m_s = n$ solutions

$$x^j e^{t_1 x} \qquad j = 0, \ldots, m_1,$$

$$\ldots$$

$$x^j e^{t_s x} \qquad j = 0, \ldots, m_s.$$

These solutions are linearly independent (try to prove), and form therefore a basis for the solution space of the equation $p(D)y = 0$.

(d) Since p has real coefficients, the conjugate $\overline{\lambda} = \alpha - i\beta$ of a root $\lambda = \alpha + i\beta$ is also a root. Furthermore, the two conjugate roots have the same multiplicity m. Prove that the $2m$ complex solutions

$$x^j e^{\lambda x},\ x^j e^{\overline{\lambda} x}, \qquad (j = 0, \ldots, m-1)$$

may be replaced in the above basis of the solution space by the $2m$ *real* solutions

$$x^j e^{\alpha x} \cos(\beta x),\ x^j e^{\alpha x} \sin(\beta x),\ j = 0, \ldots, m-1.$$

15. Use the methods of Exercise 14 to find a basis of real solutions for the solution space of the following homogeneous linear equations.
    (a) $y'' + y' - 2y = 0$.
    (b) $y'' - 6y' + 9y = 0$.
    (c) $y'' + 2y' + 2y = 0$.
    (d) $y^{(4)} + 2y'' + y = 0$.
    (e) $y^{(3)} - y'' - y' + y = 0$.
    (f) $y^{(6)} - 3y^{(4)} + 3y'' - y = 0$.
    (g) $y^{(4)} - 27\,y' = 0$.
    (h) $y^{(6)} - 64\,y = 0$. Hint: think of roots of unity.
16. Solve the Cauchy IVP
$$y'' + y' - 6y = 0, \qquad y(0) = \alpha,\ y'(0) = \beta.$$
17. Solve the Boundary Value Problem (BVP)
$$y'' + 2y' + 2y = 0, \qquad y(0) = y(\pi/2) = 1$$
in the interval $[0, \pi/2]$.
18. Find the general solution of the equation
$$y'' + y = 3\sin 2x.$$
19. Let $p(t) = \sum_j a_j t^j$ be a monic polynomial of degree n. The corresponding *Euler differential operator* is the operator $L := p(xD)$, obtained by substituting formally the operator xD instead of the variable t in $p(t)$. For $\lambda \in \mathbb{C}$, we use the binomial coefficients
$$\binom{\lambda}{k} := \frac{\lambda(\lambda-1)\cdots(\lambda-k+1)}{k!}$$
for $k \in \mathbb{N}$ (and $\binom{\lambda}{0} := 0$) to define the so-called *indicial polynomial* $q(\lambda)$ *associated with* $p(t)$:
$$q(\lambda) := \sum_{k=0}^{n} a_k\, k! \binom{\lambda}{k}.$$
Prove
    (a) $Lx^\lambda = x^\lambda q(\lambda)$, $x > 0$.
    (b) If all the roots λ_j of $q(\lambda)$ are simple roots, then $x^{\lambda_1}, \ldots, x^{\lambda_n}$ are a basis for the solution space of *Euler's equation* $Ly = 0$ in $x > 0$.
    (c) If λ_1 is a root of $q(\lambda)$ of multiplicity $m_1 \geq 2$, the functions

$$y_j(x) := \frac{\partial^j}{\partial \lambda^j} x^\lambda \Big|_{\lambda_1} \quad (= x^{\lambda_1} \log^j x),$$

$j = 0, \ldots, m_1 - 1$, are linearly independent solutions of Euler's equation in $x > 0$.

(d) Replace complex solution in the above basis by real solutions.

3.7 More on $C(I)$

Let I be a closed interval. We shall discuss in this section two important properties of the Banach space $C(I)$: the density of polynomials in $C(I)$ (the Weierstrass Approximation Theorem) and the characterization of subsets having the Bolzano-Weierstrass Property by means of equicontinuity (the Arzela-Ascoli Theorem).

We start with some elementary calculation.

Let $w = (w_0, \ldots, w_n)$ be a *weight vector*, that is, $w_i \ge 0$ and $\sum w_i = 1$. Its *k-th moment* ($k = 0, 1, 2, \ldots$) is defined by

$$\mu_k := \sum_i i^k w_i.$$

(This is the weighted mean of i^k with the weights w_i.) Set $\mu := \mu_1$ (the "mean") and

$$\sigma^2 := \sum_i (i - \mu)^2 w_i$$

(the "weighted mean square deviation from the mean"; σ is the so-called "standard deviation").

3.7.1 Proposition $\sigma^2 = \mu_2 - \mu^2$.

Proof Expanding $(i - \mu)^2$ in the definition of σ^2, we get

$$\sigma^2 = \sum i^2 w_i - 2\mu \sum i w_i + \mu^2 \sum w_i$$
$$= \mu_2 - 2\mu^2 + \mu^2 = \mu_2 - \mu^2. \qquad \square$$

We shall need the following special case of Tchebichev's inequality.

3.7.2 Proposition (discrete Tchebichev's inequality) *For any $r > 0$, set*

$$Q_r := \{i; 0 \le i \le n, \ |i - \mu| \ge r\}.$$

Then

$$\sum_{i \in Q_r} w_i \leq \frac{\sigma^2}{r^2}.$$

Proof For any $r > 0$,

$$\sum_{i \in Q_r} w_i = r^{-2} \sum_{i \in Q_r} r^2 w_i \leq r^{-2} \sum_{i \in Q_r} (i - \mu)^2 w_i \leq r^{-2} \sigma^2. \qquad \square$$

We specialize now to the *Bernoulli weights* $b_i(t)$, $i = 0, \ldots, n$ (where t is a given number in $I = [0, 1]$ and n is a fixed positive integer):

$$b_i(t) := \binom{n}{i} t^i (1 - t)^{n-i}.$$

The vector w is a weight vector, since by Newton's binomial formula

$$\sum_{i=0}^{n} b_i(t) = [t + (1 - t)]^n = 1.$$

We have the elementary binomial identity

$$i \binom{n}{i} = n \binom{n-1}{i-1} \qquad 1 \leq i \leq n. \tag{3.102}$$

Hence for $2 \leq i \leq n$,

$$(i - 1)i \binom{n}{i} = n(i - 1) \binom{n-1}{i-1} = n(n - 1) \binom{n-2}{i-2}. \tag{3.103}$$

We obtain from (3.102), writing $j = i - 1$, $m = n - 1$ (so that $n - i = m - j$)

$$\mu = \sum_{i=1}^{n} i \binom{n}{i} t^i (1 - t)^{n-i} = nt \sum_{j=0}^{m} \binom{m}{j} t^j (1 - t)^{m-j} = nt. \tag{3.104}$$

Similarly, we obtain from (3.103), writing $j = i - 2$, $m = n - 2$ (so that $n - i = m - j$)

$$\sum_{i=2}^{n} (i - 1)i \binom{n}{i} t^i (1 - t)^{n-i} = n(n - 1)t^2 \sum_{j=0}^{m} \binom{m}{j} t^j (1 - t)^{m-j} = n(n - 1)t^2. \tag{3.105}$$

Note that the index i in the sums on the left hand side of (3.104) and (3.105) can range from $i = 0$, because the added terms vanish. Adding the two equations term-by-term, we obtain a formula for μ_2:

$$\mu_2 = nt + n(n-1)t^2. \tag{3.106}$$

Hence by (3.104), (3.106), and Proposition 3.7.1,

$$\sigma^2 = \mu_2 - \mu^2 = nt + n(n-1)t^2 - n^2t^2 = nt(1-t). \tag{3.107}$$

We apply now Tchebichev's inequality to the Bernoulli weights. For any $r > 0$,

$$\sum_{i \in Q_r} b_i(t) \le \frac{nt(1-t)}{r^2}.$$

Write $\delta = r/n$ (an arbitrary positive number), and note that

$$Q_r := \{i; |i - nt| \ge r\} = \{i; |\frac{i}{n} - t| \ge \delta\}.$$

Hence

$$\sum_{\{i; |\frac{i}{n} - t| \ge \delta\}} b_i(t) \le \frac{t(1-t)}{n\delta^2} \le \frac{1}{4n\delta^2}. \tag{3.108}$$

3.7.3 Definition Let $I = [0, 1]$ and $f \in C(I)$. Fix $n \in \mathbb{N}$, and let $b_i(t)$ be the corresponding Bernoulli weights for $t \in I$ given. The n-th Bernstein polynomial for f is defined by

$$B_n(t) = \sum_{i=0}^{n} f(\frac{i}{n}) b_i(t) \qquad (t \in I).$$

Note that $B_n(\cdot)$ is a polynomial of degree n. At each point $t \in I$, its value is the weighted mean of the values of f at the points i/n, with the weights $b_i(t)$.

The Weierstrass Approximation Theorem

We are ready now to prove the Bernstein version of the Weierstrass Approximation Theorem (WAT) for continuous functions on $I = [0, 1]$.

3.7.4 Theorem *The polynomials are dense in $C(I)$. More specifically, if B_n are the Bernstein polynomials for a given $f \in C(I)$, then $B_n \to f$ in the Banach space $C(I)$.*

(Cf. Exercise 1.3.12 (11).)

In a more classical terminology, the polynomials B_n converge to f uniformly on I.

Proof Let $f \in C(I)$ and let $\epsilon > 0$. Since f is continuous on the compact set $I = [0,1]$, it is uniformly continuous on I. There exists therefore $\delta = \delta(\epsilon)$ such that

$$|f(s) - f(t)| < \frac{\epsilon}{4} \tag{3.109}$$

if $s, t \in I$ and $|s - t| < \delta$.

Fix an integer n_0 such that

$$n_0 > \frac{2||f||}{\epsilon\delta^2}. \tag{3.110}$$

Clearly n_0 depends only on the given function f and on ϵ.

Let $n > n_0$. Fix $t \in I$. We have

$$|B_n(t) - f(t)| = \Big|\sum_{i=0}^{n} \binom{n}{i} t^i (1-t)^{n-i} [f(\frac{i}{n}) - f(t)]\Big|$$

$$\leq \sum_{i=0}^{n} \binom{n}{i} t^i (1-t)^{n-i} |f(\frac{i}{n}) - f(t)| = \sum_{i \in A} + \sum_{i \in B}, \tag{3.111}$$

where

$$A := \{i;\ |\frac{i}{n} - t| < \delta\};\ \ B := \{i;\ |\frac{i}{n} - t| \geq \delta\}.$$

For $i \in A$, $|f(i/n) - f(t)| < \epsilon/4$ by (3.109), and therefore

$$\sum_{i \in A} \leq \frac{\epsilon}{4} \sum_{i \in A} b_i(t) \leq \frac{\epsilon}{4} \sum_{i=0}^{n} b_i(t) = \frac{\epsilon}{4}. \tag{3.112}$$

By (3.108) and (3.110),

$$\sum_{i \in B} \leq 2\,||f|| \sum_{i \in B} b_i(t) \leq \frac{2\,||f||}{4n\delta^2} \leq \frac{||f||}{2n_0\delta^2} < \frac{\epsilon}{4}. \tag{3.113}$$

We conclude from (3.111)–(3.113) that

$$|B_n(t) - f(t)| < \frac{\epsilon}{2}$$

for all $t \in I$. Taking the supremum over all $t \in I$, we get

$$||B_n - f|| \leq \frac{\epsilon}{2} < \epsilon. \tag{3.114}$$

Since (3.114) is valid for all $n > n_0$, we proved that $B_n \to f$ in the Banach space $C(I)$. □

For an arbitrary interval $[a, b]$, we consider the linear map

$$x(t) := a + (b - a)t : [0, 1] \to [a, b].$$

If $f \in C([a, b])$, the function $f \circ x$ is in $C(I)$ (where $I = [0, 1]$). Since $x = x(t)$ iff $t = \frac{x-a}{b-a}$ (equivalently, $1 - t = \frac{b-x}{b-a}$), the n-th Bernstein polynomial $\tilde{B}_n(t)$ for $(f \circ x)(t) = f(x(t))$ is

$$\frac{1}{(b-a)^n} \sum_{i=0}^{n} \binom{n}{i} f(a + \frac{i}{n}(b-a))(x-a)^i (b-x)^{n-i},$$

which is a polynomial $p_n(x)$ in the variable $x \in [a, b]$.

We have

$$||p_n - f||_{C([a,b])} := \sup_{x \in [a,b]} |p_n(x) - f(x)|$$

$$= \sup_{t \in I} |\tilde{B}_n(t) - (f \circ x)(t)| = ||\tilde{B}_n - f \circ x||_{C(I)} \to 0$$

as $n \to \infty$, by Theorem 3.7.4.

This proves the following general version of the WAT

3.7.5 Corollary *The polynomials are dense in* $C([a, b])$.

We stress again the meaning of this result as an approximation theorem: given a continuous function f on $[a, b]$, there exists a sequence of polynomials converging to f uniformly on $[a, b]$.

The Arzela-Ascoli Theorem

We turn next to another important property of $C(X)$, for any compact metric space X. Recall the so-called "Bolzano-Weierstrass property" (BWP) of any *bounded* subset E of $\mathbb{R}^k$: any sequence in E has a convergent subsequence. We are interested in a sufficient condition for a bounded subset $\mathbb{F}$ of the Banach space $C(X)$ to have the BWP. The Arzela-Ascoli theorem below gives *equicontinuity* of $\mathbb{F}$ as such a sufficient condition. It is also necessary (as well as the boundedness of $\mathbb{F}$), but we shall not insist here on this. In the ordinary ϵ, δ-definition of continuity, equicontinuity of a *family* $\mathbb{F}$ of functions (rather than a single function) stresses the possibility of choosing a δ which is good for *all* the functions in the family (and all the points of X).

3.7.6 Definition Let X be a compact metric space. The family $\mathbb{F}$ of real valued functions on X is equicontinuous if for every $\epsilon > 0$, there exists $\delta > 0$ such that $|f(x) - f(y)| < \epsilon$ for all $f \in \mathbb{F}$ and all $x, y \in X$ with $d(x, y) < \delta$.

Note that we have necessarily $\mathbb{F} \subset C(X)$.

We start with some general observations.

Observation 1. *A compact metric space is separable.*
(Cf. Exercise 4, Sect. 1.2.11.)

This means that if X is a compact metric space, then there exists a countable subset $Z \subset X$ which is dense in X, that is, $\overline{Z} = X$. Recall the equivalence of the statements (a)–(c) below: (a) Z is dense in X; (b) every ball in X meets Z; (c) every $x \in X$ is the limit of a sequence in Z.

Proof For each $n \in \mathbb{N}$, the family

$$\{B(x, \frac{1}{n}); \ x \in X\}$$

is an open cover of the compact set X. Let then

$$\{B(x_{j,n}, \frac{1}{n}); \ j = 1, \ldots, p(n)\} \tag{3.115}$$

be a finite subcover. Define

$$Z = \{x_{j,n}; \ j = 1, \ldots, p(n), \ n \in \mathbb{N}\}. \tag{3.116}$$

Then Z is countable. Let $B(x, r)$ be any given ball. Choose $n > 1/r$. By (3.115), there exists $x_{j,n}$ $(1 \le j \le p(n))$ such that $x \in B(x_{j,n}, 1/n)$. Hence $d(x_{j,n}, x) < 1/n < r$, i.e., $x_{j,n} \in B(x, r)$. Since $x_{j,n} \in Z$ by (3.116), we proved that $B(x, r)$ meets Z. This shows that the countable set Z is dense in X. □

Observation 2. *Let Z be a countable set, and suppose $f_n : Z \to \mathbb{R}$, $n = 1, 2, \ldots$ are uniformly bounded on Z, that is, there exists $M > 0$ such that*

$$|f_n(z)| \le M \tag{3.117}$$

for all $z \in Z$ and $n \in \mathbb{N}$. Then there exists a subsequence $\{f_n'\}$ of $\{f_n\}$ which converges pointwise on Z.

Proof This is proved by the well known Cantor diagonal process.

Write the countable set Z as a sequence $Z = \{z_i; \ i \in \mathbb{N}\}$. By (3.117), $|f_n(z_1)| \le M$ for all $n \in \mathbb{R}$. By the Bolzano-Weierstrass Theorem in $\mathbb{R}$, there exists a subsequence $\{f_{n,1}\}$ of $\{f_n\}$ such that $\{f_{n,1}(z_1)\}$ converges. Next, $|f_{n,1}(z_2)| \le M$ by (3.117). Therefore, by the BWT on $\mathbb{R}$, there exists a subsequence $\{f_{n,2}\}$ of $\{f_{n,1}\}$ which converges at the point z_2 (and at z_1). Continuing inductively, we obtain subsequences $\{f_{n,j}; \ n \in \mathbb{N}\}$, $j = 0, 1, 2, \ldots$, such that $\{f_{n,0}\} = \{f_n\}$, $\{f_{n,j}\}$ is a subsequence of $\{f_{n,j-1}\}$ for all $j \in \mathbb{N}$, and for each $j \in \mathbb{N}$, $\{f_{n,j}; \ n \in \mathbb{N}\}$ converges at the points $z_1, \ldots, z_j$.

We define the desired subsequence $\{f_n'\}$ of $\{f_n\}$ as the diagonal sequence $\{f_{n,n}\}$. For each fixed $j \in \mathbb{N}$, the subsequence $\{f_n'\}$ differs from a subsequence of $\{f_{n,j}\}$ at most for the indices $n \le j - 1$. Therefore $\{f_n'\}$ converges at the points $z_1, \ldots, z_j$.

Since $j \in \mathbb{N}$ is arbitrary, we proved that the subsequence $\{f_n'\}$ of $\{f_n\}$ converges pointwise at all the points $z_1, z_2, \ldots$ of Z. □

3.7.7 Theorem (The Arzela-Ascoli Theorem) *Let X be a compact metric space. Let $\mathbb{F} \subset C(X)$ be bounded and equicontinuous. Then $\mathbb{F}$ has the BWP: every sequence in $\mathbb{F}$ has a subsequence converging in $C(X)$.*

As mentioned before, the BWP is also called *sequential compactness*.

Proof By Observation 1 above, there exists a countable dense subset Z of X. Since $\mathbb{F}$ is a bounded set in $C(X)$, there exists $M > 0$ such that $||f|| \leq M$ for all $f \in \mathbb{F}$. Let $\{f_n\} \subset \mathbb{F}$. Then $\{f_n\}$ satisfies (3.117). By Observation 2, there exists a subsequence $\{f_n'\}$ of $\{f_n\}$ which converges pointwise on Z.

Let $\epsilon > 0$ be given. By the equicontinuity of $\mathbb{F}$, there exists $\delta > 0$ such that

$$|f(x) - f(y)| < \frac{\epsilon}{4} \tag{3.118}$$

for all $f \in \mathbb{F}$ and all $x, y \in X$ with $d(x, y) < \delta$.

Observe that

$$X = \bigcup_{z \in Z} B(z, \delta). \tag{3.119}$$

Indeed, if $x \in X$, $B(x, \delta)$ meets Z, by density of Z in X. Let $z \in B(x, \delta) \cap Z$. Then $x \in B(z, \delta)$.

By the compactness of X, it follows from (3.119) that there exist points $z_1, \ldots, z_q \in Z$ such that

$$X = \bigcup_{j=1}^{q} B(z_j, \delta). \tag{3.120}$$

Since $\{f_n'(z_j)\}$ converges for each $j = 1, \ldots, q$, there exist $n_j \in \mathbb{N}$ such that

$$|f_n'(z_j) - f_m'(z_j)| < \frac{\epsilon}{4} \tag{3.121}$$

for all $n.m > n_j$, for each $j = 1, \ldots, q$. Set $n^* := \max_{1 \leq j \leq q} n_j$. Then for all $n, m > n^*$, (3.121) is valid for all $j = 1, \ldots, q$.

Fix $x \in X$. By (3.120), there exists j, $1 \leq j \leq q$, such that $d(x, z_j) < \delta$. Hence by (3.118)

$$|f_n'(x) - f_n'(z_j)| < \frac{\epsilon}{4} \tag{3.122}$$

for all $n \in \mathbb{N}$. Let $n, m > n^*$. Then $n, m > n_j$, and therefore (3.121) is valid for the index j specified above.

By (3.121) and (3.122) with this index j, we have for all $n, m > n^*$,

$$|f_n'(x) - f_m'(x)| \leq |f_n'(x) - f_n'(z_j)| + |f_n'(z_j) - f_m'(z_j)| + |f_m'(z_j) - f_m'(x)| < \frac{3\epsilon}{4}.$$

Note that n^* was chosen independently of x. We may then take the supremum over all $x \in X$, and conclude that for all $n, m > n^*$,

$$||f_n' - f_m'|| \le \frac{3\epsilon}{4} < \epsilon.$$

Since $C(X)$ is complete, it follows that $\{f_n'\}$ converges in $C(X)$. □

Example. Let $I = [a, b] \times [\alpha, \beta]$, and let $K \in C(I)$. For each $f \in C([\alpha, \beta])$ and $x \in [a, b]$ fixed, the function $K(x, \cdot) f(\cdot)$ is continuous on $[\alpha, \beta]$, and therefore its Riemann integral is well defined on this interval. Set

$$(Tf)(x) := \int_\alpha^\beta K(x, y) f(y)\, dy.$$

The function Tf is continuous in $[a, b]$ (exercise). Thus $T : f \to Tf$ is a (linear) map of $C([\alpha, \beta])$ into $C([a, b])$.

Let $\mathbb{A}$ be a bounded subset of $C(([\alpha, \beta])$, that is, there exists $M > 0$ such that $||f|| \le M$ for all $f \in \mathbb{A}$ (all norms in the sequel are supremum norms on the relevant range of the variable).

We show that $\mathbb{B} := T\mathbb{A}$ is bounded and equicontinuous.

For all $f \in \mathbb{A}$,

$$|(Tf)(x)| \le ||K||\, ||f||\, (\beta - \alpha) \qquad (x \in [a, b]),$$

and therefore

$$||Tf|| \le ||K||\, ||f||\, (\beta - \alpha) \le ||K||\, M\, (\beta - \alpha).$$

This shows that $\mathbb{B} := T\mathbb{A}$ is a bounded subset of $C([a, b])$.

Let $\epsilon > 0$. Since K is continuous on the compact set I, it is uniformly continuous. Hence there exists $\delta > 0$ such that

$$|K(x, y) - K(x', y')| < \frac{\epsilon}{2M(\beta - \alpha)} \tag{3.123}$$

if $(x, y), (x', y') \in I$ and $d((x, y), (x', y')) < \delta$.

Let $x, x' \in [a, b]$ be such that $|x - x'| < \delta$. Then for all $y \in [\alpha, \beta]$, $d((x, y), (x', y)) = |x - x'| < \delta$, and therefore by (3.123), for all $f \in \mathbb{A}$,

$$|(Tf)(x) - (Tf)(x')| \le \int_\alpha^\beta |K(x, y) - K(x', y)|\, |f(y)|\, dy$$

$$\le \frac{\epsilon}{2M\,(\beta - \alpha)}\, M\, (\beta - \alpha) = \frac{\epsilon}{2} < \epsilon.$$

This shows that $\mathbb{B} = T\mathbb{A}$ is equicontinuous on $[a, b]$. By the AAT, $\mathbb{B}$ has the Bolzano-Weierstrass Property, that is, every sequence in $\mathbb{B}$ has a subsequence that converges in $C([a, b])$.

The Stone-Weierstrass Theorem

Back to the WAT, we now desire to extend it to the space $C(X)$, when X is a compact subset of $\mathbb{R}^k$, $k \geq 1$: polynomials (in k variables) are dense in $C(X)$. Actually, we need only some of the properties of polynomials to obtain such a density result.

The Banach space $C(X)$ has also a multiplication operation, namely, pointwise multiplication. With the vector space operations and multiplication, $C(X)$ is a unital algebra over the field $\mathbb{R}$. The adjective "unital" means that the algebra has a multiplicative unit (or identity). In the case of $C(X)$, the unit is the constant function with value 1 on X, which we denote by 1. We note that the norm of $C(X)$ has also the properties

$$||fg|| \leq ||f||\,||g|| \qquad (f, g \in C(X))$$

and $||1|| = 1$. A Banach space with a multiplication operation that gives it the structure of a unital algebra, and whose norm satisfies the above relations, is called a unital *Banach algebra* (over $\mathbb{R}$). Cf. Exercise 4, Sect. 3.6.16. We shall not need however these relations between the norm and multiplication in the present discussion.

We observe first that the set $\mathcal{A}$ of all polynomials (restricted to X) is a unital subalgebra of $C(X)$. Second, $\mathcal{A}$ *separates the points of* X, that is, if $a, b \in X$ are distinct, there exists $p \in \mathcal{A}$ such that $p(a) \neq p(b)$. Actually, given any two real numbers α, β we can choose $p \in \mathcal{A}$ to be the polynomial of degree 1

$$p(x) = \alpha + \frac{\beta - \alpha}{||b - a||^2}(b - a) \cdot (x - a) \qquad (x \in X \subset \mathbb{R}^k).$$

It turns out that the above properties of polynomials are sufficient for the density of $\mathcal{A} \subset C(X)$ in $C(X)$ in the general case where X is an arbitrary compact metric space. This is the Stone-Weierstrass Theorem stated below.

3.7.8 Theorem (The Stone-Weierstrass Theorem) *Let X be a compact metric space, and let $\mathcal{A}$ be a unital subalgebra of $C(X)$ which separates the points of X. Then $\mathcal{A}$ is dense in $C(X)$.*

Proof We begin with two properties of $\mathcal{A}$. □

Property 1. For any distinct points $a, b \in X$ and any $\alpha, \beta \in \mathbb{R}$, there exists $h \in \mathcal{A}$ such that $h(a) = \alpha$ and $h(b) = \beta$.

Indeed, since $\mathcal{A}$ separates the points of X, there exists $g \in \mathcal{A}$ such that $g(a) \neq g(b)$. Define

$$h(x) = \alpha + \frac{\beta - \alpha}{g(b) - g(a)}[g(x) - g(a)] \qquad (x \in X).$$

Since $\mathcal{A}$ is a vector space over the field $\mathbb{R}$ and $1 \in \mathcal{A}$, the function h belongs to $\mathcal{A}$, and clearly $h(a) = \alpha$ and $h(b) = \beta$.

Property 2. For all $f, g \in \mathcal{A}$, the functions $|f|$, $\max(f, g)$, and $\min(f, g)$ (defined pointwise) belong to $\overline{\mathcal{A}}$, the closure of $\mathcal{A}$ in $C(X)$.

Let $f \in \mathcal{A}$, not identically 0 without loss of generality. Define $g = \|f\|^{-1} f$. Then $g \in \mathcal{A}$ and $g(X) \subset [-1, 1]$. Consider the function $h(t) = |t|$ on $[-1, 1]$. Since it is continuous, there exists a sequence of polynomials p_n in t converging uniformly to h in $[-1, 1]$ (by the WAT). Since $\mathcal{A}$ is a unital algebra and $g \in \mathcal{A}$, $p_n \circ g$ are well defined elements of $\mathcal{A}$, and $p_n \circ g$ converge to $|g|$ uniformly on X. Hence $|g| \in \overline{\mathcal{A}}$, and therefore $|f| = \|f\|\,|g| \in \overline{\mathcal{A}}$.

Next, for any two functions $f, g \in \mathcal{A}$, we have

$$\max(f, g) = \frac{1}{2}[f + g + |f - g|] \in \overline{\mathcal{A}}$$

and

$$\min(f, g) = \frac{1}{2}[f + g - |f - g|] \in \overline{\mathcal{A}}.$$

This verifies Property 2 of $\mathcal{A}$.

We proceed now with the proof of the SWT.

Let $f \in C(X)$. We must prove that $f \in \overline{\mathcal{A}}$.

Let $\epsilon > 0$. Fix $x \in X$. For any given $y \in X$, apply Property 1 with $\alpha = f(x)$ and $\beta \le f(y) + \epsilon/2$. There exists $f_y \in \mathcal{A}$ such that

$$f_y(x) = f(x), \qquad f_y(y) = \beta \le f(y) + \frac{\epsilon}{2}. \tag{3.124}$$

By continuity of f and f_y, it follows from (3.124) that there exists a ball $B(y, r_y)$ such that

$$f_y < f + \epsilon \ on\ B(y, r_y). \tag{3.125}$$

The open cover $\{B(y, r_y);\ y \in X\}$ of the compact set X has a finite subcover, say

$$X = \bigcup_{k=1}^{n} B(y_k, r_{y_k}). \tag{3.126}$$

Define

$$g := \min(f_{y_1}, \dots, f_{y_n}).$$

By Property 2 (and induction), $g \in \overline{\mathcal{A}}$. If $z \in X$, then by (3.126) there exists k, $1 \le k \le n$, such that $z \in B(y_k, r_{y_k})$, and therefore by (3.125)

$$g(z) \le f_{y_k}(z) < f(z) + \epsilon.$$

Together with (3.124), we have

$$g(x) = f(x); \;\; g < f + \epsilon \;\; on \; X. \tag{3.127}$$

We got (3.127) for any fixed $x \in X$. We wish now to vary x. The function g constructed above depends on x, and we shall stress this fact by writing g_x instead of g. Rewrite (3.127) for $g_x \in \overline{\mathcal{A}}$ with this new notation:

$$g_x(x) = f(x); \;\; g_x < f + \epsilon \;\; on \; X. \tag{3.128}$$

By continuity of g_x and f, there exists a ball $B(x, s_x)$ such that $g_x > f - \epsilon$ on $B(x, s_x)$. The open cover $\{B(x, s_x); \; x \in X\}$ of the compact set X has a finite subcover, say

$$X = \bigcup_{j=1}^{m} B(x_j, s_{x_j}). \tag{3.129}$$

Define

$$h = \max(g_{x_1}, \ldots, g_{x_m}). \tag{3.130}$$

Then $h \in \overline{\mathcal{A}}$ by Property 2 (and induction). Let $z \in X$. By (3.129), there exists j, $1 \le j \le m$, such that $z \in B(x_j, s_{x_j})$. Hence

$$h(z) \ge g_{x_j}(z) > f(z) - \epsilon,$$

that is $h > f - \epsilon$ on X. By (3.128) and (3.130), we have also $h < f + \epsilon$ on X. Hence $||h - f|| < \epsilon$. Since $h \in \overline{\mathcal{A}}$, this proves that f belongs to the closure in $C(X)$ of the (closed) set $\overline{\mathcal{A}}$, hence $f \in \overline{\mathcal{A}}$. □

By the comments preceding the statement of the SWT, we have the following

3.7.9 Corollary (WAT in $\mathbb{R}^k$) *Let X be a compact subset of $\mathbb{R}^k$, $k \ge 1$. Then the polynomials (in k variables) are dense in $C(X)$.*

Example. Let I be a cell in $\mathbb{R}^k$, say

$$I = \prod_{i=1}^{k} [a_i, b_i].$$

Let $\mathcal{A}$ be the set of all finite linear combinations of functions $f : I \to \mathbb{R}$ of the form

$$f(x) = \prod_{i=1}^{k} f_i(x_i)$$

with $f_i \in C([a_i, b_i])$. Then $\mathcal{A}$ is a unital subalgebra of $C(I)$. If u, v are distinct points of X, there exists i, $1 \le i \le k$, such that $u_i \ne v_i$. For this index i, there exists $f_i \in C([a_i, b_i])$ such that $f_i(u_i) \neq f_i(v_i)$ (we may take f_i to be a linear function on $[a_i, b_i]$). Take $f_j = 1$ identically on $[a_j, b_j]$ for all $j \neq i$ $(1 \leq j \leq k)$. Then $f(x) := \prod_{1 \leq j \leq k} f_j(x_j) \in \mathcal{A}$ and $f(u) = f_i(u_i) \neq f_i(v_i) = f(v)$. Hene $\mathcal{A}$ separates the points of I. By the SWT, we conclude that $\mathcal{A}$ is dense in $C(I)$. Note that we could also reach this conclusion by applying Corollary 3.7.9, since $\mathcal{A}$ contains all the polynomials in the variables $x_1, \ldots, x_k$.

3.7.10 Exercises

1. Let X be a normed space, Y a Banach space, and $Z \subset X$ a dense subspace of X. Let $T \in B(Z, Y)$ (cf. Exercise 1.3.7 (5)). Prove
   (a) T has a unique extension $\tilde{T} \in B(X, Y)$;
   (b) $||\tilde{T}||_{B(X,Y)} = ||T||_{B(Z,Y)}$.
   (c) Let Z be the space of polynomials in one real variable (restricted to $I := [0, 1]$), regarded as a subspace of the Banach space $C(I)$. It is given that the operator $T \in B(Z, C(I))$ satisfies $T(x^n) = x^{n+1}/(n+1)$ for all $n = 0, 1, 2, \ldots$. Let $\tilde{T} \in B(C(I), C(I))$ be the unique extension of T (cf. WAT). Then $\tilde{T} : f(x) \to \int_0^x f(t)\,dt$.
2. Let X be a compact metric space. The space $C(X, \mathbb{C})$ (over the field $\mathbb{C}$ of complex numbers) is a unital *complex* Banach algebra, where the adjective "complex" stresses the fact that the field of scalars is $\mathbb{C}$. All operations are defined pointwise, and the norm is $||f|| := \sup_{x \in X} |f(x)|$.
   Let $\mathcal{A}$ be a unital subalgebra of $C(X, \mathbb{C})$ (over $\mathbb{C}$). We say that $\mathcal{A}$ is *selfadjoint* if the complex adjoint $\overline{f}$ of f (defined pointwise) belongs to $\mathcal{A}$ for each $f \in \mathcal{A}$. Set
   $$\mathcal{A}_r := \{f \in \mathcal{A};\ f(X) \subset \mathbb{R}\}.$$
   Prove
   (a) $\mathcal{A}_r$ is a unital subalgebra of $C(X, \mathbb{R})$ over the field $\mathbb{R}$; if $\mathcal{A}$ is selfadjoint, then $\Re\mathcal{A} := \{\Re f;\ f \in \mathcal{A}\}$ and $\Im\mathcal{A} := \{\Im f;\ f \in \mathcal{A}\}$ are subsets of $\mathcal{A}_r$ ($\Re f$ and $\Im f$ are defined pointwise);
   (b) if $\mathcal{A}$ is selfadjoint and separates the points of X, then $\mathcal{A}_r$ separates the points of X;
   (c) if $\mathcal{A}$ is a unital selfadjoint subalgebra of $C(X, \mathbb{C})$ which separates the points of X, then $\mathcal{A}$ is dense in $C(X, \mathbb{C})$.
   This is the complex version of the SWT.
3. Let $\mathbb{T}$ be the unit circle with the $\mathbb{C}$-metric:
   $$d(z, w) = |z - w| \qquad (z, w \in \mathbb{C}).$$
   Given $k \in \mathbb{N}$, consider the complex unital Banach algebra $C(\mathbb{T}^k, \mathbb{C})$ (the metric on the Cartesian product $\mathbb{T}^k$ is the metric of $\mathbb{C}^k$ properly restricted). Let $\mathcal{A}$ be the subalgebra of all polynomials in k complex variables (with complex coefficients), restricted to $\mathbb{T}^k$. Prove that $\mathcal{A}$ is dense in $C(\mathbb{T}^k, \mathbb{C})$.

Note that the complex variables z_j restricted to $\mathbb{T}$ can be written as e^{ix_j} with $x_j \in [0, 2\pi]$. Writing $n := (n_1, \ldots, n_k) \in \mathbb{N}^k$, the elements of $\mathcal{A}$ are finite sums of the form

$$\sum c_n e^{in\cdot x} \qquad x := (x_1, \ldots, x_k) \in [0, 2\pi]^k,$$

with $c_n \in \mathbb{C}$.

These sums are called *trigonometric polynomials* in k real variables, because of the well known relation $e^{it} = \cos t + i \sin t$ for $t \in \mathbb{R}$. Thus trigonometric polynomials in k real variables (with complex coefficients) are dense in $C(\mathbb{T}^k, \mathbb{C})$.

4. Let $X = \overline{B}(0, 1)$ be the closed unit ball in $\mathbb{R}^k$. Let $\mathbb{F}$ be a family of functions $f : X \to \mathbb{R}$ of class C^1, and suppose there exist positive constants L, M such that

   (a) $|f(0)| \le L$ for all $f \in \mathbb{F}$, and
   (b) $||\nabla f(x)|| \le M$ for all $x \in X$ and $f \in \mathbb{F}$.
   (The norm in (b) is the Euclidean norm on $\mathbb{R}^k$.)

   Prove:

   (i) $||f|| \le L + M$ for all $f \in \mathbb{F}$.
   (ii) $\mathbb{F}$ is equicontinuous on X.

   Therefore, by the Arzela-Ascoli Theorem, $\mathbb{F}$ has the BWP.

Chapter 4
Integration

In this chapter, we discuss Riemann integration of functions of several real variables. First we consider partial integrals, in which we fix all the variables except for one, as was done in the case of partial derivation.

4.1 Partial Integrals

4.1.1 Partial Integrals (or Integrals Dependent on Parameters) For simplicity of notation, we consider the case of two real variables. The general principle for any number of variables will be clear from it.

Consider the closed rectangle

$$I := [a, b] \times [\alpha, \beta] \subset \mathbb{R}^2,$$

and let

$$f : I \to \mathbb{R}$$

be such that $f(\cdot, y)$ is Riemann integrable in the interval $[a, b]$ for each fixed $y \in [\alpha, \beta]$. Then the relation

$$F(y) := \int_a^b f(x, y)\, dx \qquad (y \in [\alpha, \beta])$$

defines a function $F : [\alpha, \beta] \to \mathbb{R}$. We call F a partial integral of f (with respect to x), or an integral depending on the parameter y.

These notations will be fixed in this section.

The original version of this chapter was revised. An erratum to this chapter can be found at DOI 10.1007/978-3-319-27956-5_5

S. Kantorovitz, *Several Real Variables*, Springer Undergraduate Mathematics Series, DOI 10.1007/978-3-319-27956-5_4

If $f \in C(I)$, the restriction of f to the segment

$$I_y := \{(x, y);\ x \in [a, b]\} \subset I$$

is continuous for each fixed $y \in [\alpha, \beta]$, hence $f(\cdot, y)$ is integrable on $[a, b]$, and F is well-defined on $[\alpha, \beta]$.

4.1.2 Theorem *If $f \in C(I)$, then $F \in C([\alpha, \beta])$.*

Proof Let $\epsilon > 0$. Since f is continuous on the compact set I (cf. 1.2.8), it is uniformly continuous on I (cf. 1.4.11). There exists therefore $\delta > 0$ such that

$$|f(x, y) - f(x, y')| < \epsilon/(b - a) \qquad (\forall x \in [a, b])$$

when $(d((x, y),\ (x, y')) =)\ |y - y'| < \delta$. Therefore, when $|y - y'| < \delta$,

$$\begin{aligned} |F(y) - F(y')| &= |\int_a^b [f(x, y) - f(x, y')]\, dx| \\ &\le \int_a^b |f(x, y) - f(x, y')|\, dx < \frac{\epsilon}{b - a}(b - a) = \epsilon. \end{aligned}$$

□

If in addition f_y exists in I and $f_y \in C(I)$, then we have the following stronger conclusion on F:

Leibnitz' rule.

4.1.3 Theorem *Suppose $f \in C(I)$, and f_y exists in I and is continuous there. Then F is of class C^1 in $[\alpha, \beta]$ and*

$$\frac{dF}{dy}(y) = \int_a^b \frac{\partial f}{\partial y}(x, y)\, dx. \tag{4.1}$$

Formula (4.1) is Leibnitz' rule for deriving with respect to the parameter y under the integral sign.

Proof Let $\epsilon > 0$ be given. Suppose $0 \neq h \in \mathbb{R}$, and $y,\ y + h \in [\alpha, \beta]$. For each $x \in [a, b]$, $f_y(x, \cdot)$ exists in the interval $[y,\ y + h]$ (or $[y + h,\ y]$). Hence by the MVT for functions of one real variable, there exists θ, $0 < \theta < 1$, depending on $x,\ y,\ h$, such that

$$f(x, y + h) - f(x, y) = h\, f_y(x, y + \theta h). \tag{4.2}$$

Since f_y is continuous on the compact set I, it is uniformly continuous there. There exists therefore $\delta > 0$ such that

$$|f_y(x, y') - f_y(x, y)| < \frac{\epsilon}{b - a} \tag{4.3}$$

when $(d((x,y'),(x,y)) =)\ |y'-y|<\delta$, for all $x\in[a,b]$ and $y,y'\in[\alpha,\beta]$.

Let $|h|<\delta$. For x, y, h and θ as in (4.2), $|(y+\theta h)-y| = \theta|h| < |h| < \delta$, and therefore by (4.3)

$$|f_y(x,y+\theta h) - f_y(x,y)| < \frac{\epsilon}{b-a}.$$

Hence

$$\begin{aligned}
&\Big|\frac{F(y+h)-F(y)}{h} - \int_a^b f_y(x,y)\,dx\Big| \\
&= \Big|\int_a^b \Big[\frac{f(x,y+h)-f(x,y)}{h} - f_y(x,y)\Big]\,dx\Big| \\
&= \Big|\int_a^b [f_y(x,y+\theta h) - f_y(x,y)]\,dx\Big| \\
&\le \int_a^b |f_y(x,y+\theta h) - f_y(x,y)|\,dx < \frac{\epsilon}{b-a}(b-a) = \epsilon.
\end{aligned}$$

Although θ is an unknown function of x, y, h, there is no question about the integrability of the last two integrands, because they are equal, respectively, to the preceding *continuous* integrand and its absolute value. Thus

$$\lim_{h\to 0}\frac{F(y+h)-F(y)}{h} = \int_a^b f_y(x,y)\,dx.$$

This proves that F' exists in $[\alpha,\beta]$ and is given by (4.1). Since $f_y \in C(I)$, Theorem 4.1.2 implies that $F' \in C([\alpha,\beta])$, that is, F is of class C^1 in $[\alpha,\beta]$. □

We give below a version of Theorem 4.1.3 for improper integrals.

4.1.4 Theorem *Let $f,\ f_y \in C([a,\infty)\times[\alpha,\beta])$. Suppose*

(a) $F(y) := \int_a^\infty f(x,y)\,dx$ converges for each $y\in[\alpha,\beta]$, and
(b) $\int_a^\infty f_y(x,y)\,dx$ converges absolutely and uniformly with respect to y for all $y\in[\alpha,\beta]$.

Then $F\in C^1([\alpha,\beta])$ and

$$f'(y) = \int_a^\infty f_y(x,y)\,dx \qquad (y\in[\alpha,\beta]).$$

Proof Let $\epsilon>0$. By the smoothness hypothesis and Condition (b), there exists $b>a$ (which we fix) such that for all $y\in[\alpha,\beta]$

$$\int_b^\infty |f_y(x,y)|\,dx < \frac{\epsilon}{3}. \tag{4.4}$$

For $y,\ y+h \in [\alpha,\beta]$, $h \neq 0$, we have by the one-variable MVT,

$$\frac{F(y+h)-F(y)}{h} = \int_a^\infty \frac{f(x,y+h)-f(x,y)}{h}\,dx = \int_a^\infty f_y(x,y+\theta h)\,dx, \tag{4.5}$$

where θ depends on x and y. Hence

$$\left|\frac{F(y+h)-F(y)}{h} - \int_a^\infty f_y(x,y)\,dx\right| \le \int_a^\infty |f_y(x,y+\theta h) - f_y(x,y)|\,dx$$

$$\le \int_a^b |f_y(x,y+\theta h) - f_y(x,y)|\,dx + \int_b^\infty |f_y(x,y+\theta h)|\,dx + \int_b^\infty |f_y(x,y)|\,dx. \tag{4.6}$$

Let $I := [a,b]\times[\alpha,\beta]$. Since $f_y \in C(I)$ and I is compact, f_y in uniformly continuous in I. Therefore there exists $\delta > 0$ such that $|f_y(x',y') - f_y(x,y)| < \frac{\epsilon}{3(b-a)}$ whenever (x,y) and (x',y') are points of I at a distance smaller than δ. In our previous notation, if $|h| < \delta$, the distance between the points $(x, y+\theta h)$ and (x,y) is $\theta|h| < |h| < \delta$, and therefore the first integral in (4.6) is less than $\epsilon/3$. The second and the third integrals in (4.6) are less than $\epsilon/3$ by (4.4). We conclude that $F'(y)$ exists and is equal to $\int_a^\infty f_y(x,y)\,dx$. The continuity of F' in $[\alpha,\beta]$ follows from this last formula. Given $\epsilon > 0$, we fix $b > a$ and then δ as above. For $y,\ y+h \in [\alpha,\beta]$,

$$|F'(y+h) - F'(y)| \le \int_a^b |f_y(x,y+h) - f_y(x,y)|\,dx + \frac{2\epsilon}{3} < \epsilon$$

if $|h| < \delta$. □

4.1.5 Example Let

$$f(x,y) = e^{-xy}\frac{\sin x}{x}$$

for $(x,y) \in (0,\infty)\times[\alpha,\beta]$, $0 < \alpha < \beta$, and $f(0,y) = 1$ for all $y \in [\alpha,\beta]$. Clearly

$$f_y(x,y) = -e^{-xy}\sin x$$

for all (x,y) in the strip $K := [0,\infty)\times[\alpha,\beta]$. We have $f,\ f_y \in C(K)$. Since $|\sin x/x| \le 1$, both f and f_y are dominated in K by $e^{-\alpha x}$. This implies that the conditions (a)–(b) of Theorem 4.1.4 are satisfied ($a = 0$ in our example). Defining F as in the theorem, we conclude after two integrations by part that

$$F'(y) = -\int_0^\infty e^{-xy}\sin x\,dx = e^{-xy}\cos x\Big|_0^\infty + y\int_0^\infty e^{-xy}\cos x\,dx$$

$$= -1 + y e^{-xy}\sin x\Big|_0^\infty + y^2\int_0^\infty e^{-xy}\sin x\,dx$$

$$= -1 - y^2 F'(y).$$

Hence

$$F'(y) = -\frac{1}{1+y^2} \qquad y \in [\alpha, \beta].$$

Integrating over the interval $[\alpha, \beta]$, we obtain

$$F(\alpha) - F(\beta) = \arctan \beta - \arctan \alpha \tag{4.7}$$

for all $0 < \alpha < \beta < \infty$. Since $|\frac{\sin x}{x}| \leq 1$, we have

$$|F(\beta)| \leq \int_0^\infty e^{-x\beta} dx = \frac{1}{\beta} \to 0$$

as $\beta \to \infty$. Letting $\beta \to \infty$ in (4.7), we get

$$F(\alpha) = \frac{\pi}{2} - \arctan \alpha \qquad (\alpha > 0). \tag{4.8}$$

It follows in particular that

$$\lim_{\alpha \to 0+} F(\alpha) = \frac{\pi}{2}. \tag{4.9}$$

Formally, $F(0)$ is the improper integral

$$F(0) = \int_0^\infty \frac{\sin x}{x} dx. \tag{4.10}$$

The integral (4.10) converges, because for all $0 < M < N < \infty$, we see by an integration by parts that

$$\int_M^N \frac{\sin x}{x} dx = -\frac{\cos x}{x}\Big|_M^N - \int_M^N \frac{\cos x}{x^2} dx.$$

The integrated term is $= \frac{\cos M}{M} - \frac{\cos N}{N} \to 0$ as $M, N \to \infty$. The absolute value of the integral term is $\leq \int_M^N dx/x^2 = 1/M - 1/N \to 0$ as $M, N \to \infty$. Hence $\int_M^N \frac{\sin x}{x} dx \to 0$ as $M, N \to \infty$. By the Cauchy criterion for the convergence of improper integrals, we conclude that

$$F(0) := \int_0^\infty \frac{\sin x}{x} dx$$

is well-defined. This is the so-called Dirichlet integral. It can be shown (but not with the tools presently available to us) that F is continuous (from the right) at 0. We conclude from (4.9) that the value of the Dirichlet integral is $\pi/2$.

Variable Limits of Integration

The preceding results will be generalized now to the case of variable limits of integration.

4.1.6 Theorem *Let $f \in C(I)$, and define*

$$F(y, u, v) = \int_u^v f(x, y)\, dx$$

for

$$(y, u, v) \in T := [\alpha, \beta] \times [a, b] \times [a, b].$$

Then $F \in C(T)$.

Proof Let $(y, u, v) \in T$, and let $(p, q, r) \in \mathbb{R}^3$ be such that $(y, u, v) + (p, q, r) \in T$. We have

$$\Delta F := F(y + p, u + q, v + r) - F(y, u, v) = A + B + C, \qquad (4.11)$$

where

$$A := F(y + p, u + q, v + r) - F(y, u + q, v + r),$$
$$B := F(y, u + q, v + r) - F(y, u, v + r),$$
$$C := F(y, u, v + r) - F(y, u, v).$$

Let $\epsilon > 0$ be given. By the uniform continuity of f in the compact set I, there exists $\delta > 0$ such that

$$|f(x, y + p) - f(x, y)| < \frac{\epsilon}{3(b - a)} \qquad (4.12)$$

for all $x \in [a, b]$ if $|p| < \delta$.

Let $||f||$ denote the $C(I)$-norm of f (cf. 1.2.8 and 3.1.3). We may assume that $||f|| > 0$, because otherwise $f = 0$ identically in I, hence $F = 0$ identically, and the theorem is trivial.

Note that for any integrable function ϕ of x in $[a, b]$ and any $u, v \in [a, b]$, we have

$$|\int_u^v \phi(x)\, dx| \le \left| \int_u^v |\phi(x)|\, dx \right|.$$

The outer absolute value signs are needed because we may have $v < u$.

We estimate A, B, C as follows. By (4.12), if $|p| < \delta$,

$$|A| = \left| \int_{u+q}^{v+r} [f(x, y + p) - f(x, y)]\, dx \right|$$

$$\leq \frac{\epsilon}{3(b-a)}|(v+r)-(u+q)| \leq \frac{\epsilon}{3(b-a)}(b-a) = \frac{\epsilon}{3}.$$

In the above estimate, we used the fact that $u+q$ and $v+r$ are both in $[a, b]$, so that the distance between them is at most $b-a$.

$$\begin{aligned}|B| &= \left|\left(\int_{u+q}^{v+r} - \int_{u}^{v+r}\right) f(x, y)\,dx\right| \\ &= \left|\int_{u+q}^{u} f(x, y)\,dx\right| \leq \left|\int_{u+q}^{u} |f(x, y)|\,dx\right| \\ &\leq ||f||\,|u-(u+q)| = ||f||\,|q| < \frac{\epsilon}{3}\end{aligned}$$

if $|q| < \epsilon/(3||f||)$. Similarly,

$$|C| = \left|\int_{v}^{v+r} f(x, y)\,dx\right| \leq ||f||\,|r| < \frac{\epsilon}{3}$$

if $|r| < \epsilon/(3||f||)$.

Set

$$\delta^* = \min[\delta, \frac{\epsilon}{3||f||}].$$

If

$$d((y, u, v) + (p, q, r), (y, u, v)) \; (= (p^2+q^2+r^2)^{1/2}) < \delta^*,$$

then $|p| < \delta^* \leq \delta$ and $|q|$, $|r| < \delta^* \leq \epsilon/(3||f||)$; hence $|A|$, $|B|$, $|C| < \epsilon/3$, and therefore $|\Delta F| < \epsilon$ by (4.11). This proves the continuity of F at the arbitrary point $(y, u, v) \in T$. □

4.1.7 Corollary *Let $f \in C(I)$ and*

$$\phi, \; \psi : [\alpha, \beta] \to [a, b]$$

be continuous. Define

$$H(y) = \int_{\phi(y)}^{\psi(y)} f(x, y)\,dx \qquad (y \in [\alpha, \beta]). \tag{4.13}$$

Then $H \in C([\alpha, \beta])$.

Proof Let $F : T \to \mathbb{R}$ be as in Theorem 4.1.6, and let $G : [\alpha, \beta] \to T$ be defined by

$$G(y) = (y, \; \phi(y), \; \psi(y)) \qquad (y \in [\alpha, \beta]).$$

Then

$$H = F \circ G. \tag{4.14}$$

Since $F \in C(T)$ and G is continuous on $[\alpha, \beta]$ (because its components are continuous by our hypothesis on ϕ, ψ) with range in T, it follows from (4.14) and Theorem 1.4.4 that H is continuous on $[\alpha, \beta]$. □

Next we generalize Theorem 4.1.3 to the case of limits of integration depending on the parameter.

General Leibnitz' Rule

4.1.8 Theorem *Let f, $f_y \in C(I)$, and suppose ϕ, $\psi : [\alpha, \beta] \to [a, b]$ are differentiable. Define H by (4.13). Then H is differentiable in the interval $[\alpha, \beta]$ and for all y in this interval one has the identity*

$$H'(y) = \int_{\phi(y)}^{\psi(y)} f_y(x, y)\, dx + f(\psi(y), y)\psi'(y) - f(\phi(y), y)\phi'(y). \tag{4.15}$$

Moreover, if ϕ, ψ are of class C^1, then so is H (in $[\alpha, \beta]$).

Proof (Notation as in 4.1.6)
For any $u, v \in [a, b]$, we apply Theorem 4.1.3 to the sub-rectangle $[u, v] \times [\alpha, \beta]$ (or $[v, u] \times [\alpha, \beta]$ if $v < u$) of the rectangle I. Then for all $(y, u, v) \in T$

$$F_y(y, u, v) = \int_u^v f_y(x, y)\, dx. \tag{4.16}$$

Since $f_y \in C(I)$, it follows from Theorem 4.1.6 and (4.16) that $F_y \in C(T)$.

By the Fundamental Theorem of Calculus, since $f(\cdot, y)$ is continuous in $[a, b]$ for each fixed $y \in [\alpha, \beta]$, we have

$$F_u(y, u, v) = -f(u, y); \quad F_v(y, u, v) = f(v, y), \tag{4.17}$$

hence in particular F_u, $F_v \in C(T)$, because $f \in C(I)$.

We have verified up to this point that F is of class C^1 in T, hence differentiable in T, by Theorem 2.1.5.

The $\mathbb{R}^3$ valued function $G : [\alpha, \beta] \to T$ defined by

$$G(y) = (y, \phi(y), \psi(y)) \tag{4.18}$$

is differentiable, since its components are differentiable by the hypothesis on ϕ, ψ (cf. Sect. 2.1.8).

Thus the hypothesis of Theorem 2.1.9 are satisfied, and therefore the composite function $H = F \circ G$ (cf. 4.1.7) is differentiable in $[\alpha, \beta]$, and the chain rule formula in 2.1.9 gives

$$H'(y) = F_y(G(y)) + F_v(G(y))\,\psi'(y) + F_u(G(y))\,\phi'(y) \tag{4.19}$$

for all $y \in [\alpha, \beta]$. Substituting (4.16)–(4.18) in (4.19), we obtain (4.15).

Moreover, if ϕ and ψ are of class C^1 in $[\alpha, \beta]$, it follows from (4.15), Theorem 4.1.6, and Theorem 1.4.4 that H' is continuous in $[\alpha, \beta]$. □

4.1.9 Example Consider the integral

$$H(y) = \int_0^y \frac{\log(1+xy)}{1+x^2}\,dx \tag{4.20}$$

For any $L > 0$ the conditions of Theorem 4.1.8 are satisfied for $0 \le x, y \le L$. Hence

$$H'(y) = \int_0^y \frac{x}{(1+xy)(1+x^2)}\,dx + \frac{\log(1+y^2)}{1+y^2}. \tag{4.21}$$

We calculate the integral in (4.21) by decomposing the integrand into "partial fractions" in the form

$$[\frac{x+y}{1+x^2} - \frac{y}{1+xy}](1+y^2)^{-1}.$$

The integral in (4.21) is then equal to

$$[\frac{1}{2}\log(1+x^2) + y\arctan x - \log(1+xy)]\Big|_0^y (1+y^2)^{-1}$$

$$= \frac{1}{2}\Big[\frac{1}{1+y^2}\log(1+y^2) + \frac{2y}{1+y^2}\arctan y\Big] - \frac{\log(1+y^2)}{1+y^2}.$$

Hence by (4.21)

$$H'(y) = \frac{1}{2}[\arctan y\,\log(1+y^2)]'.$$

Since also $H(0) = 0$, it follows that

$$H(y) = \frac{1}{2}\arctan y\,\log(1+y^2)$$

for all $y \ge 0$ (since $L > 0$ is arbitrary). For example

$$\int_0^1 \frac{\log(1+x)}{1+x^2}\,dx\ (= H(1)) = \frac{1}{2}\arctan 1\,\log 2 = \frac{\pi \log 2}{8}.$$

We consider next integration of $F \in C([\alpha, \beta])$ for $f \in C(I)$ (cf. Theorem 4.1.2), and in analogy to the case of differentiation (cf. Theorem 4.1.3), we show that integration with respect to the parameter can be performed under the integral sign.

Changing the Order of Integration

4.1.10 Theorem *Let $f \in C(I)$. Then*

$$\int_\alpha^\beta F(y)\,dy = \int_a^b \Big(\int_\alpha^\beta f(x,y)\,dy\Big)\,dx. \tag{4.22}$$

Note that F is continuous in $[\alpha, \beta]$ by Theorem 4.1.2, so that the left hand side of (4.22) exists. Similarly, interchanging the roles of the variables, the inner integral on the right hand side defines a continuous function of x in $[a, b]$, by Theorem 4.1.2, so that the repeated integral on the right hand side exists as well. The claim of the theorem is that the two repeated integrals in (4.22) *coincide*, that is, the order of integration can be interchanged in the repeated integral of $f \in C(I)$:

$$\int_\alpha^\beta \Big(\int_a^b f(x,y)\,dx\Big)\,dy = \int_a^b \Big(\int_\alpha^\beta f(x,y)\,dy\Big)\,dx. \tag{4.23}$$

The inner brackets in (4.23) will be usually dropped for simplicity of notation.

Proof Let $\epsilon > 0$ be give. By the uniform continuity of f in the compact set I, there exists $\delta > 0$ such that

$$|f(x,y) - f(x',y')| < \frac{\epsilon}{(b-a)(\beta-\alpha)} \tag{4.24}$$

whenever $(x,y),\ (x',y') \in I$ satisfy $d((x,y),\ (x',y')) < \delta$. Let

$$a = x_0 < x_1 < \cdots < x_n = b;\ \ \alpha = y_0 < \cdots < y_m = \beta$$

be partitions of $[a, b]$ and $[\alpha, \beta]$ respectively such that the sub-rectangles of I

$$I_{ij} := [x_{i-1}, x_i] \times [y_{j-1}, y_j] \qquad (i = 1, \ldots, n;\ j = 1, \ldots, m)$$

have maximal diameter smaller than δ. Denote

$$m_{ij} = \min_{I_{ij}} f;\ \ M_{ij} = \max_{I_{ij}} f.$$

These minima and maxima are attained at points (x_{ij}, y_{ij}) and (x'_{ij}, y'_{ij}) of I_{ij} repectively (cf. Corollary 1.4.8), and since the distance between these points is no greater than the diameter of I_{ij}, hence smaller than δ, it follows from (4.24) that

$$M_{ij} - m_{ij} = f(x'_{ij}, y'_{ij}) - f(x_{ij}, y_{ij}) < \frac{\epsilon}{(b-a)(\beta-\alpha)} \tag{4.25}$$

for all $i = 1, \ldots, n$ and $j = 1, \ldots, m$.

Denote the area of I_{ij} and I by A_{ij} and A, respectively, i.e.,

$$A_{ij} := (x_i - x_{i-1})(y_j - y_{j-1}); \;\; A = (b-a)(\beta - \alpha).$$

Trivially

$$A = \sum_{ij} A_{ij}. \tag{4.26}$$

By the monotonicity property of the Riemann integral for integrable functions of one real variable, we have for all indices i and j

$$m_{ij} A_{ij} \le \int_{y_{j-1}}^{y_j} \int_{x_{i-1}}^{x_i} f(x,y)\, dx\, dy \le M_{ij} A_{ij}.$$

Summing these inequalities over all indices, we get

$$\sum_{ij} m_{ij} A_{ij} \le \int_\alpha^\beta \int_a^b f(x,y)\, dx\, dy \le \sum_{ij} M_{ij} A_{ij}. \tag{4.27}$$

Similarly,

$$\sum_{ij} m_{ij} A_{ij} \le \int_a^b \int_\alpha^\beta f(x,y)\, dy\, dx \le \sum_{ij} M_{ij} A_{ij}. \tag{4.28}$$

Denote the repeated integrals in (4.27) and (4.28) by P and Q, respectively. Since P and Q are both in the interval

$$[\sum m_{ij} A_{ij}, \sum M_{ij} A_{ij}] \tag{4.29}$$

the distance $|P - Q|$ is no greater than the length of the interval (4.29), which is

$$\sum (M_{ij} - m_{ij})\, A_{ij} < \frac{\epsilon}{(b-a)(\beta-\alpha)} \sum_{ij} A_{ij} = \epsilon,$$

where we used (4.25) and (4.26). Thus $|P - Q| < \epsilon$, and since ϵ is arbitrary, we conclude that $P = Q$. □

4.1.11 Example The function $f(x,y) = x^y$ is continuous in the rectangle $[0,1] \times [\alpha, \beta]$, for any $\beta > \alpha > 0$. Hence by Theorem 4.1.10,

$$\int_0^1 \int_\alpha^\beta x^y\, dy\, dx = \int_\alpha^\beta \int_0^1 x^y\, dx\, dy. \tag{4.30}$$

The right hand side of (4.30) is equal to

$$\int_\alpha^\beta \frac{x^{y+1}}{y+1}\Big|_{x=0}^{x=1} dy = \int_\alpha^\beta \frac{dy}{y+1} = \log\frac{\beta+1}{\alpha+1}. \tag{4.31}$$

The left hand side of (4.30) is equal to

$$\int_0^1 \frac{x^y}{\log x}\Big|_{y=\alpha}^{y=\beta} dx = \int_0^1 \frac{x^\beta - x^\alpha}{\log x}\, dx. \tag{4.32}$$

By (4.30)–(4.32), we get the following formula, for any positive numbers $\alpha,\ \beta$:

$$\int_0^1 \frac{x^\beta - x^\alpha}{\log x}\, dx = \log\frac{\beta+1}{\alpha+1}. \tag{4.33}$$

We proved (4.33) for $\beta > \alpha > 0$. Changing the roles of $\alpha,\ \beta$ when $\alpha > \beta > 0$, we get (4.33) in the latter case, since both sides of (4.33) change sign when the roles are interchanged. If $\alpha = \beta$, both sides trivially vanish.

4.1.12 Example Fix $0 < \alpha < 1$. The function $f(x, y) = (x^2 + y^2)^{-1}$ is continuous in the rectangle $I = [0, 1] \times [\alpha, 1]$ (as a rational function with non-vanishing denominator in I), and $f_y(x, y) = -2y(x^2 + y^2)^{-2} \in C(I)$ (for the same reason). We have for all $y \in [\alpha, 1]$

$$F(y) := \int_0^1 f(x, y)\, dx = \frac{1}{y} \arctan\frac{1}{y}. \tag{4.34}$$

By Theorem 4.1.3, for all such y,

$$F'(y) = \int_0^1 f_y(x, y)\, dx = \int_0^1 \frac{-2y}{(x^2 + y^2)^2} dx. \tag{4.35}$$

Hence by (4.34)

$$\int_0^1 \frac{-2y^2}{(x^2 + y^2)^2}\, dx = yF'(y) = -\frac{1}{y} \arctan\frac{1}{y} - \frac{1}{y^2+1},$$

and therefore

$$\begin{aligned} \int_0^1 \frac{x^2 - y^2}{(x^2 + y^2)^2}\, dx &= \int_0^1 \frac{(x^2 + y^2) - 2y^2}{(x^2 + y^2)^2}\, dx \\ &= \int_0^1 \frac{1}{x^2 + y^2}\, dx + \int_0^1 \frac{-2y^2}{(x^2 + y^2)^2}\, dx \\ &= \frac{1}{y} \arctan\frac{1}{y} - \frac{1}{y} \arctan\frac{1}{y} - \frac{1}{y^2+1} = -\frac{1}{y^2+1}. \end{aligned} \tag{4.36}$$

Formula (4.36) is valid for $0 < y < 1$ (since $\alpha \in (0, 1)$ is arbitrary). The right hand side of (4.36) makes sense also for $y = 0$, and can be used to extend the left hand side as a continuous function of y in $[0, 1]$. We then have

$$\int_0^1 \Big(\int_0^1 \frac{x^2 - y^2}{(x^2 + y^2)^2}\, dx\Big)\, dy = -\int_0^1 \frac{1}{y^2 + 1}\, dy = -\frac{\pi}{4}. \tag{4.37}$$

Changing the role of x, y in (4.37), we get

$$\int_0^1 \Big(\int_0^1 \frac{y^2 - x^2}{(y^2 + x^2)^2}\, dy\Big)\, dx = -\frac{\pi}{4}.$$

Changing signs, we obtain

$$\int_0^1 \Big(\int_0^1 \frac{x^2 - y^2}{(x^2 + y^2)^2}\, dy\Big)\, dx = \frac{\pi}{4}. \tag{4.38}$$

Comparing (4.37) and (4.38), we see that the change of integration order gave different values to the repeated integral. This example shows the importance of the continuity hypothesis on the integrand in Theorem 4.1.10: the integrand in (4.37)–(4.38) fails to be continuous only at the vertex $(0, 0)$ of the rectangle $[0, 1] \times [0, 1]$. Actually, it is not defined at the origin, but since its limit as $(x, y) \to (0, 0)$ does not exist, there is no way to extend it as a continuous function in the rectangle.

4.1.13 Exercises

1. Let $F(y) = \int_0^1 e^{x^2 y} dx$.
   (a) Find $F'(0)$.
   (b) For $y \neq 0$, show that F satisfies the differential equation
   $$2y\, F'(y) + F(y) = e^y.$$
2. Let $b > 0$, $f \in C([0, b])$, and $n \in \mathbb{N}$. Define $F_0 = f$ and
   $$F_n(x) = \int_0^x \frac{(x - y)^{n-1}}{(n - 1)!} f(y) dy \qquad (x \in [0, b]).$$
   Prove that F_n is of class C^n in $[0, b]$, and
   $$F_n^{(k)} = F_{n-k} \qquad (1 \le k \le n). \tag{$*$}$$
   If J denotes the *integration operator* on $C([0, b])$ defined by
   $$(Jf)(x) = \int_0^x f(y)\, dy \qquad (f \in C([0, b]),\ x \in [0, b]),$$

it follows from (*) that

$$F_n = J^n f.$$

3. Let

$$F(x) = \int_0^{x^2} \arctan(\frac{y}{x^2})\, dy.$$

Find $F'(x)$ and then express F as an elementary function.

4. Let $[\alpha, \beta] \subset (0, \infty)$, and fix $c \in [\alpha, \beta]$. For any $y \in [\alpha, \beta]$, define

$$F(y) = \int_0^{\pi/2} \log(y^2 \cos^2 x + c^2 \sin^2 x)\, dx.$$

(a) Prove that for all $y \in [\alpha, \beta]$,

$$F'(y) = \frac{\pi}{y} - \frac{2c^2}{y} G(y), \tag{4.39}$$

where

$$G(y) = \int_0^{\pi/2} \frac{\sin^2 x}{y^2 \cos^2 x + c^2 \sin^2 x}\, dx.$$

(b) Prove that

$$G(y) = \frac{1}{cy} \int_0^{\pi/2} \frac{\frac{c}{y} \sec^2 x}{1 + (\frac{c}{y} \tan x)^2}\, dx - \frac{1}{2y} F'(y). \tag{4.40}$$

(c) Conclude from (4.40) that

$$G(y) = \frac{\pi}{2cy} - \frac{1}{2y} F'(y). \tag{4.41}$$

(d) Conclude from (4.39) and (4.41) that

$$F'(y) = \frac{\pi}{y + c} \qquad (y \neq c). \tag{4.42}$$

(e) Show directly that (4.42) is also valid for $y = c$.

(f) Integrate (4.42) to obtain the identity

$$F(y) - F(c) = \pi \log \frac{y + c}{2c}. \tag{4.43}$$

(g) Show from the definition of F that $F(c) = \pi \log c$.

(h) Conclude from (4.43) and Part (g) that

$$F(y) = \pi \log \frac{y+c}{2}$$

for all $y, c > 0$.

5. Let $0 < a < b$, and

$$F(y) = \int_{a+y}^{b+y} \frac{e^{xy}}{x}\, dx \qquad (y > 0).$$

Calculate $F'(y)$ for $y > 0$.

6. Calculate the following iterated integral

$$\int_0^1 \int_0^1 x^3 e^{x^2 y} dx\, dy.$$

7. For $n \in \mathbb{N}$, calculate the iterated integral

$$\int_0^{\pi} \int_0^1 x^{2n-1} \cos(x^n y)\, dx\, dy.$$

8. (a) Calculate the iterated integral

$$\int_0^1 \int_0^1 \frac{x}{(1+x^2)(1+xy)}\, dx\, dy$$

in two different ways, and prove thereby that

$$A := \int_0^1 \frac{\log(1+x)}{1+x^2}\, dx = \frac{\pi \log 2}{8}.$$

(b) Conclude that

$$B := \int_0^1 \frac{\arctan x}{1+x}\, dx = \frac{\pi \log 2}{8}.$$

4.2 Integration on a Domain in $\mathbb{R}^k$

In this section, we shall define the Riemann (multiple) integral of a *bounded* function

$$f : \overline{D} \subset \mathbb{R}^k \to \mathbb{R}, \tag{4.44}$$

over $\overline{D}$, where the latter denotes a bounded *closed domain* in $\mathbb{R}^k$ *with content*. We first clarify the notions involved.

4.2.1 Domains in $\mathbb{R}^k$

We recall briefly some concepts we studied before. A *domain* in $\mathbb{R}^k$ is a non-empty *connected open set* D in $\mathbb{R}^k$. By Theorem 1.4.18, the open set $D \subset \mathbb{R}^k$ is connected iff it is polygonally connected, that is, iff for any two points $x, y \in D$, there exists a *polygonal path* γ_{xy} (consisting by definition of finitely many finite closed segments $\overline{x^i y^i}$, $i = 0, \ldots, n$, with $x^i = y^{i-1}$ for $i \geq 1$), such that $x^0 = x$, $y^n = y$, and $\gamma_{xy} \subset D$. Briefly, each pair of points of D can be connected by a polygonal path lying in D.

A point x is a *boundary point* of D if every neighborhood of x contains a point of D and a point of D^c (that is, x is neither interior to D nor interior to D^c). The *boundary* of D, denoted ∂D, is the set of all the boundary points of D. The set

$$\overline{D} := D \cup \partial D$$

is *closed and connected*. It is called a *closed domain*.

Next we define the *content* of a *bounded closed domain* in $\mathbb{R}^k$. For $k = 1, 2, 3$, the content is the length, the area, and the volume, respectively.

Content

Let I be the unit cell in $\mathbb{R}^k$:

$$I := [0, 1]^k := \{x \in \mathbb{R}^k;\ x_i \in [0, 1],\ i = 1, \ldots, k\}.$$

For any vector space X over $\mathbb{R}$ and $A \subset X$, recall that the translates and dilation of A are defined respectively by

$$p + A := \{p + x;\ x \in A\} \qquad p \in X$$

and

$$\rho A := \{\rho x;\ x \in A\} \qquad \rho > 0.$$

It is natural to require that if the content $S(A)$ of a set $A \subset \mathbb{R}^k$ is well-defined, then so are the contents of its dilations and of the latter's translates, and the following relations hold:

$$S(p + \rho A) = \rho^k S(A) \qquad (p \in \mathbb{R}^k,\ \rho > 0).$$

The requirement $S(I) = 1$ is the intuitively obvious normalization.

For any $p \in \mathbb{R}^k$, the cell $I_p := \Pi_i [p_i - 1, p_i]$ is a translate of I, so that $S(I_p) = 1$ and $S(\rho I_p) = \rho^k$ $(\rho > 0)$.

Fixing $\rho > 0$, we write

$$\mathbb{R}^k = \bigcup_{n \in \mathbb{Z}^k} \rho I_n.$$

Let $\overline{D}$ be a bounded closed domain in $\mathbb{R}^k$.

Call $n \in \mathbb{Z}^k$ an *interior (multi)index for D* if $\rho I_n \subset D$, and a *boundary (multi)index for D* (or for ∂D) if ρI_n meets ∂D. We set the following notation:

(a) $A_\rho = A_\rho(D) = \bigcup\{\rho I_n;\ n \text{ interior for } D\}$.
(b) $C_\rho = C_\rho(D) = \bigcup\{\rho I_n;\ n \text{ boundary for } D\}$.
(c) $B_\rho = B_\rho(D) = A_\rho \cup C_\rho$.

These are *finite* unions of cells, because of the boundedness of D. Each of these unions is not mutually disjoint in general. However the pairwise intersections have dimension smaller than k, so that their content (in $\mathbb{R}^k$) is postulated to be zero. It is then natural to define the content of the above unions of cells as the sum of the contents of the cells, that is, as ρ^k times the number of cells in the union.

By definition, we have

$$A_\rho(D) \subset D \subset \overline{D} \subset B_\rho(D) \tag{4.45}$$

and

$$S(A_\rho) \le S(B_\rho) \tag{4.46}$$

for all $\rho > 0$.

4.2.2 Definition (Notation as above.) In case the limits

$$\lim_{\rho \to 0} S(A_\rho), \quad \lim_{\rho \to 0} S(B_\rho) \tag{4.47}$$

both exist and are equal, we say that D and $\overline{D}$ *have content*, and the common value of the limits in (4.47) is defined as the content of D and $\overline{D}$, denoted $S(D)$ and $S(\overline{D})$, respectively:

$$S(D) = S(\overline{D}) = \lim_{\rho \to 0} S(A_\rho) = \lim_{\rho \to 0} S(B_\rho). \tag{4.48}$$

Note that $S(D) > 0$.

Since $S(C_\rho) = S(B_\rho) - S(A_\rho)$, a necessary condition for the domain to have content is

$$\lim_{\rho \to 0} S(C_\rho) = 0. \tag{4.49}$$

We express Condition (4.49) by saying that ∂D *has content zero*. It can be shown that this condition is also sufficient.

Condition (4.49) makes sense actually for any path in $\mathbb{R}^k$. A path satisfying this condition is said to have content zero. Note that there exist "square filling" paths in $\mathbb{R}^2$, that is, continuous maps of $[0, 1]$ *onto* $[0, 1] \times [0, 1]$. Condition (4.49) is not satisfied by such paths! On the other hand, if γ is the graph of a continuous function $y = f(x)$, $x \in [a, b]$ (or $x = g(y)$, $y \in [\alpha, \beta]$), then γ has content (=area) zero, and therefore the same is true if γ is the sum of finitely many such paths.

Let $\overline{D} \subset \mathbb{R}^k$ be a bounded closed domain with content. Suppose $\overline{D}$ is the finite union of closed sub-domains with content $\overline{D_j}$, and $D_i \cap D_j = \emptyset$ when $i \neq j$. Then it can be shown that

$$S(\overline{D}) = \sum_j S(\overline{D_j}). \tag{4.50}$$

Integration on a Bounded Closed Domain in $\mathbb{R}^k$

Our *standing hypothesis in this section* is that $\overline{D}$ is a bounded closed domain with content in $\mathbb{R}^k$, and

$$f : \overline{D} \to \mathbb{R}$$

is *bounded.*

A partition of $\overline{D}$ is a family

$$T = \{\overline{D_j};\ j = 1, \ldots, n\} \tag{4.51}$$

such that

(a) $\overline{D_j}$ are closed sub-domains with content of $\overline{D}$;
(b) $D_i \cap D_j = \emptyset$ for $i \neq j$;
(c) $\overline{D} = \bigcup_{j=1}^n \overline{D_j}$.

The maximal diameter of the sub-domains $\overline{D_j}$ in T is called the parameter of the partition T, denoted $\lambda(T)$.

A *Riemann sum* for f and the partition T is a sum

$$\sigma(f, T; x^1, \ldots, x^n) := \sum_{j=1}^n f(x^j) S(\overline{D_j}), \tag{4.52}$$

where x^j is an arbitrary point of $\overline{D_j}$, for $j = 1, \ldots, n$.

The notation (4.52) for a Riemann sum will be usually abbreviated as $\sigma(f, T)$.

4.2.3 Definition (Notation as above.) We say that f is Riemann integrable on $\overline{D}$ if f is *bounded* on $\overline{D}$ and there exists a real number J with the following Property (R):

Property (R): for every $\epsilon > 0$, there exists $\delta > 0$, such that

$$|\sigma(f, T) - J| < \epsilon$$

for all partitions T with $\lambda(T) < \delta$, and for all choices of points x^j in $\overline{D_j}$.

The number J is uniquely determined by Property (R). Indeed, if J' satisfies also Property (R) (with δ' replacing δ), then for any partition T with $\lambda(T) < \delta'' := \min[\delta, \delta']$ and any choice of x^j as above,

$$|J' - J| \le |J' - \sigma(f, T)| + |\sigma(f, T) - J| < 2\epsilon$$

Hence $J' = J$, since $|J' - J|$ does not depend on ϵ.

The uniquely determined number J is called the (multiple) integral of f on $\overline{D}$, and is denoted

$$\int_{\overline{D}} f\, dS \tag{4.53}$$

or

$$\int \cdots \int_{\overline{D}} f(x)\, dx. \tag{4.53'}$$

The symbol dx in (4.53′) is an abreviation of the usual symbol $dx_1 \dots dx_k$.

Property (R) is expressed by the equation

$$\lim_{\lambda(T) \to 0} \sigma(f, T) = J. \tag{4.54}$$

A classical physical motivation for the double (triple) integral is the evaluation of the total mass of a plane sheet (or body, respectively) $\overline{D}$ with mass density distribution on $\overline{D}$ given by the function f. The Riemann sums are approximations of the wanted quantity, and the latter's precise value is the limit of these approximations when the partitions' parameter tends to zero.

We defined above integrability and the integral by means of *limits of Riemann sums*. As in the case of one variable, we can also define the upper and lower (multiple) integrals of the bounded function f over $\overline{D}$. The following exposition omits most of the proofs, because of their similarity to the corresponding proofs for functions of one variable.

Let T be a partition of $\overline{D}$ as in (4.51). Since f is bounded in each $\overline{D_j}$, we define the real numbers m_j and M_j as the infimum and supremum (respectively) of f on $\overline{D_j}$, $j = 1, \dots, n$. If m and M are the infimum and supremum of f on $\overline{D}$, then

$$m \le m_j \le M_j \le M \qquad (j = 1, \dots, n). \tag{4.55}$$

We define the *Darboux sums* of f for the partition T by

$$\underline{\sigma}(f, T) := \sum_j m_j S(\overline{D_j}); \qquad \overline{\sigma}(f, T) := \sum_j M_j S(\overline{D_j}). \tag{4.56}$$

Since

$$S(\overline{D}) = \sum_j S(\overline{D_j})$$

(cf. (4.50)), we have by (4.55)

$$m\, S(\overline{D}) \le \underline{\sigma}(f, T) \le \sigma(f, T) \le \overline{\sigma}(f, T) \le M\, S(\overline{D}). \tag{4.57}$$

Suppose one of the sub-domains $\overline{D_j}$ is partitioned in the same manner, say

$$\overline{D_j} = \bigcup_k \overline{D_{jk}}.$$

Then as in (4.55), we have for all k

$$m_j \le m_{jk} \le M_{jk} \le M_j,$$

where m_{jk} and M_{jk} are the infimum and supremum of f on $\overline{D_{jk}}$, respectively. If T' denotes the partition of $\overline{D}$ obtained from T by partitioning the sub-domain $\overline{D_j}$ as above, then the summand $m_j S(\overline{D_j})$ in $\underline{\sigma}(f, T)$ is replaced in $\underline{\sigma}(f, T')$ by the sum

$$\sum_k m_{jk} S(\overline{D_{jk}}) \ge \sum_k m_j S(\overline{D_{jk}}) = m_j S(\overline{D_j}).$$

Hence

$$\underline{\sigma}(f, T) \le \underline{\sigma}(f, T')$$

Repeating the process, this inequality is valid whenever T' is obtained from T by partitioning as above some or all the sub-domains $\overline{D_j}$ (we call T' a *refinement* of T). Similarly,

$$\overline{\sigma}(f, T) \ge \overline{\sigma}(f, T').$$

By (4.57), the following infimum and supremum are well-defined real numbers:

$$\overline{\int}_{\overline{D}} f\, dS := \inf_T \overline{\sigma}(f, T), \tag{4.58}$$

and

$$\underline{\int}_{\overline{D}} f\, dS := \sup_T \underline{\sigma}(f, T). \tag{4.59}$$

These are called the upper and lower (multiple) integrals of f over $\overline{D}$ (respectively). We have

$$m\, S(\overline{D}) \le \underline{\int}_{\overline{D}} f\, dS \le \overline{\int}_{\overline{D}} f\, dS \le M\, S(\overline{D}).$$

The function f is integrable over $\overline{D}$ iff the lower and upper integrals of f over $\overline{D}$ *coincide* (Darboux Theorem). In that case, it is clear from (4.57) that this common value is the value of the integral $\int_{\overline{D}} f\, dS$. Another immediate consequence of Darboux' theorem is

Riemann's criterion for integrability. *f is integrable on $\overline{D}$ iff for every $\epsilon > 0$, there exists $\delta > 0$ such that*

$$\sum_j (M_j - m_j) S(\overline{D_j}) < \epsilon \tag{4.60}$$

for all partitions T with $\lambda(T) < \delta$.

The above condition is also expressed by the relation

$$\lim_{\lambda(T)\to 0} \sum_j (M_j - m_j) S(\overline{D_j}) = 0. \tag{4.61}$$

We denote by $\mathcal{R}(\overline{D})$ the set of all (Riemann) integrable real valued functions on the bounded closed domain with content $\overline{D}$. *The assumptions on $\overline{D}$ are fixed from now on*. In the next theorem, we list basic properties of $\mathcal{R}(\overline{D})$ and of the integral.

Basic Properties of $\mathcal{R}(\overline{D})$

4.2.4 Theorem *(i) $\mathcal{R}(\overline{D})$ is an algebra over the field $\mathbb{R}$ which contains the constant function* 1.

(ii) The integral over $\overline{D}$ is a positive (=monotonic) linear functional on $\mathcal{R}(\overline{D})$ and

$$\int_{\overline{D}} 1 \, dS = S(\overline{D}).$$

(iii) For any $f \in \mathcal{R}(\overline{D})$,

$$m\, S(\overline{D}) \le \int_{\overline{D}} f \, dS \le M\, S(\overline{D}).$$

(iv) $C(\overline{D}) \subset \mathcal{R}(\overline{D})$.

(v) (The Mean Value Theorem for multiple integrals). If $f \in C(\overline{D})$, then there exists a point $p \in \overline{D}$ such that

$$\int_{\overline{D}} f \, dS = f(p)\, S(\overline{D}).$$

(vi) If $\{\overline{D_j}\}$ is a partition of $\overline{D}$, and $f \in \mathcal{R}(\overline{D})$, then $f \in \mathcal{R}(\overline{D_j})$ for all j, and

$$\int_{\overline{D}} f \, dS = \sum_j \int_{\overline{D_j}} f \, dS.$$

(vii) If $f, g \in \mathcal{R}(\overline{D})$ *and* $f = g$ *on D, then*

$$\int_{\overline{D}} f\, dS = \int_{\overline{D}} g\, dS.$$

(viii) If $f \in \mathcal{R}(\overline{D})$ *and* $g \in C([m, M])$*, then* $g \circ f \in \mathcal{R}(\overline{D})$.

(ix) If $f \in \mathcal{R}(\overline{D})$*, then* $|f| \in \mathcal{R}(\overline{D})$ *and*

$$\Big|\int_{\overline{D}} f\, dS\Big| \le \int_{\overline{D}} |f|\, dS.$$

Proof *Proof of* (i)–(iii). Let $f,\ g \in \mathcal{R}(\overline{D})$ and $a, b \in \mathbb{R}$. For any partition $T = \{\overline{D_j}\}$ of $\overline{D}$ and any choice of x^j as in (4.52), we have

$$\sigma(af + bg,\ T) = a\,\sigma(f, T) + b\ \sigma(g, T).$$

Since the Riemann sums for f and g have the limits $\int_{\overline{D}} f\, dS$ and $\int_{\overline{D}} g\, dS$ respectively when $\lambda(T) \to 0$ by our hypothesis on $f,\ g$, it follows that the limit of $\sigma(af + bg,\ T)$ exists, that is, $af + bg \in \mathcal{R}(\overline{D})$, and the integral of the latter function is equal to $a \int f\, dS + b \int g\, dS$. Hence $\mathcal{R}(\overline{D})$ is a vector space over $\mathbb{R}$, and the integral is a linear functional on it. Its positivity (=monotonicity) is clear: if f is a non-negative integrable function, the Riemann sums for it are non-negative, and their limit is then non-negative. The equivalence of positivity and monotonicity is trivial for any linear functional. Since $\sigma(1,\ T) = S(\overline{D})$ for any T, it follows that 1 is integrable and $\int_{\overline{D}} 1\, dS = S(\overline{D})$.

Let $m,\ M$ be the infimum and supremum of the integrable function f over $\overline{D}$. Since $m \le f \le M$ on $\overline{D}$, the latter properties of the integral imply (iii).

Let $m_j,\ M_j$ be the infimum and supremum of $0 \le f \in \mathcal{R}(\overline{D})$ over the sub-domain $\overline{D_j}$ in the partition T. Since f is non-negative, the corresponding infimum and supremum for the function f^2 are m_j^2 and M_j^2. We have

$$\sum_j (M_j^2 - m_j^2)\, S(\overline{D_j}) = \sum_j (M_j + m_j)(M_j - m_j)\, S(\overline{D_j})$$

$$\le 2M \sum_j (M_j - m_j)\, S(\overline{D_j}) \to 0$$

as $\lambda(T) \to 0$, by the Riemann criterion of integrability for f. Hence, by the same criterion, $f^2 \in \mathcal{R}(\overline{D})$. For an arbitrary integrable function f, $f - m$ is a non-negative integrable function because $\mathcal{R}$ is a vector space over $\mathbb{R}$ which contains the constant function 1. Hence $(f - m)^2$ is integrable by what we just proved. Therefore $f^2 = (f - m)^2 - m^2 + 2mf$ is integrable. This is a special case of (viii), with $g(y) = y^2$.

Next, if f, g are integrable, then $(f+g)^2$ and $(f-g)^2$ are integrable, and therefore $fg = (1/4)[(f+g)^2 - (f-g)^2]$ is integrable. This completes the proof of (i)–(iii).

Proof of (iv)–(v). Let $f \in C(\overline{D})$, and let $\epsilon > 0$ be given. Since $\overline{D}$ is a bounded closed set in $\mathbb{R}^k$, it is compact (cf. Corollary 1.2.9), and therefore f is uniformly continuous on it (cf. Theorem 1.4.11). Let then $\delta > 0$ be such that

$$|f(p) - f(p')| < \frac{\epsilon}{S(\overline{D})} \tag{4.62}$$

whenever $p, p' \in \overline{D}$ are such that $d(p, p') < \delta$. Let $T = \{\overline{D_j}\}$ be any partition of $\overline{D}$ with $\lambda(T) < \delta$. The continuous function f attains its infimum m_j (supremum M_j, respectively) on the compact set $\overline{D_j}$ at the point p_j (p'_j, respectively), both in $\overline{D_j}$. Then $d(p_j, p'_j) \le \lambda(T) < \delta$, hence by (4.62)

$$\sum_j (M_j - m_j)\, S(\overline{D_j}) = \sum_j [f(p'_j) - f(p_j)]\, S(\overline{D_j})$$

$$< \frac{\epsilon}{S(\overline{D})} \sum_j S(\overline{D_j}) = \epsilon.$$

Thus f satisfies the Riemann Criterion for integrability, and we conclude that $f \in \mathcal{R}(\overline{D})$. By (iii),

$$S(\overline{D})^{-1} \int_{\overline{D}} f\, dS \in [m, M]. \tag{4.63}$$

Since f is continuous on the compact set $\overline{D}$, we have $m, M \in f(\overline{D})$ (cf. Corollary 1.4.8), and therefore, by the Intermediate Value Theorem for the continuous function f on the connected set $\overline{D}$, $[m, M] \subset f(\overline{D})$. In particular, there exists a point $p \in \overline{D}$ such that the expression in (4.63) is equal to $f(p)$. This proves (v).

Proof of (vi). If T_j are partitions of $\overline{D_j}$, then $T := \bigcup T_j$ is a partition of T. We have for each j

$$0 \le \overline{\sigma}(f, T_j) - \underline{\sigma}(f, T_j) \le \overline{\sigma}(f, T) - \underline{\sigma}(f, T) \to 0$$

as $\lambda(T) \to 0$, by the Riemann Criterion of integrability of f over $\overline{D}$. Hence by the same criterion, $f \in \mathcal{R}(\overline{D_j})$ for each j. Therefore, as $\lambda(T) \to 0$ (which implies $\lambda(T_j) \to 0$ for all j),

$$\int_{\overline{D}} f\, dS = \lim \sigma(f, T) = \lim \sum_j \sigma(f, T_j) = \sum_j \int_{\overline{D_j}} f\, dS.$$

(The number of summands in $\sum_j$ is fixed!)

Proof of (vii). It suffices to show that if $f = 0$ on D, then its integral over $\overline{D}$ vanishes. This follows easily from the fact that ∂D has content zero.

Proof of (viii). We may assume that g is not the zero function on $[m, M]$ (otherwise there is nothing to prove). Then $||g|| := \sup_{[m,M]} |g| > 0$. Let $\epsilon > 0$. By the uniform continuity of g on the compact set $[m, M]$, there exists $\delta > 0$ such that

$$|g(y) - g(y')| < \frac{\epsilon}{2S(\overline{D})} \tag{4.64}$$

whenever $y, y' \in [m, M]$ are such that $|y - y'| < \delta$.

By Riemann's criterion, since $f \in \mathcal{R}(\overline{D})$, there exists $\eta > 0$ such that

$$\sum_j (M_j - m_j)S(D_j) < \frac{\delta\epsilon}{4||g||} \tag{4.65}$$

for all partitions $T = \{\overline{D_j}\}$ of $\overline{D}$ with $\lambda(T) < \eta$.

Consider any such partition. Denote by m'_j, M'_j the infimum and supremum of $g \circ f$ over $\overline{D_j}$ (respectively). Write

$$\sum_j (M'_j - m'_j)\, S(\overline{D_j}) = \sum_1 + \sum_2, \tag{4.66}$$

where $\sum_1$ extends over the indices j for which $M_j - m_j < \delta$ and $\sum_2$ extends over all the remaining indices.

For the indices j in $\sum_1$, whenever $p, p' \in \overline{D_j}$, since $m_j \le f(p),\ f(p') \le M_j$, we have

$$|f(p) - f(p')| \le M_j - m_j < \delta,$$

hence by (4.64)

$$|g(f(p)) - g(f(p'))| < \frac{\epsilon}{2S(\overline{D})},$$

and therefore

$$M'_j - m'_j \le \frac{\epsilon}{2S(\overline{D})}.$$

We conclude that

$$\sum_1 \le \frac{\epsilon}{2S(\overline{D})} \sum_1 S(\overline{D_j}) \le \frac{\epsilon}{2}. \tag{4.67}$$

In $\sum_2$, we use the estimate $M'_j - m'_j \le 2||g||$. Since $M_j - m_j \ge \delta$ for the indices j in $\sum_2$, we have by (4.65)

$$\delta \sum_2 S(\overline{D_j}) \le \sum_2 (M_j - m_j)S(\overline{D_j}) \le \sum (M_j - m_j)S(\overline{D_j}) < \frac{\delta\epsilon}{4||g||},$$

since $\lambda(T) < \eta$. Hence

$$\sum_2 S(\overline{D_j}) < \frac{\epsilon}{4||g||},$$

and therefore

$$\sum_2 \leq 2||g|| \sum_2 S(\overline{D_j}) < \frac{\epsilon}{2}. \tag{4.68}$$

We now conclude from (4.66)–(4.68) that

$$\sum_j (M'_j - m'_j) S(\overline{D_j}) < \epsilon$$

for all partitions T with $\lambda(T) < \eta$. By Riemann's criterion, this proves that $g \circ f \in \mathcal{R}(\overline{D})$.

Proof of (ix). Taking $g(y) = |y|$ in (viii), we get that $|f|$ is integrable over $\overline{D}$ whenever $f \in \mathcal{R}(\overline{D})$. Since $-|f| \leq f \leq |f|$, it follows from (ii) that

$$-\int_{\overline{D}} |f|\, dS \leq \int_{\overline{D}} f\, dS \leq \int_{\overline{D}} |f|\, dS,$$

and (ix) follows. □

Multiple Integrals as Iterated Integrals

We show now that multiple integrals coincide with iterated integrals under certain conditions.

For simplicity of notation, we consider the case $k = 2$ and denote by (x, y) the points of $\mathbb{R}^2$.

4.2.5 Theorem *Let $I = [a, b] \times [\alpha, \beta]$ and $f \in C(I)$. Then*

$$\int_I f\, dS = \int_\alpha^\beta \Big(\int_a^b f(x, y)\, dx \Big)\, dy = \int_a^b \Big(\int_\alpha^\beta f(x, y)\, dy \Big)\, dx. \tag{4.69}$$

Proof By Theorem 4.1.10, the iterated integrals in (4.69) coincide. It suffices therefore to prove the first equation. Denote the double integral by A and the first iterated integral in (4.69) by B. Let $\epsilon > 0$ be given. Since f is uniformly continuous on the compact set I, there exists $\delta > 0$ such that $|f(p) - f(p')| < \epsilon / S(I)$ whenever $p, p' \in I$ satisfy $d(p, p') < \delta$. Let

$$a = x_0 < x_1 < \cdots < x_n = b; \quad \alpha = y_0 < y_1 < \cdots < y_m = \beta,$$

and denote (for $i = 1, \dots, n$ and $j = 1, \dots, m$)

$$I_{ij} = [x_{i-1}, x_i] \times [y_{j-1}, y_j]$$

and

$$M_{ij} = \sup_{I_{ij}} f; \qquad m_{ij} = \inf_{I_{ij}} f.$$

The continuous function f on I assumes its maximum M_{ij} (minimum m_{ij}) on the closed sub-rectangle I_{ij} at the point p_{ij} (p'_{ij}, respectively) in I_{ij}. We consider the partition $T = \{I_{ij}\}$ of I. If $\lambda(T) < \delta$, we have $d(p_{ij}, p'_{ij}) \le \lambda(T) < \delta$, and therefore

$$M_{ij} - m_{ij} - f(p_{ij}) - f(p'_{ij}) < \frac{\epsilon}{S(I)}. \tag{4.70}$$

By Theorem 4.2.4 (vi),

$$A = \sum_{i,j} \int_{I_{ij}} f\,dS \le \sum_{i,j} M_{ij} S(I_{ij}), \tag{4.71}$$

and similarly

$$A \ge \sum_{i,j} m_{ij} S(I_{ij}). \tag{4.71'}$$

On the other hand,

$$B = \sum_j \sum_i \int_{y_{j-1}}^{y_j} \Big(\int_{x_{i-1}}^{x_i} f(x, y)\,dx \Big)\,dy$$

$$\le \sum_j \sum_i M_{ij} S(I_{ij}),$$

and similarly $B \ge \sum_j \sum_i m_{ij} S(I_{ij})$. Therefore, if $\lambda(T) < \delta$ we have by (4.70)

$$|A - B| \le \sum_{i,j} M_{ij} S(I_{ij}) - \sum_{i,j} m_{ij} S(I_{ij})$$

$$= \sum_{i,j} (M_{ij} - m_{ij})\, S(I_{ij}) < \frac{\epsilon}{S(I)} \sum_{i,j} S(I_{ij}) = \epsilon.$$

Since ϵ is arbitrary, this proves that $A = B$. □

The next theorem generalizes Theorem 4.2.5 to functions $f \in \mathcal{R}(I)$. We keep the notation of the preceding proof.

4.2.6 Theorem *Let $f \in \mathcal{R}(I)$ be such that for each $x \in [a, b]$, the function $f(x, \cdot)$ is integrable in the interval $[\alpha, \beta]$. Define then*

$$\Phi(x) = \int_\alpha^\beta f(x, y)\, dy \qquad (x \in [a, b]). \tag{4.72}$$

Then Φ is integrable in $[a, b]$, and

$$\int_a^b \Phi(x)\, dx = \int_I f\, dS. \tag{4.73}$$

Equation (4.73) states that the double integral of f over I coincides with the iterated integral

$$\int_a^b \Big(\int_\alpha^\beta f(x, y)\, dy \Big)\, dx.$$

Proof Let $\epsilon > 0$. Since $f \in \mathcal{R}(I)$, the necessity of the Riemann Criterion implies that there exists $\delta > 0$ such that

$$\sum_{ij} (M_{ij} - m_{ij}) S(I_{ij}) < \epsilon$$

when the parameters of the partitions $\{[x_{i-1}, x_i], i = 1, \ldots, n\}$ of $[a, b]$ and $\{[y_{j-1}, y_j], j = 1, \ldots, m\}$ of $[\alpha, \beta]$ are both less than δ. Fix such a partition $\{[y_{j-1}, y_j]\}$.

For each $i = 1, \ldots, n$, we pick $t_i \in [x_{i-1}, x_i]$. By hypothesis, $f(t_i, \cdot)$ is integrable in the interval $[\alpha, \beta]$, and

$$\Phi(t_i) := \int_\alpha^\beta f(t_i, y)\, dy = \sum_{j=1}^m \int_{y_{j-1}}^{y_j} f(t_i, y)\, dy \le \sum_{j=1}^m M_{ij}(y_j - y_{j-1}).$$

Similarly

$$\Phi(t_i) \ge \sum_{j=1}^m m_{ij}(y_j - y_{j-1}).$$

Hence

$$\underline{\sigma}(f, T) := \sum_{i,j} m_{ij} S(I_{ij}) \le \sum_{i=1}^n \Phi(t_i)(x_i - x_{i-1})$$

$$\le \sum_{i,j} M_{ij} S(I_{ij}) := \overline{\sigma}(f, T). \tag{4.74}$$

On the other hand, by Eqs. (4.71) and (4.71′) (which were derived under the hypothesis that $f \in \mathcal{R}(I)$ only!),

$$\underline{\sigma}(f, T) \le \int_I f\, dS \le \overline{\sigma}(f, T).$$

Hence

$$\left|\sum_{i=1}^{n} \Phi(t_i)(x_i - x_{i-1}) - \int_I f\, dS\right| \le \overline{\sigma}(f, T) - \underline{\sigma}(f, T) = \sum_{i,j} (M_{ij} - m_{ij})\, S(I_{ij}) < \epsilon \tag{4.75}$$

whenever the parameter of the partition $\{[x_{i-1}, x_i]\}$ of $[a, b]$ is less than δ. This proves that the Riemann sums for Φ on the left hand side of (4.75) converge to $\int_I f\, dS$ when the parameter of the partitions $a = x_0 < x_1 < \cdots < x_n = b$ of $[a, b]$ tends to zero. □

In the next theorem, we generalize Theorem 4.2.5 to more general domains. We need first the following

4.2.7 Lemma *Let $\overline{D} \subset \mathbb{R}^2$ be a bounded closed domain with area, and let $f \in C(\overline{D})$. Let $I = [a, b] \times [\alpha, \beta]$ contain $\overline{D}$ in its interior, and extend the definition of f to I by setting $f = 0$ on $I \setminus \overline{D}$. Then f is integrable on I, and $\int_{\overline{D}} f\, dS = \int_I f\, dS$.*

Proof We assume f is not the zero function (otherwise there is nothing to prove). Then $||f|| > 0$, where the norm is the supremum norm on $C(\overline{D})$. Let $\epsilon > 0$. Since ∂D has area zero (cf. 4.2.2), we may fix $r > 0$ such that

$$S(C_r) < \frac{\epsilon}{4||f||}. \tag{4.76}$$

By uniform continuity of f on the compact set $\overline{D}$, there exists $\delta > 0$ such that

$$|f(p) - f(p')| < \frac{\epsilon}{2S(\overline{D})} \tag{4.77}$$

whenever $d(p, p') < \delta$, $p, p' \in \overline{D}$.

We may also assume that δ is such that whenever $T = \{\overline{D_j}\}$ is a partition of I with $\lambda(T) < \delta$, then all the subdomains $\overline{D_j}$ are contained in squares belonging to the fixed mesh chosen above.

With notation as before, we have

$$\sum_j (M_j - m_j) S(\overline{D_j}) = \sum_1 + \sum_2 + \sum_3, \tag{4.78}$$

where $\sum_1$ extends over the indices j such that $\overline{D_j} \subset D$, $\sum_2$ extends over the indices j such that $\overline{D_j}$ meets ∂D, and $\sum_3$ extends over the remaining indices. Clearly $\sum_3 = 0$, because $f = 0$ on $I \backslash \overline{D}$.

By (4.77), when $\lambda(T) < \delta$,

$$0 \leq \sum_1 < \frac{\epsilon}{2S(\overline{D})} \sum_1 S(\overline{D_j}) \leq \frac{\epsilon}{2}.$$

By (4.76),

$$\sum_2 \leq 2||f|| \sum_2 S(\overline{D_j}) \leq 2||f|| S(C_r) < \frac{\epsilon}{2}.$$

Hence by (4.78)

$$\sum_j (M_j - m_j) S(\overline{D_j}) < \epsilon$$

whenever $\lambda(T) < \delta$. This proves that $f \in \mathcal{R}(I)$.

Consider the partition $\{\overline{D},\ I \backslash D\}$ of I into closed subdomains with area. By Theorem 4.2.4 (vi),

$$\int_I f\, dS = \int_{\overline{D}} f\, dS + \int_{I \backslash D} f\, dS = \int_{\overline{D}} f\, dS$$

because $f = 0$ on $I \backslash \overline{D}$ (cf. 4.2.4 (vii)). □

Normal Domains

4.2.8 Terminology Let $\phi,\ \psi \in C([a, b])$, with $\phi < \psi$ on (a, b), and define

$$\overline{D} = \{(x, y) \in \mathbb{R}^2;\ \phi(x) \leq y \leq \psi(x);\ x \in [a, b]\}.$$

Then $\overline{D}$ is a bounded closed domain with area; its area is actually given by the known Calculus formula

$$S(\overline{D}) = \int_a^b [\psi(x) - \phi(x)]\, dx.$$

We shall refer to such a domain as an *x-type normal domain*. Interchanging the roles of the variables, we get similarly a *y-type normal domain*. A domain may be of both types.

4.2.9 Theorem *Let $\overline{D}$ be an x-type normal domain, determined by the functions $\phi,\ \psi \in C([a, b])$ ($\phi < \psi$ in (a, b)). Then for each $f \in C(\overline{D})$,*

$$\int_{\overline{D}} f\, dS = \int_a^b \Big(\int_{\phi(x)}^{\psi(x)} f(x, y)\, dy \Big)\, dx. \tag{4.79}$$

By Theorem 4.2.4 (iv), the left hand side of (4.79) is well defined; by Corollary 4.1.7, the integral in parenthesis on the right hand side defines a continuous function on $[a, b]$, so that the right hand side is also well defined. The point of the present theorem is that the double integral and the repeated integral in (4.79) coincide.

Proof Since $\phi, \psi \in C([a, b])$, they are bounded on $[a, b]$. We may then choose real numbers α, β such that $\alpha < \phi < \psi < \beta$ in (a, b). Then

$$\overline{D} \subset I := [a, b] \times [\alpha, \beta].$$

Extend the definition of f to I by setting $f = 0$ on $I \backslash \overline{D}$. By Lemma 4.2.8, $f \in \mathcal{R}(I)$ and

$$\int_{\overline{D}} f\, dS - \int_I f\, dS. \tag{4.80}$$

For each $x \in [a, b]$, $f(x, \cdot)$ is continuous on $[\alpha, \beta]$ except possibly at the two points $\phi(x)$ and $\psi(x)$; in addition, $f(x, \cdot)$ is bounded in $[\alpha, \beta]$, since it is continuous in the closed subinterval $[\phi(x),\ \psi(x)]$ (hence bounded there) and vanishes on its complement in $[\alpha, \beta]$. Therefore $f(x, \cdot)$ is integrable on $[\alpha, \beta]$, and

$$\int_\alpha^\beta f(x, y)\, dy = \int_{\phi(x)}^{\psi(x)} f(x, y)\, dy. \tag{4.81}$$

Formula (4.79) follows now from (4.80), (4.81), and Theorem 4.2.6. □

Corollary *Let* $\overline{D}$ *be a normal domain of both x-type and y-type in* $\mathbb{R}^2$*, say*

$$\begin{aligned}\overline{D} &= \{(x, y) \in \mathbb{R}^2;\ \phi(x) \le y \le \psi(x),\ x \in [a, b]\}\\ &= \{(x, y) \in \mathbb{R}^2;\ \alpha(y) \le x \le \beta(y),\ y \in [c, d]\},\end{aligned}$$

where $\alpha, \beta, \phi, \psi$ *are continuous functions in the respective intervals. Then for all* $f \in C(\overline{D})$*,*

$$\begin{aligned}\int_a^b \Big(\int_{\phi(x)}^{\psi(x)} f(x, y)\, dy\Big)\, dx &= \int_c^d \Big(\int_{\alpha(y)}^{\beta(y)} f(x, y)\, dx\Big)\, dy\\ &= \int_{\overline{D}} f\, dS.\end{aligned}$$

4.2.10 Examples (a) Let $\overline{D}$ be the triangle

$$\{(x, y);\ 0 \le y \le x;\ x \in [0, 1]\}.$$

Applying Theorem 4.2.9, we evaluate the double integral $\int_{\overline{D}} y e^{x^3}\, dS$ as the repeated integral

$$\int_0^1 \Big(\int_0^x y e^{x^3}\, dy\Big)\, dx = \int_0^1 \frac{x^2}{2} e^{x^3}\, dx = \frac{1}{6} e^{x^3}\Big|_0^1 = \frac{e-1}{6}.$$

(b) With $\overline{D}$ as in Example (a), consider the function $f : \overline{D} \to \mathbb{R}$ defined by

$$f(x, y) = \frac{x^2}{x^2 + y^2} e^{x^2} \qquad (x, y) \neq (0, 0),$$

and $f(0, 0) = 0$. The function f is bounded on $\overline{D}$ (by e), and continuous in $\overline{D} \setminus \{(0, 0)\}$. Theorem 4.2.9 is still valid for bounded functions with a finite number of discontinuities. We then have

$$\int_{\overline{D}} f\, dS = \int_0^1 \int_0^x \frac{x^2}{x^2 + y^2}\, dy\, e^{x^2}\, dx$$

$$= \int_0^1 [x \arctan(\frac{y}{x})\Big|_0^x e^{x^2}\, dx = \frac{\pi}{8} \int_0^1 e^{x^2}\, d(x^2) = \frac{\pi}{8}(e-1).$$

(c) Let

$$\overline{D} = \{(x, y) \in \mathbb{R}^2;\ 0 \leq x \leq y,\ 1 \leq y \leq 2\}.$$

Then $\overline{D}$ is normal of y-type. It is also normal of x-type, since

$$\overline{D} = \{(x, y) \in \mathbb{R}^2;\ \phi(x) \leq y \leq 2,\ 0 \leq x \leq 2\},$$

where $\phi(x) = 1$ for $x \in [0, 1]$ and $\phi(x) = x$ for $x \in [1, 2]$. It follows then from the above corollary that for all $f \in C(\overline{D})$,

$$\int_1^2 [\int_0^y f(x, y)\, dx]\, dy = \int_0^2 [\int_{\phi(x)}^2 f(x, y)\, dy]\, dx$$

$$= \int_0^1 [\int_1^2 f(x, y)\, dy] dx + \int_1^2 [\int_x^2 f(x, y)\, dy]\, dx.$$

Change of Variables

The next theorem gives a change of variable formula for the multiple integral.

4.2.11 Theorem *Let $\overline{D}$, $\overline{D'}$ be bounded closed domains with content in $\mathbb{R}^k$, and let $x(\cdot) : \overline{D'} \to \overline{D}$ be a bijective map of class C^1. Then for all $f \in C(\overline{D})$,*

$$\int_{\overline{D}} f(x)\,dx = \int_{\overline{D'}} f(x(u))\,|J(u)|\,du, \tag{4.82}$$

where J denotes the Jacobian $\frac{\partial x}{\partial u}$.

A proof may be found in [Taylor-Mane].

Since $x(\cdot)$ is bijective, we have necessarily $J \neq 0$ on $\overline{D'}$.

Examples: formula for the content of $\overline{D}$. The special case $f = 1$ identically on $\overline{D}$ gives a useful integral formula for the content $S(\overline{D})$:

$$S(\overline{D}) = \int_{\overline{D'}} |J(u)|\,du. \tag{4.82'}$$

4.2.12 Example

1. The one-dimensional case. Let $x(\cdot) : [\alpha, \beta] \to [a, b]$. The Jacobian $J = x'$ is continuous and non-vanishing (by the hypothesis of the theorem) and therefore it has a constant sign in the interval $[\alpha, \beta]$. If it is positive there, $x(\cdot)$ is increasing, so that $x(\alpha) = a$ and $x(\beta) = b$, and (4.82) is the usual change of variables formula

$$\int_a^b f(x)\,dx = \int_\alpha^\beta f(x(u))\,x'(u)\,du.$$

If x' is negative throughout the interval $[\alpha, \beta]$, $x(\cdot)$ is decreasing there, so that $x(\alpha) = b$ and $x(\beta) = a$. Then the usual change of variables formula gives

$$\int_a^b f(x)\,dx = \int_\beta^\alpha f(x(u))\,x'(u)\,du$$

$$= \int_\alpha^\beta f(x(u))\,[-x'(u)]\,du = \int_\alpha^\beta f(x(u))\,|x'(u)|\,du,$$

in accordance with (4.82).

2. Linear maps.

The simplest map $x(\cdot)$ is a linear map of $\mathbb{R}^k$ onto itself, $x = Au$, where $A = (a_{ij})$ is a non-singular matrix over $\mathbb{R}$, and we wrote the vectors x, u as columns. We have $x_i = \sum_j a_{ij} u_j$, hence $\frac{\partial x_i}{\partial u_j} = a_{ij}$, that is, the Jacobian matrix of the map coincides with A, and so $J = \det A$. If we simplify the notation by writing $E = \overline{D'}$, we obtain in this case

$$\int_{AE} f(x)\,dx = |\det A| \int_E f(Au)\,du. \tag{4.83}$$

In particular, for $f = 1$ identically, we get the following formula relating the contents of E and AE:

$$S(AE) = |\det A|\, S(E). \tag{4.84}$$

Formula (4.84) can be proved directly for the so-called *elementary matrices*. Since any non-singular matrix is a product of elementary matrices, and the determinant of a product of matrices is the product of their determinants, the general case of (4.84) follows. Then we use (4.84) to prove (4.83), by means of the definition of the integral as a limit of Riemann sums. The general formula (4.82) is obtained by using the *local* linear approximation of the map $x(\cdot)$ by means of its differential, which is a linear map whose matrix is (locally) the Jacobian matrix of the given map $x(\cdot)$.

Example. The closed domain $\overline{D}$ in $\mathbb{R}^k$ is said to be x_i-symmetric for a given index i, $1 \le i \le k$, if the reflection (linear) map L_i, which replaces x_i by $-x_i$ and leaves all other coordinates of x unchanged, is mapping

$$\overline{D^-} := \{x \in \overline{D};\ x_i \le 0\}$$

onto

$$\overline{D^+} := \{x \in \overline{D};\ x_i \ge 0\}.$$

The Jacobian of the map L_i has absolute value 1. Let $f \in C(\overline{D})$ be *even* with respect to x_i, that is, $f(L_i x) = f(x)$ for all $x \in \overline{D}$. Since $\{\overline{D^+}, \overline{D^-}\}$ is a partition of $\overline{D}$, we have

$$\int_{\overline{D}} f(x)\, dx = \int_{\overline{D^+}} f(x)\, dx + \int_{\overline{D^-}} f(x)\, dx.$$

The reflection $x = L_i x'$ shows that the second integral on the right hand side is equal to the first, and therefore

$$\int_{\overline{D}} f(x)\, dx = 2 \int_{\overline{D^+}} f(x)\, dx.$$

The same argument shows that if $\overline{D}$ is x_i-symmetric and $f \in C(\overline{D})$ is *odd* with respect to x_i (that is, $f(L_i x) = -f(x)$ for all $x \in \overline{D}$), then

$$\int_{\overline{D}} f(x)\, dx = 0.$$

3. Polar coordinates in $\mathbb{R}^2$. The map involved is

$$(r, \phi) \in [0, \infty) \times [0, 2\pi] \to (r \cos \phi, r \sin \phi) \in \mathbb{R}^2.$$

We calculate that

$$\frac{\partial(x, y)}{\partial(r, \phi)} = r.$$

Although the Jacobian vanishes at the origin, the formula in Theorem 4.2.11 remains valid for f continuous on the domain.

Examples. The closed disc $\overline{D} = \overline{B}(0, R)$ corresponds through the above map to the closed rectangle $\overline{D'} = I := [0, R] \times [0, 2\pi]$. Therefore by Theorems 4.2.11 and 4.2.5, we have for any f continuous on $\overline{B}(0, R)$,

$$\int_{\overline{B}(0,R)} f(x, y)\, dx\, dy = \int_I f(r \cos\phi, r \sin\phi) r\, dr\, d\phi$$

$$= \int_0^{2\pi} \left(\int_0^R f(r \cos\phi, r \sin\phi) r\, dr \right) d\phi.$$

We consider below some special examples.

(a) Taking $f = 1$ identically in the above formula, we get

$$S(\overline{B}(0, R)) = \int_0^{2\pi} \int_0^R r\, dr\, d\phi = \pi R^2.$$

(b) Let $\overline{D}$ be defined by

$$\overline{D} = \{(x, y);\ \frac{x^2}{a^2} + \frac{y^2}{b^2} \le 1\},$$

where $a, b > 0$ are constants (D is the interior of the ellipse with semi-axis a, b and with center at the origin). The non-singular linear function $(x, y) = (au, bv)$ maps $\overline{D'} := \overline{B}(0, 1)$ onto $\overline{D}$, and has the Jacobian ab. Hence

$$S(\overline{D}) = ab\, S(\overline{B}(0, 1)) = \pi ab.$$

(c) Let

$$\overline{D} = \{(x, y) \in \mathbb{R}^2;\ x,\ \sqrt{x},\ y,\ \sqrt{y} \in [0, \infty),\ \sqrt{x} + \sqrt{y} \le 1\}.$$

We wish to calculate the double integral

$$A := \int_{\overline{D}} \sqrt{\sqrt{x} + \sqrt{y}}\, dS.$$

We use first the map $(u, v) \to (x, y) = (u^4, v^4)$. We have

$$\overline{D'} = \{(u, v);\ u^2 + v^2 \le 1\}$$

and $J = 16u^3v^3$. Therefore

$$A = 16 \int_{u^2+v^2 \le 1} \sqrt{u^2+v^2}\, |u^3v^3|\, du\, dv.$$

By symmetry,

$$A = 64 \int_K \sqrt{u^2+v^2}\, u^3v^3\, du\, dv,$$

where

$$K = \{(u, v);\ u, v \ge 0,\ u^2+v^2 \le 1\}.$$

Next we use the (polar) map $(r, \phi) \to (u, v) = (r\cos\phi, r\sin\phi) \in K$. The domain for (r, ϕ) is the rectangle $[0, 1] \times [0, \pi/2]$. Hence

$$A = 64 \int_0^{\pi/2} [\int_0^1 r^8 \cos^3\phi \sin^3\phi\, dr]\, d\phi = \frac{64}{9} \int_0^{\pi/2} \cos^3\phi \sin^3\phi\, d\phi$$

$$= \frac{64}{9} \int_0^{\pi/2} (1-\sin^2\phi) \sin^3\phi\, d(\sin\phi) = \frac{64}{9} (\frac{\sin^4\phi}{4} - \frac{\sin^6\phi}{6})\Big|_0^{\pi/2} = \frac{16}{27}.$$

(d) Consider the double integral

$$A := \int_{\overline{D}} \sqrt{1-x^2-y^2}\, dS$$

over the closed domain

$$\overline{D} = \{(x, y);\ x \ge 0,\ x^2 - y^2 \ge (x^2+y^2)^2\}.$$

The domain of integration is y-symmetric and the integrand is even with respect to y. By Example 2, it follows that $A = 2A^+$, where A^+ is the integral over $\overline{D^+} := \{(x, y) \in \overline{D};\ y \ge 0\}$. The closed domain corresponding to $\overline{D^+}$ under the polar map $(r, \phi) \to (x, y) = (r\cos\phi, r\sin\phi)$ is the set

$$E := \{(r, \phi);\ r^2(\cos^2\phi - \sin^2\phi) \ge r^4,\ \sin\phi,\ \cos\phi \ge 0,\ r \ge 0,\ \phi \in [0, 2\pi]\}.$$

Thus (r, ϕ) (with $r \ge 0$ and $\phi \in [0, 2\pi]$ belongs to E iff both $\sin\phi$ and $\cos\phi$ are non-negative, and $\cos 2\phi \ge r^2$, that is, iff $\phi \in [0, \pi/4]$, and $r^2 \le \cos 2\phi$. Therefore

$$A = 2A^+ = 2 \int_E \sqrt{1-r^2}\, r\, dr\, d\phi$$

$$= \int_0^{\pi/4} \int_{\cos^{1/2} 2\phi}^0 \sqrt{1-r^2}\, d(1-r^2)\, d\phi = \frac{2}{3} \int_0^{\pi/4} (1-r^2)^{3/2}\Big|_{\cos^{1/2} 2\phi}^0\, d\phi$$

$$= \frac{2}{3}\int_0^{\pi/4} [1 - (1 - \cos 2\phi)^{3/2}]\, d\phi = \frac{2}{3}[\frac{\pi}{4} + 2^{3/2}\int_0^{\pi/4} \sin^3\phi\, d\phi]$$

$$= \frac{\pi}{6} + \frac{2^{5/2}}{3}\int_{\pi/4}^{0} (1 - \cos^2\phi)\, d\cos\phi = \frac{\pi}{6} + \frac{2^{5/2}}{3}[\cos\phi - \frac{\cos^3\phi}{3}]\Big|_{\pi/4}^{0}$$

$$= \frac{\pi}{6} + \frac{2}{9}(4\sqrt{2} - 5).$$

4. Polar coordinates in $\mathbb{R}^3$. The map from the polar coordinates (also called *spherical coordinates*) (r, θ, ϕ) of a point in $\mathbb{R}^3$ to its Cartesian coordinates (x, y, z) is given by

$$(x, y, z) = (r \sin\theta \cos\phi,\ r \sin\theta \sin\phi,\ r \cos\theta),$$

where (r, θ, ϕ) varies in

$$[0, \infty) \times [0, \pi] \times [0, 2\pi].$$

We calculate that

$$\frac{\partial(x, y, z)}{\partial(r, \theta, \phi)} = r^2 \sin\theta.$$

Note that the Jacobian vanishes on the z-axis, but the conclusion of Theorem 4.2.11 is still valid in this situation.

The cell

$$\overline{D'} = I := [0, R] \times [0, \pi] \times [0, 2\pi]$$

is mapped onto the closed ball

$$\overline{D} = \overline{B}(0, R).$$

By Theorems 4.2.11 and 4.2.5 (in its $\mathbb{R}^3$-version!), we then have for any f continuous on the ball $\overline{B}(0, R)$

$$\int_{\overline{B}(0,R)} f\, dS$$

$$= \int_0^{2\pi} \Big[\int_0^{\pi} \Big(\int_0^R f(r \sin\theta \cos\phi,\ r \sin\theta \sin\phi,\ r \cos\theta)\, r^2\, dr\Big) \sin\theta\, d\theta\Big]\, d\phi.$$

Examples.

(a) We calculate the volume $vol(\overline{B}(0, R))$ of the ball $\overline{B}(0, R)$ in $\mathbb{R}^3$ by taking $f = 1$ identically:

$$vol(\overline{B}(0, R)) = \int_0^{2\pi} \Big[\int_0^{\pi} \Big(\int_0^R r^2 dr\Big) \sin\theta\, d\theta\Big]\, d\phi$$

$$= 2\pi \frac{R^3}{3} \int_0^{\pi} \sin\theta \, d\theta = \frac{4\pi R^3}{3}.$$

(b) Let $0 < a < b < \infty$ and

$$\overline{D} = \{x \in \mathbb{R}^3;\ a \le ||x|| \le b,\ x_i \ge 0\}.$$

For $p \in \mathbb{R}$, we wish to calculate the triple integrals

$$A_i := \int_{\overline{D}} x_i ||x||^p dx.$$

By symmetry with respect to the index i, we clearly have $A_1 = A_2 = A_3$. We then calculate A_3, since it has the simplest expression in spherical coordinates. The closed domain $\overline{D}'$ corresponding to $\overline{D}$ in the transformation to spherical coordinates is $[a, b] \times [0, \pi/2] \times [0, \pi/2]$. Hence for $i = 1, 2, 3$

$$A_i = A_3 = \int_0^{\pi/2} \int_0^{\pi/2} \int_a^b r\cos\theta \, r^p r^2 \sin\theta \, dr \, d\theta \, d\phi$$

$$= \frac{\pi}{2} \frac{\sin^2\theta}{2}\Big|_0^{\pi/2} \frac{r^{p+4}}{p+4}\Big|_a^b = \frac{\pi}{4} \frac{b^{p+4} - a^{p+4}}{p+4}$$

for $p \neq -4$, and $A_i = \frac{\pi}{4} \log \frac{b}{a}$ for $p = -4$.

(c) Let

$$\overline{D} = \{(x, y, z) \in \mathbb{R}^3;\ x^2 + y^2 + z^2 \le 2x + 4y + 4z\}.$$

We shall calculate the triple integral

$$I := \int_{\overline{D}} z \, dV.$$

Observe first that

$$\overline{D} = \{(x, y, z) \in \mathbb{R}^3;\ (x-1)^2 + (y-2)^2 + (z-2)^2 \le 9\} = \overline{B}(p, 3)$$

(the closed ball of radius 3 centred at the point $p = (1, 2, 2)$). We apply the translation map

$$(x, y, z) = p + (x', y', z') = (x' + 1, y' + 2, z' + 2).$$

The Jacobian is equal to 1 and the closed domain corresponding to $\overline{D}$ is

$$\overline{D'} = \{(x', y', z') \in \mathbb{R}^3;\ (x')^2 + (y')^2 + (z')^2 \le 9\}.$$

Then

$$I = \int_{\overline{D'}} (z' + 2)\, dx'dy'dz'. \tag{4.85}$$

We apply next the polar map

$$(x', y', z') = (r\cos\phi\sin\theta,\ r\sin\phi\sin\theta,\ r\cos\theta).$$

We have $\frac{\partial(x',y',z')}{\partial(r,\phi,\theta)} = r^2\sin\theta$ and the closed domain corresponding to $\overline{D'}$ is the cell $[0, 3] \times [0, 2\pi] \times [0, \pi]$. Therefore

$$\begin{aligned}
I &= 2\pi \int_0^{\pi} [\int_0^3 (r\cos\theta + 2)r^2 \sin\theta\, dr]\, d\theta \\
&= 2\pi \int_0^{\pi} \Big[\frac{r^4}{4}\Big|_0^3 \cos\theta\sin\theta + 2\frac{r^3}{3}\Big|_0^3 \sin\theta\Big]\, d\theta \\
&= \frac{81\pi}{2} \int_0^{\pi} \sin\theta\, d(\sin\theta) + 36\pi\, \cos\theta\Big|_{\pi}^{0} = 72\pi.
\end{aligned}$$

We can reach this result with almost no calculation. By (4.85),

$$I = \int_{\overline{D'}} z' dV + 2\, vol(\overline{D'}). \tag{4.86}$$

Since $\overline{D'}$ is z'-symmetric and the function $(x', y', z') \to z'$ is odd with respect to z', the integral on the right hand side of (4.86) is equal to 0, by Example 2. The volume of the closed ball $\overline{D'}$ is $(4/3)\pi 3^3 = 36\pi$. We conclude from (4.86) that $I = 72\pi$.

5. Cylindrical coordinates in $\mathbb{R}^3$. The function mapping the cylindrical coordinates (r, ϕ, z) of a point in $\mathbb{R}^3$ to its Cartesian coordinates is given by

$$(r, \phi, z) \in [0, \infty) \times [0, 2\pi] \times \mathbb{R} \to (x, y, z) = (r\cos\phi,\ r\sin\phi,\ z).$$

We calculate that the Jacobian of the map is r. The cell $\overline{D'} := [0.R] \times [0, 2\pi] \times [0, L]$ corresponds to the cylinder

$$\overline{D} := \{(x, y, z) \in \mathbb{R}^3;\ x^2 + y^2 \le R^2,\ z \in [0, L]\}.$$

Examples.

(a) Taking $f = 1$ identically on the cylinder $\overline{D}$, we get

$$vol(\overline{D}) = \int_0^L \int_0^{2\pi} \int_0^R r\, dr\, d\phi\, dz = \pi R^2 L.$$

(b) The ball

$$\overline{D} = \overline{B}(0, R) = \{(x, y, z);\ x^2 + y^2 + z^2 \le R^2\}$$

corresponds to the domain

$$\overline{D'} = \{(r, \phi, z);\ -\sqrt{R^2 - r^2} \le z \le \sqrt{R^2 - r^2},\ 0 \le r \le R,\ 0 \le \phi \le 2\pi\}.$$

Hence

$$vol(\overline{D}) = \int_0^{2\pi} \int_0^R \int_{-\sqrt{R^2-r^2}}^{\sqrt{R^2-r^2}} dz\, r\, dr\, d\phi$$

$$= 2\pi \int_0^R \sqrt{R^2 - r^2}\, 2r dr = 2\pi \frac{2}{3} (R^2 - r^2)^{3/2}\Big|_R^0 = \frac{4\pi R^3}{3}.$$

(c) Let

$$f : z \in [a, b] \to y = f(z)$$

be continuous and positive, except possibly at the endpoints. Rotate the graph of f about the z-axis to obtain the closed domain $\overline{D} \subset \mathbb{R}^3$ given by

$$\overline{D} := \{(x, y, z) \in \mathbb{R}^3; x^2 + y^2 \le f(z)^2,\ z \in [a, b]\}.$$

The corresponding domain in cylindrical coordinates is given by

$$\overline{D'} = \{(r, \phi, z);\ 0 \le r \le f(z),\ z \in [a, b],\ \phi \in [0, 2\pi]\}.$$

Therefore

$$vol(\overline{D}) = \int_0^{2\pi} [\int_a^b (\int_0^{f(z)} r\, dr)\, dz]\, d\phi$$
$$= \pi \int_a^b f(z)^2 dz.$$

This is the formula for the *volume of a solid of revolution*.

(d) We apply the formula in Example (c) to calculate the volume of the ball $\overline{B}(0, R)$. The relevant function f is

$$y = f(z) = \sqrt{R^2 - z^2}, \qquad z \in [-R, R].$$

Therefore

$$vol(\overline{B}(0, R)) = \pi \int_{-R}^R (R^2 - z^2)\, dz$$

$$= \pi(R^2 z - \frac{z^3}{3})\Big|_{-R}^{R} = 2\pi(R^3 - \frac{R^3}{3}) = \frac{4\pi}{3} R^3.$$

(e) We apply again the formula of Example (c) to calculate the volume of the torus, which is the closed domain in $\mathbb{R}^3$ obtained by revolving the closed disc of radius a in the (y, z)-plane, centred at $(R, 0)$, where $0 < a < R$, about the z-axis. The said disc

$$E := \{(y, z);\ R - \sqrt{a^2 - z^2} \le y \le R + \sqrt{a^2 - z^2},\ z \in [-a, a]\}$$

is considered as the set difference $E = E_1 \setminus E_2$, where

$$E_1 := \{(y, z);\ 0 \le y \le R + \sqrt{a^2 - z^2},\ z \in [-a, a]\},$$

and

$$E_2 := \{(y, z);\ 0 \le y \le R - \sqrt{a^2 - z^2},\ z \in [-a, a]\}.$$

Correspondingly, the torus is the set difference of the respective solids of revolution, and the volume V of the torus is then given by

$$V = \pi \int_{-a}^{a} \{[R + \sqrt{a^2 - z^2}]^2 - [R - \sqrt{a^2 - z^2}]^2\}\, dz$$
$$= \pi \int_{-a}^{a} 4R\sqrt{a^2 - z^2}\, dz.$$

The change of variable $z = a \cos\theta$ yields

$$V = 4\pi R\, a^2 \int_0^{\pi} \sin^2\theta\, d\theta = 2\pi^2 R\, a^2.$$

(f) Let $a, b, c > 0$, and let $\overline{D}$ be the closed domain in the upper half-space

$$\mathbb{R}^3_+ := \{(x, y, z);\ x, y \in \mathbb{R},\ z \ge 0\},$$

bounded by the plane with equation $z = c$ and the elliptic cone with equation $(x/a)^2 + (y/b)^2 - (z/c)^2 = 0$. We wish to calculate the triple integral

$$I := \int_{\overline{D}} z\, dV.$$

We have

$$\overline{D} = \{(x, y, z);\ x, y \in \mathbb{R},\ \frac{x^2}{a^2} + \frac{y^2}{b^2} \le \frac{z^2}{c^2},\ 0 \le z \le c\}.$$

We apply first the linear map

$$(x, y, z) = (ax', by', cz').$$

The Jacobian of the map is equal to abc, and the corresponding domain is

$$\overline{D'} = \{(x', y', z');\ x', y' \in \mathbb{R},\ x'^2 + y'^2 \le z'^2,\ 0 \le z' \le 1\}.$$

Hence

$$I = abc^2 \int_{\overline{D'}} z' dx' dy' dz'.$$

We apply next the cylindrical map

$$(x', y', z') = (r \cos \phi,\ r \sin \phi,\ z').$$

The Jacobian of the map is r and the corresponding domain is

$$\{(r, \phi, z');\ 0 \le r \le z',\ 0 \le z' \le 1,\ 0 \le \phi \le 2\pi\}.$$

Therefore

$$I = abc^2 \int_0^{2\pi} [\int_0^1 z' (\int_0^{z'} r\, dr)\, dz']\, d\phi$$
$$= 2\pi abc^2 \int_0^1 z' \frac{r^2}{2}\Big|_0^{z'} dz' = \pi abc^2 \int_0^1 z'^3 dz' = \frac{\pi}{4} abc^2.$$

(g) Let $I = \int_{\overline{D}} z|y|\, dV$, where

$$\overline{D} = \{(x, y, z) \in \mathbb{R}^3;\ x^2 + y^2 \le 2x,\ 0 \le z \le 1\}.$$

Clearly

$$\overline{D} = \{(x, y, z);\ (x-1)^2 + y^2 \le 1,\ 0 \le z \le 1\}.$$

This closed circular cylinder is y-symmetric, and the integrand $z|y|$ is even with respect to y. Therefore, by Example 2,

$$I = 2 \int_{\overline{D^+}} zy\, dV,$$

where

$$\overline{D^+} = \{(x, y, z);\ (x-1)^2 + y^2 \le 1,\ y \ge 0,\ 0 \le z \le 1\}.$$

The translation map $(x, y, z) = (x' + 1, y, z)$ has Jacobian equal to 1, and the closed domain corresponding to $\overline{D^+}$ is

$$\overline{K} := \{(x', y, z);\ (x')^2 + y^2 \leq 1,\ y \geq 0,\ 0 \leq z \leq 1\}.$$

Hence

$$I = 2 \int_{\overline{K}} zy\, dx'dydz.$$

We apply now the cylindrical map $(x', y, z) = (r \cos \phi,\ r \sin \phi,\ z)$. The closed domain corresponding to $\overline{K}$ is the cell

$$\overline{K'} = \{(r, \phi, z);\ 0 \leq r \leq 1,\ 0 \leq \phi \leq \pi,\ 0 \leq z \leq 1\}.$$

Hence

$$I = 2 \int_0^1 [\int_0^{\pi} (\int_0^1 r^2 dr) \sin \phi\, d\phi] z\, dz$$
$$= 2 \frac{r^3}{3}\Big|_0^1 \cos \phi|_{\pi}^0 \frac{z^2}{2}\Big|_0^1 = \frac{2}{3}.$$

Integration on Unbounded Domains in $\mathbb{R}^k$

In this section, we study briefly Riemann integration on *unbounded* domains. Denote by $\overline{B}(0, R)$ the closed ball in $\mathbb{R}^k$ with radius R and center at the origin.

4.2.13 Definition Let $\overline{D}$ be a closed domain in $\mathbb{R}^k$, such that

$$\overline{D}_R := \overline{D} \cap \overline{B}(0, R) \tag{4.87}$$

has content for each $R > 0$.

Let $f \in C(\overline{D})$ (hence the integrals $J_R := \int_{\overline{D}_R} f\, dS$ exist for all R; cf. Theorem 4.2.4). If the limit $\lim J_R$ exists as $R \to \infty$, we say that the *improper* integral $\int_{\overline{D}} f\, dS$ *converges*, and its value is defined as that limit:

$$\int_{\overline{D}} f\, dS := \lim_{R \to \infty} \int_{\overline{D}_R} f\, dS. \tag{4.88}$$

The Gauss integral.

4.2.14 Example Let $\overline{D}$ be the closed first quadrant in $\mathbb{R}^2$:

$$\overline{D} = \{(x, y);\ x \geq 0,\ y \geq 0\}.$$

We consider the integral J of the positive continuous function $e^{-x^2-y^2}$ on that domain. For each $R > 0$, the closed domain $\overline{D}_R$ is the closed "quarter disc" $\{(x, y);\ x^2 + y^2 \le R^2,\ x \ge 0,\ y \ge 0\}$, which corresponds under the plane polar map to the closed rectangle $I := \overline{D}'_R = [0, R] \times [0, \pi/2]$. By Theorem 4.2.5,

$$J_R := \int_{\overline{D}_R} e^{-x^2-y^2}\,dx\,dy = \int_I e^{-r^2} r\,dr\,d\phi$$

$$= \int_0^{\pi/2} \Big(\int_0^R e^{-r^2} r\,dr \Big)\,d\phi = \frac{\pi}{4}\,(1 - e^{-R^2}). \tag{4.89}$$

Hence $J_R \to \pi/4$ as $R \to \infty$. This proves that J converges, and its value is $\pi/4$.

We clearly have the inclusions

$$\overline{D}_R \subset [0, R] \times [0, R] \subset \overline{D}_{R\sqrt{2}}.$$

Since $e^{-x^2-y^2} > 0$, it follows that

$$J_R \le \int_{[0,R]\times[0,R]} e^{-x^2-y^2}\,dx\,dy \le J_{R\sqrt{2}}. \tag{4.90}$$

By Theorem 4.2.5, the middle integral in (4.90) is equal to

$$\Big(\int_0^R e^{-x^2}\,dx \Big)^2. \tag{4.91}$$

Hence by (4.89)

$$\frac{\sqrt{\pi}}{2}\,(1 - e^{-R^2})^{1/2} \le \int_0^R e^{-x^2}\,dx \le \frac{\sqrt{\pi}}{2}\,(1 - e^{-2R^2})^{1/2}.$$

Therefore the integral $\int_0^\infty e^{-x^2}\,dx$ converges, and its value is $\sqrt{\pi}/2$. By symmetry, $\int_{-\infty}^0 e^{-x^2}\,dx$ converges to the same value, and consequently

$$\int_{\mathbb{R}} e^{-x^2}\,dx = \sqrt{\pi}.$$

The integral on the left hand side is the *Gauss integral*, which plays an important role in Probability Theory.

4.2.15 Exercises

1. Let $\overline{D}$ be the closed domain bounded by the parabola $x = y^2$ and the line $x = y$. Calculate the double integral
$$\int_{\overline{D}} \sin \pi \frac{x}{y} \, dS.$$

2. Let $\overline{D}$ be the closed ring with radii $0 < a < b$. Calculate the double integral
$$\int_{\overline{D}} \arctan \frac{y}{x} \, dS.$$

3. Let $\overline{D}$ be the closed domain in $\mathbb{R}^2$ bounded by the parabola $y = x^2$ and the lines $x = 1$ and $y = 0$. Calculate
$$\int_{\overline{D}} y \, e^{x^5} \, dS.$$

4. Let $\overline{D}$ be the closed domain in $\mathbb{R}^2$ bounded by the circles $x^2 + y^2 = 1$ and $x^2 + y^2 = 4$, and the lines $y = x$ and $y = x/\sqrt{3}$. Calculate the double integral of $(x^2 + y^2 - 1)^{1/2}$ over $\overline{D}$.
5. (a) Let $\overline{D}$ be the closed disc $\overline{B}(0, R)$ in $\mathbb{R}^2$, let F be a real function of class C^1 on $[0, R^2]$, and denote $f := F'$. Show that
$$\int_{\overline{D}} f(x^2 + y^2) \, dS = \pi[F(R^2) - F(0)].$$

   (b) Calculate the double integrals

   (i) $\int_{\overline{D}} \cos(x^2 + y^2) \, dS$.
   (ii) $\int_{\overline{D}} \exp(-x^2 - y^2) \, dS$.

6. Find the area of the closed domain bounded by the two parabolas $y^2 = ax$ and $y^2 = bx$ $(0 < a < b)$, and the two hyperbolas $xy = \alpha$ and $xy = \beta$ $(0 < \alpha < \beta)$. Hint: use the map
$$(x, y) \to (u, v) := (\frac{y^2}{x}, xy).$$

7. Let
$$\overline{D} = \{(x, y); x, \ y \geq 0, \ x^{1/2} + y^{1/2} \leq 1\}.$$

   Calculate the double integral of $(x^{1/2} + y^{1/2})^3$ over the domain. Hint: use the map
$$(r, \phi) \to (x, y) = r^4 \, (\cos^4 \phi, \ \sin^4 \phi).$$

8. Calculate the double integral of $|\cos(x+y)|$ over the closed square $[0,\pi] \times [0,\pi]$.
9. Calculate $\int_K xy\, e^{x^2-y^2}\, dS$, where $K = \{(x,y);\ 1 \le x^2 - y^2 \le 9,\ 0 \le x \le 4,\ y \ge 0\}$.
10. Calculate $\int_K \sqrt{1-x^2-y^2}\, dS$, where K is the closed disc $\{(x,y); x^2+y^2 \le x\}$.
11. Let
$$\overline{D} := \{x \in \mathbb{R}^3;\ x_i \ge 0,\ a \le ||x|| \le b\},$$
where $0 < a < b < \infty$. For $p, q \in \mathbb{R}$, $q \ge 0$, calculate the (triple) integrals
$$A_i = \int_{\overline{D}} x_i^q ||x||^p dx.$$
12. Find the volume of the closed domain $\overline{D}$ in $\mathbb{R}^3$ bounded by the elliptic paraboloid
$$\frac{x^2}{a^2} + \frac{y^2}{b^2} + \frac{z}{c} = 1$$
$(a, b, c > 0)$ and the x, y-plane.
13. Calculate the triple integral $\int\int\int_{\overline{D}} z\, dx\, dy\, dz$ for the closed domain $\overline{D}$ in $\mathbb{R}^3$ bounded by the one-sheeted hyperboloids
$$x^2 + y^2 - z^2 = 1, \qquad x^2 + y^2 - z^2 = 4$$
and the planes $z = 0$ and $z = 1$.
14. Calculate the triple integral of $||x||$ over the closed domain $\overline{D} = \{x \in \mathbb{R}^3;\ ||x||^2 \le x_3\}$.
15. Calculate the volume of the closed domain in $\mathbb{R}^3$ bounded by the surfaces $z = x^2 + y^2$, $y = x^2$, $y = 1$, and $z = 0$.

4.3 Line Integrals

In this section, we shall generalize the partial integrals considered in Sect. 4.1, which are in fact integrals over line segments parallel to the axis of coordinates, to line integrals over more general curves. We start with concepts related to curves in $\mathbb{R}^k$.

4.3.1 Curves in $\mathbb{R}^k$ A *path* in $\mathbb{R}^k$ is an element $x(\cdot)$ of $C([a,b], \mathbb{R}^k)$. The natural order on $[a,b]$ induces an *orientation* in the range of $x(\cdot)$: if $t < s$ in $[a,b]$, then $x(t) \prec x(s)$ ($x(t)$ precedes $x(s)$). In particular, $x(a)$ and $x(b)$ are the *initial point* and the *final point* of the path, respectively; these are the path's *end points*. The path is *closed* if its end points coincide. If $x(t) = x(s)$ for some $t \ne s$ in $(a,b]$ (or $[a,b)$), we say that the path intersects itself; otherwise, the path is *simple*.

Let $x(\cdot) : [a, b] \to \mathbb{R}^k$ and $x'(\cdot) : [c, d] \to \mathbb{R}^k$ be given paths. We say that $x'(\cdot)$ is equivalent to $x(\cdot)$ if there exists an increasing continuous function h mapping $[c, d]$ onto $[a, b]$ such that $x' = x \circ h$. The meaning of this equivalence is that the two paths have the same range and the same orientation.

The relation we just defined is an equivalence relation in the sense of Set Theory. The corresponding cosets γ are called *curves*, and each path in a given coset is a *representative of the curve* or a *parametrization of the curve*. It is common practice to identify a curve with anyone of its parametrizations. This will be done implicitly whenever convenient.

The *inverse* γ^{-1} of the curve γ with parametrization $x(\cdot) : [a, b] \to \mathbb{R}^k$ is defined by the parametrization

$$t \in [-b, -a] \to x(-t).$$

The curve and its inverse have the same range, but have reversed orientation.

If the curves γ and η are such that the final point of γ coincides with the initial point of η, we define the *composition* $\gamma\eta$ as the curve whose range is the union of the ranges of γ, η with their given orientation, and with the points of γ preceding the points of η

Exercise Write a parametrization of the curve $\gamma\eta$ in terms of given parametrizations of the curves γ and η.

The set of all closed curves with a common point x^0 is a group under the above operation. The identity of the group is the singleton curve $\{x^0\}$, parametrized by the constant path $x(t) = x^0$.

Length of curve.

Let $x(\cdot) : [a, b] \to \mathbb{R}^k$ be a parametrization of the curve γ. Let

$$T : a = t_0 < t_1 < \cdots < t_n = b$$

be a partition of $[a, b]$, and denote $\lambda(T) = \max_i (t_i - t_{i-1})$.

The points $x(t_i)$ on γ are the vertices of a polygonal curve γ_T depending on T, with the same initial point and final point as γ. The length $L(\gamma_T)$ of γ_T is defined naturally as the sum of the lengths of the line segments $\overline{x(t_{i-1})\, x(t_i)}$, which are themselves defined as the Euclidean distance between their end points, that is, $||x(t_i) - x(t_{i-1})||$. Thus

$$L(\gamma_T) := \sum_{i=1}^{n} ||x(t_i) - x(t_{i-1})||. \tag{4.92}$$

4.3.2 Definition (Notation as above.)
We say that γ *has length* (or is *rectifiable*) if

$$L(\gamma) := \sup_T L(\gamma_T) < \infty. \tag{4.93}$$

When this is the case, the number $L(\gamma)$ is called *the length of* γ.

It can be shown that the definition of the length does not depend on the parametrization. We shall prove this below in the special case of smooth curves. The concept of rectifiability for the vector function $x(\cdot)$ is equivalent to the property that all the real-valued component functions $x_j(\cdot)$ $(j = 1, \ldots, k)$ are functions of *bounded variation*.

4.3.3 Theorem *If the limit*

$$\lim_{\lambda(T)\to 0} L(\gamma_T) \tag{4.94}$$

exists and is equal to the real number L, *then* γ *has length and* $L(\gamma) = L$.

Proof Let $T = \{t_i\}$ be any partition of $[a, b]$. It follows from the triangle inequality for the norm that if the partition T' is a *refinement* of T, that is, if T' is obtained from T by adding finitely many points, then

$$L(\gamma_{T'}) \geq L(\gamma_T). \tag{4.95}$$

Let $\{T_0, T_1, \ldots\}$ be a sequence of partitions of $[a, b]$ such that $T_0 = T$, T_j is a refinement of T_{j-1}, and $\lambda(T_j) \to 0$. By (4.95),

$$L(\gamma_T) \leq L(\gamma_{T_j})$$

for all j. By our hypothesis, the limit L in (4.94) exists; hence, letting $j \to \infty$, we get

$$L(\gamma_T) \leq L \tag{4.96}$$

for all partitions T. Hence γ has length and $L(\gamma) \leq L$.

Given $\epsilon > 0$, it follows from (4.96) and the hypothesis (4.94) that there exists $\delta > 0$ such that

$$0 \leq L - L(\gamma_T) < \epsilon \tag{4.97}$$

if $\lambda(T) < \delta$. Taking any such partition T_0, we have by (4.97)

$$L(\gamma_{T_0}) > L - \epsilon. \tag{4.98}$$

By (4.96) and (4.98), $L = \sup_T L(\gamma_T) := L(\gamma)$. □

Smooth Curves

4.3.4 Definition The curve γ is *smooth* if it has a parametrization of class C^1. If there exist smooth curves $\gamma_1, \ldots, \gamma_n$ such that $\gamma = \gamma_1 \ldots \gamma_n$, the curve γ is said to be *piecewise smooth.*

Since $x'(t)$ is a tangent vector to the curve at the point $x(t)$, the geometric meaning of smoothness is the continuous change of the tangent. For example, a circle is smooth, but the boundary of a rectangle is piecewise smooth but not smooth, since the tangent is not continuous at the vertices.

For smooth or piecewise smooth curves, we agree that the only permissible changes of parametrization *are by means of* C^1 *functions* h between the parameter intervals (with $h' > 0$, since h is required to be increasing).

4.3.5 Theorem *Let* γ *be a smooth curve, and let then* $x(\cdot) : [a, b] \to \mathbb{R}^k$ *be a parametrization of class* C^1 *for* γ*. Then* γ *has length, and*

$$L(\gamma) = \int_a^b ||x'(t)||\, dt.$$

The conclusion of the theorem is valid for piecewise smooth curves as well. The additional argument to show this is left as an exercise.

Proof (Notation as above.) Let $T = \{t_i\}$ be a partition of $[a, b]$. Since the derivatives x_j' exist in $[a, b]$ for all $j = 1, \ldots, k$, it follows from the MVT for functions of one variable that there exist points $t_{ij} \in (t_{i-1}, t_i)$ such that

$$x_j(t_i) - x_j(t_{i-1}) = (t_i - t_{i-1})x_j'(t_{ij}) \qquad i = 1, \ldots, n;\ j = 1, \ldots, k. \tag{4.99}$$

Hence

$$L(\gamma_T) = \sum_i \Big(\sum_j [x_j(t_i) - x_j(t_{i-1})]^2\Big)^{1/2}$$

$$= \sum_i \Big(\sum_j (t_i - t_{i-1})^2 x_j'(t_{ij})^2\Big)^{1/2} = \sum_i (t_i - t_{i-1}) \Big(\sum_j x_j'(t_{ij})^2\Big)^{1/2}$$

$$= \sum_i (t_i - t_{i-1}) u(t_{i1}, \ldots, t_{ik}), \tag{4.100}$$

where $u : [a, b]^k \to \mathbb{R}$ is defined by

$$u(s) := \Big(\sum_{j=1}^k x_j'(s_j)^2\Big)^{1/2}.$$

Since x_j are assumed to be of class C^1 in $[a, b]$, u is continuous on the compact set $[a, b]^k$, hence uniformly continuous there. Given $\epsilon > 0$, there exists then $\delta_1 > 0$ such that

$$|u(s) - u(s')| < \frac{\epsilon}{2(b-a)} \tag{4.101}$$

whenever $s, s' \in [a, b]^k$ and $||s - s'|| < \delta_1$.

Suppose $\lambda(T) < \delta_1/\sqrt{k}$. Since $t_{ij} \in (t_{i-1}, t_i)$, we have

$$|t_{ij} - t_i| \leq (t_i - t_{i-1}) \leq \lambda(T) < \frac{\delta_1}{\sqrt{k}},$$

and therefore

$$||(t_{i1}, \dots, t_{ik}) - (t_i,\ t_i, \dots, t_i)|| = \Big(\sum_{j=1}^{k} (t_{ij} - t_i)^2\Big)^{1/2}$$

$$< [k\frac{\delta_1^2}{k}]^{1/2} = \delta_1.$$

It then follows from (4.101) that (when $\lambda(T) < \delta_1/\sqrt{k}$)

$$|u(t_{i1}, \dots, t_{ik}) - u(t_i, \dots, t_i)| < \frac{\epsilon}{2(b-a)}. \tag{4.102}$$

Consider the Riemann sums $\sigma(T) := \sigma(||x'(\cdot)||,\ T)$ given by

$$\sigma(T) = \sum_{i=1}^{n} (t_i - t_{i-1})||x'(t_i)||.$$

By (4.100) and (4.102),

$$|L(\gamma_T) - \sigma(T)| = \Big|\sum_i (t_i - t_{i-1})[u(t_{i1}, \dots, t_{ik}) - u(t_i, \dots, t_i)]\Big|$$

$$\leq \sum_i (t_i - t_{i-1})|u(t_{i1}, \dots, t_{ik}) - u(t_i, \dots, t_i)|$$

$$< \frac{\epsilon}{2(b-a)} \sum_i (t_i - t_{i-1}) = \frac{\epsilon}{2}, \tag{4.103}$$

whenever $\lambda(T) < \delta_1/\sqrt{k}$. On the other hand, since $||x'(\cdot)||$ is continuous on $[a, b]$,

$$\lim_{\lambda(T)\to 0} \sigma(T) = \int_a^b ||x'(t)||\, dt,$$

that is, there exists $\delta_2 > 0$ such that

$$|\sigma(T) - \int_a^b ||x'(t)||\,dt| < \frac{\epsilon}{2} \tag{4.104}$$

if $\lambda(T) < \delta_2$.

Let $\delta = \min(\delta_1/\sqrt{k},\ \delta_2)$. Then if $\lambda(T) < \delta$, both (4.103) and (4.104) are valid, and therefore

$$\left|L(\gamma_T) - \int_a^b ||x'(t)||\,dt\right| \leq \left|L(\gamma_T) - \sigma(T)\right| + \left|\sigma(T) - \int_a^b ||x'(t)||\,dt\right| < \epsilon.$$

This shows that

$$\lim_{\lambda(T)\to 0} L(\gamma_T) = \int_a^b ||x'(t)||\,dt,$$

and the conclusion of the theorem follows from Theorem 4.3.3. □

Remarks 1. For smooth curves, we can see from the theorem that the length of the curve does not depend on the parametrization. Recalling that the only permissible changes of parameter in this case are of the form $t = h(s)$ with h *of class* C^1 and $h' > 0$ in $[\alpha, \beta]$ (cf. Sect. 4.3.4), we have by the change of variable formula for integrals and the chain rule

$$\begin{aligned}\int_a^b ||x'(t)||\,dt &= \int_\alpha^\beta ||x'(h(s))||\,h'(s)\,ds \\ &= \int_\alpha^\beta ||x'(h(s))\,h'(s)||\,ds = \int_\alpha^\beta ||(x\circ h)'(s)||\,ds.\end{aligned}$$

By Theorem 4.3.5, this shows that the length of γ given by the parametrization $x(\cdot)$ is the same as its length given by the equivalent parametrization $x \circ h$.

2. The independence of the integral in the theorem on the parametrization motivates some authors to consider only piecewise smooth curves, and to *define* their length by means of the integral formula in Theorem 4.3.5.

Examples. 1. Let γ be the circle with radius R. Choosing its centre as the origin, we have the parametrization

$$x(\phi) = (R\cos\phi,\ R\sin\phi) \qquad \phi \in [0, 2\pi].$$

Then

$$||x'(\phi)|| = ||(-R\sin\phi,\ R\cos\phi)|| = R.$$

Hence by Theorem 4.3.5

$$L(\gamma) = \int_0^{2\pi} R\, d\phi = 2\pi R.$$

2. Let γ be the *cycloid* arc, parametrized by

$$x(t) = a(t - \sin t,\ 1 - \cos t), \qquad t \in [0, 2\pi],$$

where a is a positive constant. We calculate

$$||x'(t)||^2 = a^2||(1 - \cos t,\ \sin t)||^2 = 2a^2(1 - \cos t) = 4a^2 \sin^2 \frac{t}{2}.$$

Hence by Theorem 4.3.5

$$L(\gamma) = 2a \int_0^{2\pi} \sin \frac{t}{2}\, dt = 4a \cos \frac{t}{2}\Big|_{2\pi}^{0} = 8a.$$

Line Integrals

Let γ be a piecewise smooth curve in $\mathbb{R}^k$, and let $x(\cdot) : [a, b] \to \mathbb{R}^k$ be a parametrization for it. Suppose $f \in C(\gamma, \mathbb{R}^k)$. Such a vector valued function is also called a vector field on γ. *In the present context, we shall use the symbol dx to denote the symbolic vector* $(dx_1, \ldots, dx_k)$, so that the symbolic inner product $f \cdot dx$ denotes the so-called *first order differential form*

$$f \cdot dx := f_1 dx_1 + \cdots + f_k dx_k. \tag{4.105}$$

4.3.6 Definition The line integral on γ of the differential form $f \cdot dx$ is defined by the equation

$$\int_\gamma f \cdot dx := \int_a^b f(x(t)) \cdot x'(t)\, dt. \tag{4.106}$$

The integrand on the right is piecewise continuous on $[a, b]$, so that the integral exists. It depends only on f and γ, and not on the particular parametrization of γ. We show this in the case of a smooth curve (the additional argument needed for the general case is left as an exercise). Let $h : [\alpha, \beta] \to [a, b]$ be of class C^1 with $h' > 0$. Then by the change of variables formula for integrals,

$$\int_a^b f(x(t)) \cdot x'(t)\, dt = \int_\alpha^\beta f(x(h(s))) \cdot x'(h(s))\, h'(s)\, ds$$

$$= \int_\alpha^\beta f(\tilde{x}(s)) \cdot \tilde{x}'(s)\, ds,$$

where $\tilde{x} := x \circ h : [\alpha, \beta] \to \mathbb{R}^k$ is the new parametrization of γ. This proves the independence of the integral on the right hand side of (4.106) on the parametrization.

The physical motivation of the line integral is the concept of work. If f is a force field, the physical meaning of its line integral along γ is the work done by moving along the curve in the force field.

If p is a *real valued* continuous function on γ, the line integrals $\int_\gamma p\,dx_i$ are defined similarly by the formula

$$\int_\gamma p\,dx_i := \int_a^b p(x(t))\,x_i'(t)\,dt \qquad (i = 1, \ldots, k),$$

where $x(\cdot) : [a, b] \to \mathbb{R}^k$ is a piecewise C^1 parametrization of γ. As before, the value of these integrals is invariant under a permissible change of parametrization. It is obvious from the definitions that

$$\int_\gamma f \cdot dx = \sum_{i=1}^k \int_\gamma f_i\,dx_i$$

for any continuous vector field f on γ.

An important parametrization of piecewise smooth curves uses the arc length

$$s(t) := \int_a^t ||x'(\tau)||\,d\tau$$

as the parameter for the curve

$$\gamma : t \in [a, b] \to x(t) \in \mathbb{R}^k.$$

If f is a real continuous function on γ, the line integral $\int_\gamma f\,ds$ is defined by

$$\int_\gamma f\,ds := \int_a^b f(x(t))\,||x'(t)||\,dt.$$

(Formally, $ds := ||x'(t)||\,dt$.) Clearly, for $f = 1$ identically on γ, the line integral $\int_\gamma ds$ is equal to the length of γ.

The elementary properties of the line integral are stated in the following

Proposition *(a) The map* $f \to \int_\gamma f \cdot dx$ *is a linear functional on* $C(\gamma, \mathbb{R}^k)$.
(b) $\int_{\gamma_1\gamma_2} f \cdot dx = \int_{\gamma_1} f \cdot dx + \int_{\gamma_2} f \cdot dx$ *for any* $f \in C(\gamma_1\gamma_2, \mathbb{R}^k)$.
(c) $\int_{\gamma^{-1}} f \cdot dx = -\int_\gamma f \cdot dx$ *for any* $f \in C(\gamma, \mathbb{R}^k)$.
(d) Let $M := \max_\gamma ||f||$ *and* $L := L(\gamma)$ *be the length of the curve* γ. *Then*

$$|\int_\gamma f\cdot dx| \le ML. \tag{4.107}$$

Proof We verify only Property (d) (for smooth curves). Note first that γ is a compact set, as the image of the compact interval $[a, b]$ by the continuous function $x(\cdot)$. Since the norm on a normed space is continuous, the function $||f||$ is a continuous real valued function on the *compact* set γ, and assumes therefore its maximum M on γ. Applying Theorem 4.2.4 (ix) and Schwarz' inequality, we obtain from (4.106)

$$|\int_\gamma f\cdot dx| \le \int_a^b |f(x(t))\cdot x'(t)|\, dt$$

$$\le \int_a^b ||f(x(t))||\, ||x'(t)||\, dt \le M\int_a^b ||x'(t)||\, dt = ML$$

by Theorem 4.3.5. □

Conservative Fields

4.3.7 Definition Let D be a domain in $\mathbb{R}^k$. The continuous vector valued function (or "vector field") $f : D \to \mathbb{R}^k$ is said to be *conservative* in D if for *every* piecewise smooth curve $\gamma \subset D$, the line integral $\int_\gamma f\cdot dx$ does not depend on γ, but *only on its endpoints*.

The above condition is equivalent to the following:
For every *closed* piecewise smooth curve $\gamma \subset D$,

$$\int_\gamma f\cdot dx = 0. \tag{4.108}$$

Indeed, suppose f satisfies the condition in Definition 4.3.7, and let $\gamma \subset D$ be a closed curve. Let $x(\cdot) : [a, b] \to \mathbb{R}^k$ be a parametrization of γ. We have $x(a) = x(b) := x^0$, since the endpoints of a *closed* curve coincide. The curve $\gamma' = \{x^0\}$, that is, the coset of the constant path $y(t) = x^0$, has the same end points as γ, and therefore

$$\int_\gamma f\cdot dx = \int_{\gamma'} f\cdot dx = \int_a^b f(y(t))\cdot y'(t)\, dt = 0,$$

since $y'(t) = 0$ identically.

On the other hand, suppose (4.108) is satisfied for every closed piecewise smooth curve $\gamma \subset D$. Let γ_1, γ_2 be piecewise smooth curves lying in D with the same initial point and the same final point. Then $\gamma := \gamma\gamma^{-1}$ is a *closed* piecewise smooth curve contained in D, and therefore by (4.108) and the proposition in Sect. 4.3.6,

$$0 = \int_{\gamma} f \cdot dx = \int_{\gamma_1} f \cdot dx - \int_{\gamma_2} f \cdot dx.$$

This shows that the field f is conservative in D.

The next theorem gives a characterization of continuous conservative fields.

4.3.8 Theorem *The continuous field f in the domain $D \subset \mathbb{R}^k$ is conservative in D if and only if there exists a C^1 function $u : D \to \mathbb{R}$ such that $f = \nabla u$.*

Proof Suppose there exists a C^1 function $u : D \to \mathbb{R}$ such that $f = \nabla u$. Let γ be a smooth curve in D, with the parametrization $x(\cdot)$ over the parameter interval $[a, b]$. By the Chain Rule (Theorem 2.1.10),

$$(u \circ x)'(t) = \nabla u\Big|_{x(t)} \cdot x'(t) = f(x(t)) \cdot x'(t).$$

Integrating over $[a, b]$, we get by the Fundamental Theorem of Calculus and the definition of the line integral

$$u(x(b)) - u(x(a)) = \int_{\gamma} f \cdot dx. \tag{4.109}$$

Equation (4.109) shows that the line integral of f does not depend on the curve, but only on its end points $x(a)$ and $x(b)$. Hence the vector field f is conservative in D.

The additional argument needed in the case of a *piecewise* smooth curve is left as an exercise.

Next, suppose that the continuous vector field f is conservative in D. Fix $x^0 \in D$, and let x be an arbitrary point of D. Since D is connected, there exists a polygonal path $\gamma(x) \subset D$ with initial point x^0 and final point x. Since f is continuous on $\gamma(x)$, and the latter is piecewise smooth, we may define

$$u(x) := \int_{\gamma(x)} f \cdot dx \qquad (x \in D). \tag{4.110}$$

The right hand side of (4.110) depends only on the endpoint x of $\gamma(x)$, since the field is conservative in D; hence u is well defined on D. We shall verify that u is of class C^1 and $f = \nabla u$ on D.

Fix $x \in D$. Since D is open, there exists a ball $B(x, r)$ contained in D. Let $h \in (-r, r)$. Then for each $i = 1, \dots, k$, the line segment $\gamma_i(h) := \{x + the^i;\ t \in [0, 1]\}$ lies in $B(x, r)$, because $d(x + the^i, x) = ||the^i|| = t|h| < r$. Hence $\gamma_i(h) \subset D$, and therefore the polygonal path $\gamma(x)\gamma_i(h)$ connecting x^0 to $x + he^i$ can be chosen as $\gamma(x + he^i)$, since it lies in D. Hence for $0 < |h| < r$,

$$h^{-1}[u(x + he^i) - u(x)] = h^{-1}\Big[\int_{\gamma(x)\gamma_i(h)} f \cdot dx - \int_{\gamma(x)} f \cdot dx\Big]$$

$$= h^{-1} \int_{\gamma_i(h)} f \cdot dx = h^{-1} \int_0^1 f(x + the^i) \cdot he^i \, dt = \int_0^1 f_i(x + the^i) \, dt. \tag{4.111}$$

Since $\phi(t) := f_i(x + the^i) \in C([0, 1])$, it follows from the Mean Value Theorem for integrals of real continuous functions of one variable that there exists $\theta \in [0, 1]$ such that the last integral in (4.111) is equal to $f_i(x + \theta he^i)$. Therefore, by (4.111) and the continuity of f at x,

$$\lim_{h \to 0} h^{-1}[u(x + he^i) - u(x)] = f_i(x) \qquad i = 1, \ldots, k,$$

that is, for all i and $x \in D$,

$$\frac{\partial u}{\partial x_i}(x) = f_i(x), \tag{4.112}$$

i.e., $f = \nabla u$. Since f is continuous in D, we read from (4.112) that u is of class C^1 in D. □

4.3.9 Exact Differential Form and Potential

If $f = \nabla u$ for some u of class C^1 in the domain $D \subset \mathbb{R}^k$, the differential form $f \cdot dx$ is said to be *exact*. The latter is also said to be the *total differential* of u. This terminology does not stictly conform with the one used in this text, since differentials were defined here as linear functionals and not as differential forms.

If v is also a C^1 function such that $f = \nabla v$ on D, then $\nabla w = 0$ on D for $w := v - u$. If the segment $\overline{x\,x'}$ lies in D, it follows from the Mean Value Theorem in $\mathbb{R}^k$ that there exists $\theta \in (0, 1)$ such that

$$w(x') = w(x) + (x' - x) \cdot \nabla w\Big|_{x + \theta(x' - x)} = w(x).$$

Next, fixing $x^0 \in D$, *every point* $x \in D$ can be connected to x^0 by a polygonal path $\gamma(x)$ lying in the *domain* D. If x^j are the latter's vertices ($j = 0, \ldots, n$, $x^n = x$), then by our preceding conclusion, $w(x^j) = w(x^{j-1})$ for all j. Consequently $w(x) = w(x^0)$. This shows that w is constant on D, that is, the functions u, v differ only by a constant on D. On the other hand, if $v = u + c$ on D for some constant c, we trivially have $\nabla v = f$. In other words, the most general C^1 function v with gradient equal to f on D is $v = u + c$. We refer to this fact by saying that u is *uniquely determined on D up to an additive constant*. The function u is called a *potential* for the vector field f. Theorem 4.3.8 states that a continuous vector field is conservative on a domain D if and only if it is the gradient field of a potential on D.

Necessary Condition

4.3.10 Theorem *Let f be a conservative field of class C^1 in the domain $D \subset \mathbb{R}^k$. Then the Jacobian matrix $\left(\frac{\partial f}{\partial x}\right)$ is symmetric in D.*

Proof By Theorem 4.3.8, f has a potential u of class C^1 in D. Thus $f_i = u_{x_i}$ in D for $i = 1, \ldots, k$. By hypothesis, f_i has partial derivatives in D with respect to x_j (for all $j = 1, \ldots, k$), that is, $u_{x_i x_j}$ exists and equals $(f_i)_{x_j}$, hence is continuous in D, by our hypothesis. Therefore $u_{x_i x_j} = u_{x_j x_i}$, that is,

$$\frac{\partial f_i}{\partial x_j} = \frac{\partial f_j}{\partial x_i}$$

in D, i.e., the Jacobian matrix $\left(\frac{\partial f}{\partial x}\right)$ is symmetric in D. □

The symmetry of the Jacobian matrix is *not sufficient* in general for the field to be conservative. This is shown in the following example.

4.3.11 Example Let

$$D := \{(x, y) \in \mathbb{R}^2;\ 0 < x^2 + y^2 < 1\},$$

and

$$f(x, y) := \left(\frac{-y}{x^2 + y^2}, \frac{x}{x^2 + y^2}\right).$$

The components of f are rational functions with non-vanishing denominator in D, hence are of class C^1 in D, and we have there

$$\frac{\partial f_1}{\partial y} = \frac{y^2 - x^2}{(x^2 + y^2)^2} = \frac{\partial f_2}{\partial x},$$

so that *the Jacobian matrix is symmetric in D*.

However the field f *is not conservative*. Indeed, consider the circle γ with radius $0 < r < 1$ and centre at the origin. Then $\gamma \subset D$ is a closed smooth curve, but we show below that $\int_\gamma f \cdot dx \neq 0$.

A parametrization of γ is

$$(x, y)(t) = r\,(\cos t,\ \sin t) \qquad (t \in [0, 2\pi]).$$

Hence

$$\int_\gamma f \cdot dx = \int_0^{2\pi} \left(\frac{-r \sin t}{r^2}, \frac{r \cos t}{r^2}\right) \cdot r\,(-\sin t,\ \cos t)\, dt$$

$$= \int_0^{2\pi} ||(-\sin t,\ \cos t)||^2 dt = \int_0^{2\pi} dt = 2\pi \neq 0.$$

Sufficient Condition

In contrast to Example 4.3.11, the symmetry of the Jacobian matrix is also sufficient for the field to be conservative *in suitable domains*. The simplest domains for which this is true are the so-called *star-like domains*. The domain D is star-like if there exists a point $x^0 \in D$ such that, *for every $x \in D$, the segment* $\gamma(x) := \overline{x^0 x}$ *lies in* D.

Convex domains, that is, domains which contain the segment joining any pair of points in them, are an important sub-class of star-like domains. The plane domain whose boundary is given by the equation $x^{2/3} + y^{2/3} = 1$ is star-like but not convex.

4.3.12 Theorem *Let D be a star-like domain in $\mathbb{R}^k$. Then the C^1 vector field f in D is conservative if and only if the Jacobian matrix $\left(\frac{\partial f}{\partial x}\right)$ is symmetric in D.*

Proof By Theorem 4.3.10, we have only to prove the sufficiency of the symmetry condition. Without loss of generality, we may assume $x^0 = 0$, since we may translate the space if needed. For any $x \in D$, the segment $\gamma(x) := \overline{0x}$ lies in D. A parametrization for it is

$$x(t) = tx \qquad (t \in [0, 1]).$$

Note that $x'(t) = x$. Define

$$u(x) := \int_{\gamma(x)} f \cdot dx = \int_0^1 f(tx) \cdot x \, dt.$$

For each fixed $i = 1, \ldots, k$, the integrand in the last integral is of class C^1 with respect to (x_i, t) in a suitable rectangle $[x_i - r, x_i + r] \times [0, 1]$. Hence by Theorem 4.1.3,

$$\begin{aligned} u_{x_i}(x) &= \int_0^1 \Big[f(tx) \cdot x \Big]_{x_i} dt \\ &= \int_0^1 \left[\frac{\partial f(tx)}{\partial x_i} \cdot x + f(tx) \cdot \frac{\partial x}{\partial x_i} \right] dt. \end{aligned} \tag{4.113}$$

By the symmetry of the Jacobian matrix, the first summand of the integrand is equal to

$$t \sum_{j=1}^{k} x_j \frac{\partial f_j}{\partial x_i}(tx) = t \sum_j x_j \frac{\partial f_i}{\partial x_j}(tx).$$

By the chain rule, the last expression is equal to $t\, \frac{df_i(tx)}{dt}$. Since $\frac{\partial x}{\partial x_i} = e^i$, the second summand of the integrand is equal to $f_i(tx)$. We then conclude that the integrand is equal to $(d/dt)[t\, f_i(tx)]$. Therefore by (4.113)

$$u_{x_i}(x) = \int_0^1 \frac{d}{dt}[t\, f_i(tx)]\, dt = [t\, f_i(tx)]\Big|_{t=0}^{t=1} = f_i(x),$$

for all $i = 1, \ldots, k$. This shows that $\nabla u = f$ on D. Therefore the field f is conservative in D, by Theorem 4.3.8. □

Examples.

(a) Let γ be the ellipse with equation $(x/a)^2 + (y/b)^2 = 1$. Consider the vector field

$$f(x, y) = (\frac{y^2}{1+x^2}, 2y \arctan x) \qquad (x, y) \in \mathbb{R}^2.$$

We wish to calculate the line integral $\int_\gamma f \cdot dx$. The vector field is of class C^1 in $\mathbb{R}^2$ and

$$\frac{\partial f_1}{\partial y} = \frac{2y}{1+x^2} = \frac{\partial f_2}{\partial x}$$

in $\mathbb{R}^2$ (which is trivially a star-like domain). Therefore $\int_\gamma f \cdot dx = 0$, by Theorem 4.3.12.

(b) Let γ be the cycloid arc given by the parametrization

$$t \in [0, 2\pi] \to x(t) = a\,(t - \sin t,\ 1 - \cos t) \in \mathbb{R}^2,$$

where a is a positive constant. Note that the end points of γ are $(0, 0)$ and $(2\pi a, 0)$.
Let $f : \mathbb{R}^2 \to \mathbb{R}^2$ be the vector field

$$f(x, y) = \frac{2}{1+x^2+y^2}(x, y).$$

We are asked to calculate the line integral $\int_\gamma f \cdot dx$. We first observe that the field f is of class C^1 in $\mathbb{R}^2$ since its components are rational functions with non-vanishing denominator. We have also

$$\frac{\partial f_1}{\partial y} = -\frac{4xy}{(1+x^2+y^2)^2} = \frac{\partial f_2}{\partial x}$$

for all $(x, y) \in \mathbb{R}^2$. By Theorem 4.3.12, the field f is conservative in $\mathbb{R}^2$, and therefore

$$\int_\gamma f \cdot dx = \int_{\gamma'} f \cdot dx,$$

where γ' is the line segment from the initial point $(0, 0)$ to the final point $(2\pi a, 0)$ of γ. A parametrization of γ' is

$$t \in [0, 2\pi] \to (at, 0).$$

Hence

$$\int_\gamma f \cdot dx = \int_0^{2\pi} f_1(at, 0)\, a\, dt$$
$$= \int_0^{2\pi} \frac{2a^2 t}{1 + a^2 t^2}\, dt = \log(1 + a^2 t^2)\Big|_0^{2\pi} = \log(1 + 4\pi^2 a^2).$$

Locally Conservative Fields

The symmetry of the Jacobian matrix is necessary and sufficient for the vector field to be *locally conservative* in the domain D.

4.3.13 Definition A vector field is *locally conservative* in a domain D if each point of D has a neighborhood in which the field is conservative.

4.3.14 Theorem *A C^1 vector field in a domain D is locally conservative in D if and only if its Jacobian matrix is symmetric in D.*

Proof If the C^1 field f is locally conservative in D, each $x \in D$ has a neighborhood $B(x, r)$ in which f is conservative, and therefore its Jacobian matrix $M := (\partial f/\partial x)$ is symmetric in $B(x, r)$ by Theorem 4.3.10. Since this is true for each $x \in D$, M is symmetric in D. Conversely, suppose M is symmetric in D, and let $x \in D$. Since D is open, there exists $r > 0$ such that $B(x, r) \subset D$. The latter ball is a star-like domain, and f is of class C^1 with symmetric Jacobian matrix in it; hence f is consevative in $B(x, r)$, by Theorem 4.3.12. □

Remark It can be proved that if the domain D is *simply connected*, then a C^1 vector field is conservative in D if (and only if) its Jacobian matrix is symmetric in D.

We say that the domain D is simply connected if every closed curve in D can be continuously contracted into a singleton in D. We shall not elaborate on this topological concept and on the proof of the above statement.

4.3.15 Exercises

1. Let γ be the *helix*

$$\gamma: \quad x(t) = (a\cos t,\ a\sin t,\ bt) \qquad (0 \le t \le T).$$

$(a, b > 0)$.

(a) Find the arc length $s(t)$ of the arc $\{x(\tau);\ 0 \le \tau \le t\}$.
(b) Find the length of one turn of the helix γ.
(c) Let f be the vector field $(-y, x, z)$ in $\mathbb{R}^3$. Calculate the line integral $\int_\gamma f \cdot dx$ over one turn of the helix γ.
(d) Calculate $\int_\gamma (x^2 + y^2 + z^2)^{-1/2} ds$ when γ is one turn of the above helix.

2. Find the length of the curve

$$\gamma: \quad x(t) = (e^t \cos t,\ e^t \sin t,\ e^t) \qquad 0 \le t \le 1.$$

3. The curve $\gamma \subset \mathbb{R}^2$ is given in polar coordinates by the C^1 function

$$r = g(\theta) \qquad \theta \in [a, b].$$

(a) Show that the arc length function on γ is given by

$$s(\theta) = \int_a^\theta [g'(\tau)^2 + g(\tau)^2]^{1/2} d\tau.$$

(b) Let γ be the curve in $\mathbb{R}^2$ given in polar coordinates by

$$\gamma = \{(r, \theta); r = 1 - \cos\theta, \qquad 0 \le \theta \le 2\pi\}.$$

Find the length of γ.

4. Let γ be the circle centred at the origin with radius r. Calculate $\int_\gamma f \cdot dx$ for the following vector field f:

(a) $f(x) = (x_1 - x_2,\ x_1 + x_2)$;
(b) $f(x) = \nabla u$, where $u(x) = \arctan(x_2/x_1)$.
(c) Explain your results in (a)–(b) in light of the theory.

5. Let γ be the cycloid arc given by the parametrization

$$x(t) = a(t - \sin t,\ 1 - \cos t) \qquad 0 \le t \le 2\pi,$$

where a is a positive constant.

(a) Find the arc length function on γ.
(b) Find the length of the arc γ.
(c) Let $f : \mathbb{R}^2 \to \mathbb{R}^2$ be the vector field

$$f(x, y) = \frac{4}{1 + 2x^2 + y^4}(x,\ y^3).$$

Calculate $\int_\gamma f \cdot dx$.

6. Let γ be the ellipse with equation $(x/a)^2 + (y/b)^2 = 1$, described counterclockwise. Let f be the vector field $f(x, y) = (y^2/(1 + x^2),\ 2y \arctan x)$. Calculate $\int_\gamma f \cdot dx$.

7. For γ as in Exercise 6 and f the vector field

$$f(x, y) = e^x (\sin y, \cos y),$$

calculate the following line integrals

(a) $\int_\gamma f \cdot dx$.
(b) $\int_{\gamma_1} f \cdot dx$, where γ_1 is the part of γ in the first quadrant.
(c) $\int_{\gamma_2} f \cdot dx$, where γ_2 is the line segment from $(a, 0)$ to $(0, b)$.

4.4 Green's Theorem in $\mathbb{R}^2$

Green's theorem gives an identity relating a double integral on a bounded closed domain with area $\overline{D} \subset \mathbb{R}^2$ with a line integral on the boundary ∂D. We start with some restriction on the boundary.

4.4.1 Definition A closed curve $\gamma \subset \mathbb{R}^2$ with parametrization $x(\cdot) : [a, b] \to \mathbb{R}^2$ is *regular* if $x(\cdot)$ is piecewise smooth and $x'(\cdot) \neq 0$ in each sub-interval where $x(\cdot)$ is smooth.

The closed domain $\overline{D}$ is *regular* if its boundary ∂D is a finite union of mutually disjoint regular curves γ_i. Each γ_i is oriented so that D is *on the left of a point moving in the positive direction on the curve.*

Example. Let $\overline{D}$ be the ring

$$\{(x, y) \in \mathbb{R}^2;\ r \leq x^2 + y^2 \leq R^2\}$$

(where $0 < r < R$). Then ∂D is the disjoint union of the circle γ_1 with radius R with counter-clockwise orientation, and the circle γ_2 with radius r with clockwise orientation (both centred at the origin). The parametrization $x(t) = R\,(\cos t,\ \sin t)$ of γ_1 with $t \in [0, 2\pi]$) is of class C^1, and $||x'(t)|| = ||(-R \sin t, R \cos t)|| = R \neq 0$ for all t; the same is true for γ_2. Thus the closed domain $\overline{D}$ is regular.

If f is a continuous vector field defined on the boundary ∂D of the regular domain D, we define the line integral of f over ∂D by

$$\int_{\partial D} f \cdot dx := \sum_i \int_{\gamma_i} f \cdot dx, \tag{4.114}$$

where γ_i are the mutually disjoint regular curves with union equal to ∂D.

We shall first establish Green's theorem for *normal* domains (cf. Sect. 4.2.8). Note that bounded closed normal domains have area; therefore double integration over them makes sense.

Green's Theorem for Normal Domains

4.4.2 Theorem *Let $\overline{D}$ be a bounded closed domain in $\mathbb{R}^2$, which is regular and normal. Let f be a vector field of class C^1 on $\overline{D}$. Then*

$$\int_{\partial D} f \cdot dx = \int_{\overline{D}} \Big(\frac{\partial f_2}{\partial x_1} - \frac{\partial f_1}{\partial x_2}\Big)\, dS. \tag{4.115}$$

Relation (4.115) is called *Green's formula for the field f on the domain $\overline{D}$.*

Proof We prove the theorem when the domain is normal of x-type. The case of a y-type domain is similar. For simplicity of notation, we write (x, y) instead of (x_1, x_2) and (p, q) instead of (f_1, f_2). Since the domain is also regular, there exist continuous piececewise C^1 real functions ϕ, ψ on $[a, b]$ such that $\phi < \psi$ on (a, b), and

$$\overline{D} = \{(x, y);\ \phi(x) \le y \le \psi(x), \quad x \in [a, b]\}. \tag{4.116}$$

We now use Theorem 4.2.9 for the continuous function $-p_y$:

$$\int_{\overline{D}} (-p_y)\, dS = \int_a^b \Big(\int_{\phi(x)}^{\psi(x)} (-p_y)\, dy\Big)\, dx$$

$$= \int_a^b [p(x, \phi(x)) - p(x, \psi(x))]\, dx. \tag{4.117}$$

On the other hand,

$$\partial D = \gamma_1\gamma_2\gamma_3^{-1}\gamma_4^{-1} \tag{4.118}$$

where $\gamma_1 := \{(x, \phi(x));\ x \in [a, b]\}$.
$\gamma_3 := \{(x, \psi(x));\ x \in [a, b]\}$.
$\gamma_2 := \{(b, y);\ y \in [\phi(b), \psi(b)]\}$.
$\gamma_4 := \{(a, y);\ y \in [\phi(a), \psi(a)]\}$.
Therefore

$$\int_{\partial D} p\, dx := \int_{\gamma_1} p\, dx + \int_{\gamma_2} p\, dx - \int_{\gamma_3} p\, dx - \int_{\gamma_4} p\, dx. \tag{4.119}$$

In γ_1 and γ_3, the parameter is x, so

$$\int_{\gamma_1} p\, dx = \int_a^b p(x, \phi(x))\, dx, \tag{4.120}$$

and

$$\int_{\gamma_3} p\,dx = \int_a^b p(x,\,\psi(x))\,dx. \tag{4.121}$$

On γ_2 and γ_4, taking the parametrizations $y \to (b,\,y)$ ($y \to (a,\,y)$, respectively), we have $dx/dy = 0$ identically, and therefore

$$\int_{\gamma_2} p\,dx = \int_{\phi(b)}^{\psi(b)} p(b,\,y)\,\frac{dx}{dy}\,dy = 0,$$

and similarly $\int_{\gamma_4} p\,dx = 0$.

We then conclude from (4.117) and (4.119)–(4.121) that

$$\int_{\partial D} p\,dx = \int_a^b p(x,\,\phi(x))\,dx - \int_a^b p(x,\,\psi(x))\,dx = \int_{\overline{D}} (-p_y)\,dS. \tag{4.122}$$

Next, applying again Theorem 4.2.9, we have

$$\int_{\overline{D}} q_x dS = \int_a^b \Big(\int_{\phi(x)}^{\psi(x)} q_x(x,\,y)\,dy\Big)\,dx. \tag{4.123}$$

Let

$$F(x) := \int_{\phi(x)}^{\psi(x)} q(x,\,y)\,dy. \tag{4.124}$$

We apply Theorem 4.1.8, assuming that $\phi,\ \psi$ are C^1. The general case of piecewise C^1 functions follows easily from this special case. Thus

$$F'(x) = \int_{\phi(x)}^{\psi(x)} q_x(x,\,y)\,dy + q(x,\,\psi(x))\,\psi'(x) - q(x,\,\phi(x))\,\phi'(x),$$

that is,

$$\int_{\phi(x)}^{\psi(x)} q_x(x,\,y)\,dy = F'(x) + q(x,\,\phi(x))\,\phi'(x) - q(x,\,\psi(x))\,\psi'(x).$$

Hence by (4.123)

$$\int_{\overline{D}} q_x dS = F(b) - F(a) + \int_a^b q(x,\,\phi(x))\,\phi'(x)\,dx - \int_a^b q(x,\,\psi(x))\,\psi'(x)\,dx. \tag{4.125}$$

Now by (4.124)

$$F(b) = \int_{\phi(b)}^{\psi(b)} q(b,\,y)\,dy = \int_{\gamma_2} q\,dy,$$

and similarly

$$F(a) = \int_{\gamma_4} q\, dy.$$

Also

$$\int_{\gamma_1} q\, dy = \int_a^b q(x, \phi(x))\, \phi'(x)\, dx,$$

and a similar formula on γ_3.

We conclude from (4.125) that

$$\begin{aligned}\int_{\overline{D}} q_x dS &= \Big(\int_{\gamma_2} - \int_{\gamma_4} + \int_{\gamma_1} - \int_{\gamma_3}\Big) q\, dy \\ &= \int_{\partial D} q\, dy.\end{aligned}$$

Together with (4.122), this gives

$$\int_{\partial D} (p\, dx + q\, dy) = \int_{\overline{D}} (-p_y + q_x)\, dS,$$

which is Green's formula (4.115). □

In the proof of the following general statement of Green's theorem in $\mathbb{R}^2$, we shall use without proof the fact that *every bounded closed regular domain* $\overline{D} \subset \mathbb{R}^2$ *can be partitioned into finitely many domains* $\overline{D_j}$ *of the same kind, that are in addition normal of either* x*-type or* y*-type*. Since bounded closed normal domains in $\mathbb{R}^2$ have area, it follows from this partitioning that every regular bounded closed domain in $\mathbb{R}^2$ has area, and consequently double integrals over them are well-defined, and are equal to the sum of the double integrals over the normal sub-domains $\overline{D_j}$.

Example. The ring domain in the example of Sect. 4.4.1 is not normal, but it can be partitioned into the four (regular) normal domains $\overline{D_j}$, $j = 1, \ldots, 4$, where

$$\overline{D_1} := \{(x, y);\ r \le y \le \sqrt{R^2 - x^2}, \quad -\sqrt{R^2 - r^2} \le x \le \sqrt{R^2 - r^2}\}$$

and $\overline{D_3} := -\overline{D_1}$ are (normal) of x-type, while

$$\overline{D_2} := \{(x, y);\ \sqrt{r^2 - y^2} \le x \le \sqrt{R^2 - y^2}, \quad -r \le y \le r\}$$

and $\overline{D_4} := -\overline{D_2}$ are (normal) of y-type.

General Green's Theorem in $\mathbb{R}^2$

4.4.3 Theorem *Let $\overline{D}$ be a bounded closed regular domain in $\mathbb{R}^2$, and let f be a C^1 vector field on $\overline{D}$. Then Green's formula for the field f on the domain $\overline{D}$ is valid.*

Proof We partition $\overline{D}$ into a finite union of regular, normal, closed domains $\overline{D_j}$, with $D_i \cap D_j = \emptyset$ for $i \neq j$. Applying Theorem 4.4.2 to the *normal* domains D_j, we get

$$\int_{\overline{D}} \left(\frac{\partial f_2}{\partial x_1} - \frac{\partial f_1}{\partial x_2} \right) dS = \sum_j \int_{\overline{D_j}} \left(\frac{\partial f_2}{\partial x_1} - \frac{\partial f_1}{\partial x_2} \right) dS$$
$$= \sum_j \int_{\partial D_j} f \cdot dx. \tag{4.126}$$

If the curve γ is part of the boundaries ∂D_i and ∂D_j for $i \neq j$, the integral $\int_\gamma f \cdot dx$ is included twice in the sum (4.126) of line integrals, with *opposite signs*. Indeed, if one domain is on the left of a point moving on γ in its positive direction, the other domain is necessarily on its right. Therefore the above sum includes only line integrals over curves γ that are parts of the boundary of D, and their composition is precisely ∂D. Hence the sum of line integrals in (4.126) is equal to $\int_{\partial D} f \cdot dx$. □

4.4.4 Corollary *Let $\overline{D}$ be a bounded regular closed domain in $\mathbb{R}^2$. Then*

$$S(\overline{D}) = \frac{1}{2} \int_{\partial D} (-y dx + x dy).$$

Proof Apply Green's formula to the field $f(x, y) = (1/2)(-y, x)$. □

Examples. 1. Let $\overline{D}$ be the bounded closed domain whose boundary is the ellipse with half-axis a, b. We may assume the centre of the ellipse is at the origin, so a parametrization for ∂D is $(x, y) = (a \cos t,\ b \sin t)$, $t \in [0, 2\pi]$. Hence by Corollary 4.4.4,

$$S(\overline{D}) = \frac{1}{2} \int_0^{2\pi} \Big[(-b \sin t)(-a \sin t) + (a \cos t)(b \cos t) \Big] dt = \pi ab.$$

2. Let $\overline{D}$ be the closed domain in the first quadrant $\{(x, y) \in \mathbb{R}^2; x, y \geq 0\}$ bounded by the coordinates axis and the ellipse arc

$$\gamma := \{(x, y); x, y \geq 0,\ (x/a)^2 + (y/b)^2 = 1\},$$

where $a, b > 0$ are constants. We calculate the double integral

$$J := \int_{\overline{D}} xy\, dS$$

directly, as a repeated integral:

$$J = \int_0^a x \int_0^{b\sqrt{1-(x/a)^2}} y\, dy\, dx = \frac{b^2}{2} \int_0^a (x - \frac{x^3}{a^2})\, dx = \frac{a^2 b^2}{8}.$$

Next, we calculate J by using Green's theorem.

Consider the vector field $f : \mathbb{R}^2 \to \mathbb{R}^2$ defined by

$$f(x, y) = (0, \frac{x^2 y}{2}).$$

We have

$$J = \int_{\overline{D}} (\frac{\partial f_2}{\partial x} - \frac{\partial f_1}{\partial y})\, dS = \int_{\partial D} (f_1 dx + f_2 dy)$$

$$= \int_{\partial D} \frac{x^2 y}{2}\, dy.$$

The integrand vanishes on the coordinates axis, so that the above line integral reduces to the line integral on the arc γ (with counter-clockwise orientation). A parametrization of the latter is

$$t \in [0, \frac{\pi}{2}] \to (a \cos t,\ b \sin t).$$

Hence

$$J = \frac{1}{2} \int_0^{\pi/2} a^2 b\, \cos^2 t\ \sin t\, b \cos t\, dt = \frac{a^2 b^2}{2} \int_0^{\pi/2} \cos^3 t\ \sin t\, dt = \frac{a^2 b^2}{8},$$

in accordance with the above result.

4.4.5 Exercises

Find the area $S(\overline{D})$ of the closed domains $\overline{D} \subset \mathbb{R}^2$ with the given boundary $\partial D = \gamma$.

1. γ is the *cardioid*

$$\gamma : t \in [0,\ 2\pi] \to (1 - \cos t)\, (\cos t,\ \sin t) \in \mathbb{R}^2.$$

2. γ is the *hypocycloid*

$$\gamma = \{(x, y);\ (\frac{x}{a})^{2/3} + (\frac{y}{b})^{2/3} = 1\}, \qquad a, b > 0.$$

Hint: use a trigonometric parametrization.

3. γ has the parametrization

$$t \in [-1, 1] \to (t^2 - 1)\,(1, t) \in \mathbb{R}^2.$$

4.5 Surface Integrals in $\mathbb{R}^3$

Let

$$f : \overline{D} \subset \mathbb{R}^2 \to \mathbb{R}^3 \tag{4.127}$$

be a parametrization of the surface $M \subset \mathbb{R}^3$, where $\overline{D}$ is a bounded closed domain with area in $\mathbb{R}^2$ and f is continuous (cf. Sect. 3.5).

Recall that M is the range of f in $\mathbb{R}^3$

$$M = \{f(s, t);\ (s, t) \in \overline{D}\}, \tag{4.128}$$

and is a compact set, as the continuous image of the compact set $\overline{D}$. We say that the parametrization f of M is *smooth* if f is of class C^1 and the normal vector field

$$n := f_s \times f_t \tag{4.129}$$

does not vanish on D.

The present notation is fixed in the sequel.

Surface Area

4.5.1 Definition The *area* $\sigma(M)$ of the surface M with the smooth parametrization f is defined by the double integral

$$\sigma(M) := \int_{\overline{D}} ||n||\, dS = \int\int_{\overline{D}} ||n(s, t)||\, ds\, dt.$$

Geometric motivation. Let $\{I_{ij}\}$ be a mesh as in the definition of the double integral (cf. Sects. 4.2.2 and 4.2.3). The square I_{ij} with lower left corner (s_0, t_0) is mapped by the linear approximation of f onto the parallelogram on the tangent plane to M at the point $x^0 := f(s_0, t_0)$, determined by the tangent vectors $rf_s(s_0, t_0)$ and $rf_t(s_0, t_0)$ (r is the length of the squares' edges). The area of the parallelogram is equal to the length of the cross product of these vectors, which is $r^2||n(s_0, t_0)||$ by (4.129).

The sum of these areas over all the indices of the interior squares I_{ij} is then an approximation of the area of M, and the approximation is better for smaller values of r. On the other hand, the sum is a Riemann sum for the double integral of the continuous function $||n||$ on the bounded closed domain with area $\overline{D}$, with summands corresponding to boundary squares omitted. We verify easily that the sum converges to $\int_{\overline{D}} ||n||\, dS$ as $r \to 0$.
It can be shown that the definition of $\sigma(M)$ is invariant under *permissible* change of parametrization (we shall not elaborate on the precise meaning of this statement).

4.5.2 Example A parametrization of the sphere with radius R is given by the vector field

$$f(\theta, \phi) = R\Big(\sin\theta\,\cos\phi,\ \sin\theta\,\sin\phi,\ \cos\theta\Big), \tag{4.130}$$

with

$$(\theta, \phi) \in \overline{D} := [0, \pi] \times [0.2\pi].$$

We calculate

$$n := f_\theta \times f_\phi = R^2\Big(\sin^2\theta\,\cos\phi,\ \sin^2\theta\,\sin\phi,\ \sin\theta\,\cos\theta\Big), \tag{4.131}$$

and

$$||n|| = R^2 \sin\theta. \tag{4.132}$$

Note that the condition $n \neq 0$ is satisfied on the ball, except at the North and South Poles, i.e., except for $\theta = 0,\ \pi$. However finitely many points have no influence on the definition.

We have by (4.132)

$$\sigma(M) = R^2 \int_0^{2\pi} \int_0^{\pi} \sin\theta\, d\theta\, d\phi = 4\pi R^2.$$

Surface Integral

4.5.3 Definition Let M be a surface in $\mathbb{R}^3$ with the smooth parametrization f (cf. Sect. 4.5.1 for terminology and notation). Suppose $U \subset \mathbb{R}^3$ is a domain containing M, and $g : U \to \mathbb{R}$ is continuous. The surface integral of g on M, denoted $\int_M g\, d\sigma$, is defined by

$$\begin{aligned}\int_M g\, d\sigma &:= \int_{\overline{D}} g \circ f\ ||n||\, dS \\ &= \int\!\!\int_{\overline{D}} g(f(s,t))\ ||n(s,t)||\, ds\, dt.\end{aligned} \tag{4.133}$$

By our assumptions, the integrand on the right hand side of (4.133) is continuous on $\overline{D}$, so that the double integral exists. It can be shown to be invariant under permissible change of parametrization, so that the surface integral in (4.133) depends only on M and g, and not on the particular parametrization.

Physical motivation. If we think of g as a mass (or electric charge) distribution, then for each interior square I_k of the r-mesh $\{I_{ij}\}$, $g(f(s_k, t_k))\sigma(f(I_k))$ with $(s_k, t_k) \in I_k$ is an approximation of the mass of the part M_k of M determined by the condition $(s, t) \in I_k$. Summing over all the interior squares, we get an approximation of the total mass (or electric charge) of M. The latter is the limit of these sums as $r \to 0$. On the other hand, by Definition 4.5.1 and the Mean Value Theorem for double integrals (cf. Theorem 4.2.4)

$$\sigma(f(I_k)) = \int_{I_k} ||n||\, dS = ||n(s_k', t_k')||\, S(I_k),$$

where $(s_k', t_k') \in I_k$. Choosing these points as the points (s_k, t_k) above, the sum discussed before is a Riemann sum for the integral on the right hand side of (4.133), with the omission of summands corresponding to boundary squares. Their limit as $r \to 0$ is the double integral of the *continuous* function $g \circ f\, ||n||$ on $\overline{D}$. The surface integral defined in (4.133) is interpreted in this physical context as the total mass (or electric charge) of the surface M with mass distribution (or electric charge distribution, respectively) given by the function g.

Example. Let M be the upper hemisphere with radius 1 centred at 0. We use the parametrization (4.130) (with $\theta \in [0, \pi/2]$). By (4.132), $||n(\theta, \phi)|| = \sin\theta$. Let $g(x, y, z) = (x^2 + y^2)z$, defined on the domain $U = \mathbb{R}^3 \supset M$. Then

$$\begin{aligned}\int_M g\, d\sigma &= \int_0^{2\pi} \Big(\int_0^{\pi/2} [\sin^2\theta \cos^2\phi + \sin^2\theta \sin^2\phi] \cos\theta\ \sin\theta\, d\theta\Big)\, d\phi \\ &= 2\pi \int_0^{\pi/2} \sin^3\theta \cos\theta\, d\theta = \frac{\pi}{2} \sin^4\theta\Big|_0^{\pi/2} = \frac{\pi}{2}.\end{aligned}$$

Flux of a Vector Field Through a Surface

Notation and assumptions on M are as in the preceding sections.

4.5.4 Definition Let $g : U \to \mathbb{R}^3$ be a continuous vector field. Let $\tilde{n}$ be the unit normal field on M, that is, $\tilde{n} := \frac{n}{||n||}$ (recall that $n \neq 0$ on M). The *flux* $\Phi_M(g)$ *of the vector field* g *through the surface* M is defined as the surface integral

$$\Phi_M(g) := \int_M g \cdot \tilde{n}\, d\sigma. \tag{4.134}$$

It follows from (4.133) that

$$\Phi_M(g) = \int\int_{\overline{D}} g(f(s,t)) \cdot n(s,t)\, ds\, dt. \tag{4.135}$$

Physical motivation. If g is the velocity field of a fluid flow, $\Phi_M(g)$ is the instantaneous amount of fluid going through the surface M.

Example. Let M be the upper hemisphere as in the example of Sect. 4.5.3. Consider the continuous vector field

$$g(x,y,z) = (x^2+y^2+z^2)^{-1}(1,1,1) : \mathbb{R}^3 \setminus \{(0,0,0)\} \to \mathbb{R}^3.$$

Clearly $g \circ f = (1,1,1)$. Hence by (4.135)

$$\Phi_M(g) = \int_0^{2\pi} \Big[\int_0^{\pi/2} (1,1,1) \cdot (\sin^2\theta\, \cos\phi,\, \sin^2\theta\, \sin\phi,\, \sin\theta\, \cos\theta)\, d\theta\Big]\, d\phi$$

$$= J_1 + J_2 + J_3,$$

where

$$J_1 := \int_0^{2\pi} \Big(\int_0^{\pi/2} \sin^2\theta\, d\theta\Big) \cos\phi\, d\phi = \int_0^{\pi/2} \sin^2\theta\, d\theta \int_0^{2\pi} \cos\phi\, d\phi = 0,$$

and similarly

$$J_2 := \int_0^{2\pi} \Big(\int_0^{\pi/2} \sin^2\theta\, d\theta\Big) \sin\phi\, d\phi = 0,$$

while

$$J_3 := \int_0^{2\pi} \Big(\int_0^{\pi/2} \sin\theta\, \cos\theta\, d\theta\Big)\, d\phi = \pi\, \sin^2\theta\Big|_0^{\pi/2} = \pi.$$

Hence $\Phi_M(g) = \pi$.

A generalization of Green's formula to three dimensions is a formula relating a triple integral on a *regular* bounded closed domain $\overline{V}$ in $\mathbb{R}^3$ to a *surface integral* on ∂V. Such a formula is given in the following Divergence Theorem (or Gauss Theorem).

If the partial derivatives $(g_i)_{x_i}$ of the vector field g exist at some point p^0 for all i, the *divergence* of g at p^0, $\operatorname{div} g|_{p^0}$, is defined as their sum, that is, $\operatorname{div} g|_{p^0} = \nabla \cdot g|_{p^0}$. Let g be a C^1 vector field in some domain $U \subset \mathbb{R}^3$. Then $\operatorname{div} g$ is a continuous real valued function on U, and consequently the integral of $\operatorname{div} g$ on any bounded closed domain with content $\overline{V} \subset U$ exists.

We shall not elaborate on the precise meaning of *regularity* regarding $\overline{V}$ or its boundary surface ∂V; it is a staightforward adaptation of the two-dimensional concept. A bounded closed regular domain has content. It has a partition into sub-domains of the same kind, each of which is normal of x_i-type for some $i = 1, 2, 3$ (with i depending on the sub-domain).

The Divergence Theorem

4.5.5 Theorem *Let g be a C^1 vector field in a domain $U \subset \mathbb{R}^3$. Let $\overline{V}$ be a bounded closed regular domain contained in U, whose boundary ∂V has a non-vanishing continuous normal field n. Then*

$$\Phi_{\partial V}(g) := \int_{\partial V} g \cdot \tilde{n}\, d\sigma = \int_{\overline{V}} \nabla \cdot g\, dV. \tag{4.136}$$

With the additional assumption that V is normal of x_i-type *for all* $i = 1, 2, 3$, we obtain (4.136) by applying the Fundamental Theorem of the Differential and Integral Calculus with respect to each one of the variables x_i. When V is normal of x_i-type for *some* i, we need some additional effort as in the proof of Green's formula. The general case follows then from the partition of V into subdomains of the same kind that are also normal of type-x_i for some i (see comments preceding 4.5.5).

The Divergence Theorem is also called Gauss' theorem.

Example 1. The divergence of the vector field $g(x, y, z) = (1/3)(x, y, z)$ is equal indentically to 1. We then get from (4.136)

$$\begin{aligned} \mathrm{vol}(\overline{V}) = \Phi_{\partial V}(g) &= \frac{1}{3}\int_{\partial V} (x, y, z) \cdot \tilde{n}\, d\sigma \\ &= \frac{1}{3}\int_{\partial V} (x\tilde{n}_1 + y\tilde{n}_2 + z\tilde{n}_3)\, d\sigma. \end{aligned} \tag{4.137}$$

Example 2. Let $\overline{V}$ be the closed ball with radius R centred at 0. We shall apply (4.137) to find the volume of $\overline{V}$. Using the parametrization (4.130), we get

$$\mathrm{vol}(\overline{V}) = \frac{1}{3}\int_0^{\pi}\Big[\int_0^{2\pi} R\,(\sin\theta\, \cos\phi,\ \sin\theta \sin\phi,\ \cos\theta)\cdot$$

$$R^2(\sin^2\theta\, \cos\phi,\ \sin^2\theta\, \sin\phi,\ \sin\theta\, \cos\theta)\, d\phi\Big]\, d\theta.$$

The inner product above reduces to $R^3 \sin\theta$. Hence

$$\mathrm{vol}(\overline{V}) = \frac{1}{3}\, R^3\, 2\pi \int_0^{\pi} \sin\theta\, d\theta = \frac{4\pi R^3}{3}.$$

Example 3. Let $F : \mathbb{R}^3 \to \mathbb{R}^3$ be the vector field

$$F(x, y, z) = (x^3,\ y^3,\ z^3).$$

We calculate the flux of F through the positively oriented upper sphere

$$M := \{(x, y, z) \in \mathbb{R}^3;\ x^2 + y^2 + z^2 = R^2,\ z \geq 0\}$$

of radius $R > 0$.

We apply the Divergence Theorem to the upper closed half ball

$$\overline{V} := \{(x, y, z); x^2 + y^2 + z^2 \leq R^2;\ z \geq 0\}.$$

We have

$$\partial V = M \cup M',$$

where M' is the (oriented) disc

$$M' := \{(x, y, 0); x^2 + y^2 \leq R^2\}.$$

On M', we have $F \cdot \tilde{n} = -F_3 = 0$ identically. Therefore

$$\Phi_M(F) := \int_M F \cdot \tilde{n}\, d\sigma = \int_{\partial V} F \cdot \tilde{n}\, d\sigma$$

$$= \int_{\overline{V}} \operatorname{div} F\, dV = 3 \int_{\overline{V}} (x^2 + y^2 + z^2)\, dx\, dy\, dz.$$

Using spherical coordinates, we get

$$\Phi_M(F) = 3 \int_0^{2\pi} \int_0^{\pi/2} \int_0^R r^4 dr \sin\theta\, d\theta\, d\phi = \frac{6\pi}{5} R^5.$$

Next, we calculate $\Phi_M(F)$ directly from its definition as a surface integral. We parametrize M as in (4.130)

$$(\theta, \phi) \in [0, \pi/2] \times [0, 2\pi] \to (R \sin\theta \cos\phi,\ R \sin\theta \sin\phi,\ R \cos\theta).$$

Recall that the normal field of M is given by

$$n = R^2(\sin^2\theta \cos\phi, \sin^2\theta \sin\phi, \sin\theta \cos\theta).$$

(Cf. (4.131)) Hence

$$\Phi_M(F) = R^5 \int_0^{2\pi} \left(\int_0^{\pi/2} [\sin^5\theta(\cos^4\phi + \sin^4\phi) + \cos^4\theta \sin\theta]\, d\theta \right) d\phi$$

$$= J_1 + J_2 + J_3.$$

Clearly

$$J_3 = -2\pi\, R^5 \int_0^{\pi/2} \cos^4\theta\, d(\cos\theta) = 2\pi\, R^5 \frac{\cos^5\theta}{5}\Big|_{\pi/2}^{0} = \frac{2\pi}{5} R^5.$$

The evaluations of J_1 and of J_2 are similar. We go through the details for J_1; an appropriate change of signs gives J_2. We have

$$J_1 = R^5 \int_0^{\pi/2} \sin^5\theta\, d\theta \int_0^{2\pi} \cos^4\phi\, d\phi.$$

The first integral on the right hand side is equal to

$$-\int_0^{\pi/2} (1-\cos^2\theta)^2 d(\cos\theta) = \int_0^1 (1-u^2)^2 du = \frac{8}{15}.$$

The second integral can be evaluated by using the trigonometric identity

$$8\cos^4\phi = 3 + 4\cos 2\phi + \cos 4\phi.$$

(We have a similar identity for $\sin^4\phi$, with minus signs replacing the plus signs.) Integrating over $[0, 2\pi]$, we get

$$\int_0^{2\pi} \cos^4\phi\, d\phi = \int_0^{2\pi} \sin^4\phi\, d\phi = \frac{3\pi}{4}.$$

We conclude that

$$J_1 = J_2 = J_3 = \frac{2\pi}{5} R^5,$$

and $\Phi_M(F) = \frac{6\pi}{5} R^5$, in accordance with our previous calculation.

Stokes' Formula

Green's theorem may also be generalized by replacing the plane surface $\overline{D}$ by an arbitrary surface M with smooth parametrization (cf. Sect. 4.5.1). Recall that the normal field n does not vanish on M, by definition. Therefore it makes sense to define the positive orientation of the boundary ∂M as the counter-clockwise orientation relative the direction of the normal field. In the special case where M is a plane domain $\overline{D}$ (in the xy-plane, to fix the ideas), we have the trivial parametrization

$$(x, y) \in \overline{D} \to (x, y, 0) \in M \subset \mathbb{R}^3.$$

The normal field is the constant field $n = e^3 \neq 0$, so that the above parametrization is smooth, and the positive orientation mentioned above is the usual counter-clockwise orientation.

Suppose g is a C^1 vector field in a domain $U \subset \mathbb{R}^3$ containing M. The generalization of Green's formula for the line integral $\int_{\partial M} g \cdot dx$ is the so-called *Stokes' formula* (4.138) below.

4.5.6 Stokes' Theorem *Under the preceding hypothesis,*

$$\int_{\partial M} g \cdot dx = \int_M (\nabla \times g) \cdot \tilde{n}\, d\sigma := \Phi_M(\nabla \times g). \tag{4.138}$$

The field $\nabla \times g$ is called the *curl* of the vector field g, denoted curl g.

Example 1. In the special case $M = \overline{D}$, we have $\tilde{n} = e^3$, and the integral on the right hand side of (4.138) reduces to

$$\int_{\overline{D}} (\nabla \times g)_3 dS = \int_{\overline{D}} [(g_2)_x - (g_1)_y]\, dS,$$

that is, Stokes' formula reduces in the above special case to Green's formula in $\mathbb{R}^2$. The general case itself follows from Green's formula and the chain rule. We omit the details.

Example 2. Consider the field

$$g(x, y, z) = (y^2, z^2, x^2)$$

defined on $\mathbb{R}^3$. Let M be the positively oriented surface

$$M := \{(x, y, z);\ x^2 + y^2 + z^2 = 4,\ z \geq 1\}.$$

The boundary of M is the circle

$$\gamma = \{(x, y, 1);\ x^2 + y^2 = 3\}$$

described counter-clockwise. We calculate the line integral $\int_\gamma g \cdot dx$ directly. Let us parametrize γ by

$$t \in [0, 2\pi] \to (\sqrt{3} \cos t, \sqrt{3} \sin t, 1).$$

Then

$$\int_\gamma g \cdot dx = \int_0^{2\pi} [3 \sin^2 t\, \sqrt{3}(-\sin t) + \sqrt{3} \cos t]\, dt$$

$$= \sqrt{3}[3 \cos t - \cos^3 t + \sin t]\Big|_0^{2\pi} = 0.$$

Next, we calculate the same line integral by using Stokes' formula:

$$\int_\gamma g \cdot dx = \Phi_M(\nabla \times g).$$

We have

$$(\nabla \times g)(x, y, z) = -2(z, x, y).$$

We parametrize M as in (4.130):

$$(\theta, \phi) \in [0, \pi/3] \times [0, 2\pi] \to 2(\sin\theta\cos\phi,\ \sin\theta\sin\phi,\ \cos\theta).$$

Then (cf. Sect. 4.5.2)

$$\int_\gamma g \cdot dx = -2\int_M (z, x, y) \cdot \tilde{n}\, d\sigma$$

$$= -16\int_0^{2\pi} \cos\phi\, d\phi \int_0^{\pi/3} \sin^2\theta\, \cos\theta\, d\theta$$

$$-16\int_0^{2\pi} \sin\phi\, d(\sin\phi) \int_0^{\pi/3} \sin^3\theta\, d\theta$$

$$-16\int_0^{2\pi} \sin\phi\, d\phi \int_0^{\pi/3} \sin^2\theta\, d(\sin\theta).$$

All three integrals with respect to ϕ vanish trivially. Therefore $\int_\gamma g \cdot dx = 0$, as we obtained above by direct calculation of the line integral.

4.5.7 Exercises

1. Let M be the upper hemisphere

$$M : x^2 + y^2 + z^2 = 1, \quad z \geq 0.$$

Calculate the surface integral $\int_M (x^2 + y^2) z\, d\sigma$ using the following parametrization of M:

$$(r, \phi) \in \overline{D} := [0, 1] \times [0, 2\pi] \to (r\cos\phi,\ r\sin\phi,\ \sqrt{1 - r^2}).$$

(Cf. Example in Sect. 4.5.3.)
Note that n *vanishes* on a subset of $\overline{D}$ of area zero, so this fact can be disregarded.

2. Let M be the portion of the unit sphere in $\mathbb{R}^3$ above the closed domain

$$\overline{D} := \{(x, y) \in \mathbb{R}^2;\ \frac{1}{4} \le x^2 + y^2 \le 1,\ x, y \ge 0\}.$$

Evaluate the surface integral $\int_M \arctan(y/x)\, d\sigma$. Remark as in Exercise 1. Note also that the integrand is not defined on the y-axis, but is bounded otherwise, so this fact can be disregarded as well.

3. (a) The surface M in $\mathbb{R}^3$ is given as the graph of the C^1 function

$$z = h(x, y) \qquad (x, y) \in \overline{D},$$

where $\overline{D}$ is a closed bounded domain with area in the x, y-plane. Let U be an open set in $\mathbb{R}^3$ containing M, and let $g : U \to \mathbb{R}$ be continuous. Express the surface integral $\int_M g\, d\sigma$ as a double integral over $\overline{D}$.
(b) Let M be the part of the paraboloid $z = x^2 + y^2$ in the first octant $(x, y, z \ge 0)$ below the plane $z = 1$. Calculate the surface integral

$$\int_M xy\,(1 + 4z)^{1/2} d\sigma.$$

4. Let $F = (xy,\ 0,\ -z^2)$, let $V = [0, 1]^3$, and $M = \partial V$ (oriented so that the normal points outwards). Calculate the flux $\Phi_M(F)$ in the following two ways:

   (a) by applying the Divergence Theorem;
   (b) directly.

5. Same as Exercise 4 for the vector field $F = (y^2,\ x^3,\ z)$ and V the closed domain in $\mathbb{R}^3$ bounded by the cylinder $x^2 + y^2 = R^2$ and the planes $z = 0$ and $z = R$ $(R > 0)$.
6. A smooth curve in the yz-plane is given by the parametrization

$$\gamma : t \in [a, b] \to (0,\ y(t),\ z(t)).$$

The surface M is obtained by revolving γ about the z-axis.

   (a) Show that the surface area of M is given by

$$\sigma(M) = 2\pi \int_a^b |y(t)|\,[y'(t)^2 + z'(t)^2]^{1/2} dt.$$

   Observe that if $s(\cdot)$ is the curve length function on γ, the right hand side is the line integral $2\pi \int_\gamma |y|\, ds$.
   (b) Take γ to be the circle centred at $(0,\ R,\ 0)$ with radius r $(0 < r < R)$ in the yz-plane, so that M is a torus. Find the area of the torus M.

7. Let γ be the closed curve in $\mathbb{R}^3$ given by

$$\gamma : t \in [0, 2\pi] \to (\cos t,\ \sin t,\ \cos 2t).$$

Let M be the portion of the hyperbolic paraboloid $S : z = x^2 - y^2$ with boundary γ (note that γ lies on S!). Calculate the line integral $\int_\gamma F \cdot dx$ for the vector field $F(x, y, z) = (x^2 + z^2,\ y,\ z)$ in the following two ways:

(a) using Stokes' formula;
(b) directly.

Erratum to: Several Real Variables

Shmuel Kantorovitz

S. Kantorovitz, *Several Real Variables*, Springer Undergraduate Mathematics Series, DOI 10.1007/978-3-319-27956-5

This book was inadvertently published without updating the following corrections:

Abbreviations:

p. = page
l. = line
fb = from bottom
→ = should be replaced by

Corrections:

p.6, l. before 1.1.6: omit "The conceptual importance"
p.12, (v): omit "(Fig. 1.1)"
p.12, (vi): omit "(Fig. 1.2)"
p.12, l.6 fb: omit "(Fig. 1.3)"
p.12, l.5 fb: omit "(Fig. 1.4)"
p.73, l.17: "(2.17) in Theorem 2.1.5" → (2.10)
p.75, l.13 fb: omit "in Theorem 2.1.12"

The updated original online version for these chapters can be found at
DOI 10.1007/978-3-319-27956-5_1
DOI 10.1007/978-3-319-27956-5_2
DOI 10.1007/978-3-319-27956-5_3
DOI 10.1007/978-3-319-27956-5_4
DOI 10.1007/978-3-319-27956-5

S. Kantorovitz (✉)
Bar-Ilan University, Ramat Gan, Israel
e-mail: shmuel.kantorovitz@gmail.com

S. Kantorovitz, *Several Real Variables*, Springer Undergraduate Mathematics Series, DOI 10.1007/978-3-319-27956-5_5

p.106, line before 3.2.6: (2,−1) → (3,−2)
p.134, l.11 fb: 1.3.3 → "Proposition in Sect. 1.3.7"
p.137, l.8 fb: 3.6.1 → 3.6.2
p.162, last line: "Arcoli-Arzela" → "Arzela-Ascoli"
p.169, l.9 fb: (4.7) → (4.11)
p.171, l. 3–4: "Example 4.1.5" → "Theorem 4.1.6"
p.194, l.8: "Example" → "Examples"
p.246, l.1: "$d(x, v)$" → "$d(x, y)$"
p.256, l.14: omit "below"
p.266, l.16: "$|Tu\|$" → "$\|Tu\|$"
p.290, lines 8–12: omit entirely.
p.290, l.7: add the following: "letting $\epsilon \to 0+$, we conclude that the value of the given integral is $1/2\pi - 2/\pi^3$."

Appendix A
Solutions

In order to facilitate the use of this book as a self-study text, solutions of some of the "exercises" are provided in the present section. The reader is encouraged to work his own solution before looking at the one presented below. A certain amount of repetition is allowed in order to provide mostly self-contained solutions.

Section 1.1.15

In this section, X is a metric space, with the metric d.

1. *Define*

$$d'(x, y) = \frac{d(x, y)}{1 + d(x, y)} \qquad (x, y \in X).$$

Show that d' is a metric on X.

Solution.
Clearly, $d'(x, y) = 0$ iff $d(x, y) = 0$, hence iff $x = y$.

Let b, c be non-negative numbers. Since

$$1 + b + c \le (1 + b)(1 + c),$$

$$\frac{2 + b + c}{(1 + b)(1 + c)} \le \frac{2 + b + c}{1 + b + c},$$

that is,

$$\frac{1}{1 + b} + \frac{1}{1 + c} \le 1 + \frac{1}{1 + b + c} \le 1 + \frac{1}{1 + a}$$

The original version of this chapter was revised. An erratum to this chapter can be found at DOI 10.1007/978-3-319-27956-5_5

S. Kantorovitz, *Several Real Variables*, Springer Undergraduate Mathematics Series, DOI 10.1007/978-3-319-27956-5

if $0 \le a \le b + c$. Adding 1 to both sides of the inequality and rearranging, we get

$$1 - \frac{1}{1+a} \le (1 - \frac{1}{1+b}) + (1 - \frac{1}{1+c}),$$

that is,

$$\frac{a}{1+a} \le \frac{b}{1+b} + \frac{c}{1+c} \tag{A.1}$$

for any non-negative numbers a, b, c such that $a \le b + c$.

For any $x, y, z \in X$, the non-negative numbers $a = d(x, y)$, $b = d(x, z)$, and $c = d(z, y)$ satisfy the rquirement $a \le b + c$ by the triangle inequality for the metric d. Therefore (A.1) gives the triangle inequality for $d'(x, y) := \frac{d(x,y)}{1+d(x,y)}$, and we conclude that d' is a metric on X.

2. *Let $Y \subset X$ be considered as a metric space with the metric d restricted to $Y \times Y$. Then $E \subset Y$ is open in Y iff there exists an open set V in X such that $E = V \cap Y$.*

Solution.
Let E be an open subset of Y. Then E is a union of balls in Y, say

$$E = \bigcup_y B_Y(y, r_y),$$

where

$$B_Y(y, r_y) := \{z \in Y;\ d(z, y) < r_y\} = B_X(y, r_y) \cap Y.$$

(Cf. Example 4, Sect. 1.1.10.) Hence

$$E = \bigcup_y B_X(y, r_y) \cap Y = V \cap Y,$$

where $V := \bigcup_y B_X(y, r_y)$ is an open set in X (as a union of balls in X). This proves that every open set E in Y is of the form $V \cap Y$ with V open in X.

Conversely, let $E = V \cap Y$ for V open in X. Let $x \in E$. Then $x \in Y$, and since x belongs to the open set V in X, there exists a ball $B_X(x, r)$ contained in V. Then

$$x \in B_Y(x, r) = B_X(x, r) \cap Y \subset V \cap Y = E,$$

which proves that x is an interior point of E as a subset of the metric space Y.

3. (Cf. Examples 3 and 6, Sect. 1.1.10, and Example 7, Sect. 1.1.11.)
Let $E \subset X$. Prove

(a) $\overline{E} = E \cup E'$.

Solution.
If $E \subset F$ and F is closed, then $E' \subset F' \subset F$. Hence

$$E' \subset \bigcap_{E \subset F,\, F closed} F := \overline{E}.$$

Since $E \subset \overline{E}$, we obtain the inclusion

$$E \cup E' \subset \overline{E}. \tag{A.2}$$

On the other hand, since $E'' \subset E'$, we have

$$(E \cup E')' = E' \cup E'' \subset E' \subset E \cup E',$$

that is, $E \cup E'$ is closed. Since this set contains E, it follows that $\overline{E} \subset E \cup E'$. Together with (A.2), we obtain the desired equality.

(b) $x \in \overline{E}$ *iff every ball* $B(x, r)$ *meets* E.
Solution.
Let $x \in \overline{E} = E \cup E'$ (cf. Part (a)). If $x \in E$, every ball $B(x, r)$ meets E at x. If $x \in E'$, every ball $B(x, r)$ meets E at some point $y \neq x$. In any case, every ball $B(x, r)$ meets E.

Conversely, suppose $B(x, r) \cap E \neq \emptyset$ for all $r > 0$. If for all r the intersection contains some $y \neq x$, then $x \in E'$. If there exists r for which the intersection contains only the point x, then $x \in E$. Therefore, in any case, $x \in E \cup E' = \overline{E}$ (by Part (a)).

4. *Let* $E \subset X$. *Denote by* E° *the union of all the open subsets of* X *contained in* E. *Prove*

(a) E° *is the maximal open subset of* X *contained in* E.
Solution.
If U is an open subset of X contained in E, then

$$U \subset \bigcup_{V \subset E,\, V open} V := E^\circ.$$

On the other hand, as the union of the open sets contained in E, E° is an open set contained in E.

The above statements are the meaning of the proposition: E° *is the maximal open set in* X *which is contained in* E.

If E is open (trivially contained in E), the maximality property above implies that $E \subset E^\circ$. Since $E^\circ \subset E$ for any set $E \subset X$, it follows that $E = E^\circ$. On the other hand, the latter equality implies that E is open, since E° is open for any E.

(b) $x \in E^\circ$ *iff* x *is an interior point of* E.
Solution.
If $x \in E^\circ$, then since E° is open, there exists a ball $B(x, r)$ contained in $E^\circ \subset E$, hence $B(x, r) \subset E$. This shows that x is an interior point of E.

Conversely, if x is an interior point of E, there exists a ball $B(x, r)$ contained in E. This ball is an open set contained in E, hence $B(x, r) \subset E^\circ$ by the maximality property of E°. In particular $x \in E^\circ$. We conclude that x is an interior point of E iff $x \in E^\circ$.

(c) *How does the operation $E \to E^\circ$ on subsets of X relate to the union of two sets?*
Solution.
Note first that if $A \subset B \subset X$, then $A^\circ \subset B^\circ$. Indeed, by Part (b), if $x \in A^\circ$, then x is an interior point of A. Hence there exists $r > 0$ such that $B(x, r) \subset A$. But then $B(x, r) \subset B$, hence x is an interior point of B, that is, $x \in B^\circ$ by Part (b).

Let E, F be subsets of X. Since $E \subset E \cup F$, we have $E^\circ \subset (E \cup F)^\circ$ by the preceding remark, and similarly $F^\circ \subset (E \cup F)^\circ$. Hence

$$E^\circ \cup F^\circ \subset (E \cup F)^\circ. \tag{A.3}$$

In general, we do not have equality in (A.3). As a simple example, take $E = [0, 1]$ and $F = [1, 2]$ in the metric space $\mathbb{R}$. The left hand side of (A.3) is $(0, 1) \cup (1, 2)$, while the right hand side of (A.3) is $(0, 2)$.

(d) *Denote $\partial E = \overline{E} \setminus E^\circ$. Then $x \in \partial E$ iff every ball $B(x, r)$ meets both E and E^c.*
Solution.
The statement $x \in \partial E := \overline{E} \setminus E^\circ$ is (trivially) equivalent to the validity of both statements (i)–(ii) below:

(i) $x \in \overline{E}$; and
(ii) $x \notin E^\circ$.

By Part (b) of Exercise 3, (i) is valid iff every ball $B(x, r)$ meets E.

By Part (b) above, (ii) is valid iff x is not an interior point of E, that is, iff every ball $B(x, r)$ meets E^c.

We conclude that $x \in \partial E$ iff every ball $B(x, r)$ meets both E and E^c.

5. *For $E, F \subset X$, define*

$$d(E, F) := \inf_{x \in E,\, y \in F} d(x, y).$$

In particular, $d(x, F) := d(\{x\}, F)$. Prove

(a) *$x \in \overline{E}$ iff $d(x, E) = 0$.*
Solution.
(Cf. Exercise 3(b).)
Let $x \in \overline{E}$. Then for each $n \in \mathbb{N}$, the ball $B(x, 1/n)$ meets E, say $x_n \in B(x, 1/n) \cap E$. Then

$$0 \le d(x, E) \le d(x, x_n) < \frac{1}{n}$$

for all n. Letting $n \to \infty$, we get $d(x, E) = 0$.

Conversely, if $d(x, E) = 0$ (i.e., $\inf_{y \in E} d(x, y) = 0$), then for every $r > 0$, there exists $y_r \in E$ such that $d(x, y_r) < r$. Equivalently, every ball $B(x, r)$ meets E "at y_r" (i.e., $y_r \in B(x, r) \cap E$). This shows that $x \in \overline{E}$.

(b) *The subsets* $E = \{(x, 0);\ x \geq 1\}$ *and* $F = \{(x, 1/x);\ x \geq 1\}$ *of* $\mathbb{R}^2$ *are closed, disjoint, non-empty subsets with* $d(E, F) = 0$.
Solution.
Let $p = (p_1, p_2) \in \overline{E}$. Then for all $n \in \mathbb{N}$, the ball $B(p, 1/n)$ meets E, say $(x_n, 0) \in B(p, 1/n)$, where $x_n \geq 1$. Then

$$|p_2| \leq \sqrt{(x_n - p_1)^2 + p_2^2} = d((x_n, 0),\ p) < \frac{1}{n}$$

for all n, and therefore $p_2 = 0$ and $p = (p_1, 0)$. If $p_1 < 1$ then $B(p, 1 - p_1)$ does not meet E, so that $p \notin \overline{E}$, contradicting our hypothesis. We conclude that $p = (p_1, 0)$ with $p_1 \geq 1$, that is, $p \in E$. We proved that $\overline{E} \subset E$, hence $E = \overline{E}$, and E is closed.

A similar argument shows that F is closed: if $p \in \overline{F}$, the balls $B(p, 1/n)$ meet F at points $(x_n, 1/x_n)$ with $x_n \geq 1$. Then

$$|p_1 - x_n| \leq d(p, (x_n, \frac{1}{x_n})) < \frac{1}{n}$$

for all $n \in \mathbb{N}$. Hence $p_1 = \lim x_n$, and consequently $p_1 \geq 1$. Similarly, $p_2 = \lim \frac{1}{x_n} = \frac{1}{p_1}$. Hence $p = (p_1, 1/p_1)$ with $p_1 \geq 1$, that is, $p \in F$. We proved that $\overline{F} \subset F$, and we conclude that F is closed.

Trivially, $E \cap F = \emptyset$, because $1/x \neq 0$ for all $x \geq 1$.

For all $n \in \mathbb{N}$, $(n, 0) \in E$ and $(n, 1/n) \in F$; hence

$$0 \leq d(E, F) \leq d((n, 0),\ (n, \frac{1}{n})) = \frac{1}{n}.$$

Letting $n \to \infty$, we get $d(E, F) = 0$.

6. *Denote by* $B(X)$ *the set of all bounded functions* $f : X \to \mathbb{R}$. *Define* $||f|| := \sup_{x \in X} |f(x)|$. *Prove*

(a) $B(X)$ *is a normed space for the pointwise operations and the norm* $||\cdot||$.
Solution.
By definition, the function $f : X \to \mathbb{R}$ belongs to $B(X)$ iff $||f|| < \infty$.

If $f \in B(X)$, then for all $c \in \mathbb{R}$, $|cf(x)| = |c|\,|f(x)|$, hence $||cf|| = |c|\,||f|| < \infty$. Thus $cf \in B(X)$.

If $f, g \in B(X)$, then for all $x \in X$, $|f(x)| \leq ||f||$ and $|g(x)| \leq ||g||$. Therefore $|f(x) + g(x)| \leq |f(x)| + |g(x)| \leq ||f|| + ||g||$. Taking the supremum over all $x \in X$, we get $||f + g|| \leq ||f|| + ||g||$. Hence $f + g \in B(X)$, and $||\cdot||$ is a norm on $B(X)$ (the definiteness of $||\cdot||$ is trivial). We conclude that $B(X)$ with the pointwise operations and the norm $||\cdot||$ is a normed space.

(b)–(c) *Fix $p \in X$, and for each $x \in X$, define*

$$f_x(v) := d(v, x) - d(v, p) \qquad (v \in X).$$

Prove that $f_x \in B(X)$, $|f_x(v) - f_y(v)| \le d(x, y)$, and $||f_x - f_y|| = d(x, y)$ for all $x, y, v \in X$.
Solution.
By the triangle inequality,

$$f_x(v) \le d(v, p) + d(p, x) - d(v, p) = d(p, x)$$

and

$$-f_x(v) = d(v, p) - d(v, x) \le d(v, x) + d(x, p) - d(v, x) = d(x, p).$$

Hence $|f_x(v)| \le d(x, p)$ for all $v \in X$, and consequently

$$||f_x|| \le d(x, p).$$

This shows that $f_x \in B(X)$. Since $f_x(p) = d(x, p)$, we have actually

$$||f_x|| = d(x, p). \tag{A.4}$$

If $x, y \in X$, then for all $v \in X$ we have by the triangle inequality

$$f_x(v) - f_y(v) = d(v, x) - d(v, y) \le d(x, y) + d(y, v) - d(y, v) = d(x, y). \tag{A.5}$$

Interchanging the roles of x and y, we get

$$-[f_x(v) - f_y(v)] \le d(x, y).$$

Hence $|f_x(v) - f_y(v)| \le d(x, y)$ for all $v \in X$. Therefore $||f_x - f_y|| \le d(x, y)$. However, taking $v = y$ in (A.5), we get $(f_x - f_y)(y) = d(x, y)$. Therefore $||f_x - f_y|| = d(x, y)$.

7. *Let $X^k := \{x = (x_1, \dots, x_k);\ x_i \in X\}$. For any $p \in [1, \infty)$, define*

$$d_p(x, y) = \Big(\sum_i d(x_i, y_i)^p\Big)^{1/p} \qquad (x, y \in X^k).$$

Prove

(a) *d_p is a metric on X^k.*
Solution.

The positive definiteness and symmetry of d_p are trivial. We verify the triangle inequality. Let $x, y, z \in X^k$, and denote $d(x_i, y_i) := a_i$, $d(x_i, z_i) := b_i$, and

$d(z_i, y_i) := c_i$. By the triangle inequality in X, $a_i \le b_i + c_i$ for all $i = 1, \ldots, k$. Let $a, b, c \in \mathbb{R}^k$ be defined by $a := (a_1, \ldots, a_k)$, etc. Clearly

$$d_p(x, y) = ||a||_p,\ d_p(x, z) = ||b||_p,\ d_p(z, y) = ||c||_p.$$

Since the function t^r is increasing in $[0, \infty)$ for any exponent $r > 0$, we have by Minkowski's inequality in $\mathbb{R}^k$

$$d_p(x, y) = ||a||_p = \Big(\sum_i a_i^p\Big)^{1/p} \le \Big(\sum_i (b_i + c_i)^p\Big)^{1/p}$$

$$= ||b + c||_p \le ||b||_p + ||c||_p = d_p(x, z) + d_p(z, y).$$

(b) *The metrics d_p are equivalent to one another and to the metric*

$$d_\infty(x, y) = \max_i d(x_i, y_i) \qquad (x, y \in X^k).$$

Solution.
Using the notation of Part (a), we have for all $i = 1, \ldots, k$

$$a_i \le b_i + c_i \le d_\infty(x, z) + d_\infty(z, y),$$

hence $d_\infty(x, y) \le d_\infty(x, z) + d_\infty(z, y)$. The positive definiteness and symmetry of d_∞ are obvious.

By the equivalence of the norms $||\cdot||_p$ on $\mathbb{R}^k$ for all $p \in [1, \infty]$, given any $p, p' \in [1, \infty]$, there exist positive constants L, M such that $L\,||v||_p \le ||v||_{p'} \le M\,||v||_p$ for all $v \in \mathbb{R}^k$. Taking $v = a$ (cf. Part (a)), we get

$$L\,d_p(x, y) \le d_{p'}(x, y) \le M\,d_p(x, y)$$

for all $x, y \in X^k$, that is, the metrics d_p and $d_{p'}$ on X^k are equivalent.

(c) *If X is a normed space, then X^k is a normed space for the componentwise vertor space operations and anyone of the equivalent norms*

$$||x||_p := \Big(\sum_i ||x_i||^p\Big)^{1/p} \qquad (x \in X^k),$$

$1 \le p < \infty$, *and* $||x||_\infty := \max_i ||x_i||$.
Solution.
The function $||\cdot||_p$ on X^k is obviously positive definite and homogeneous. The triangle inequality follows from Parts (a)–(b): for all $x, y \in X^k$ and $p \in [1, \infty]$,

$$||x + y||_p = d_p(x + y, 0) \le d_p(x + y, y) + d_p(y, 0) = ||x||_p + ||y||_p.$$

8. *Prove the equivalence of the following statements about X:*

(a) *X is connected.*
(b) *X is not the union of two disjoint non-empty closed sets.*
(c) *The only subsets of X that are both open and closed are $\emptyset$ and X.*
(d) *The boundary of any proper subset of X is non-empty.*

Solution.
Denote the *negation* of Statement (a) by (a′), etc. We shall prove the implications

(a′) $\Rightarrow$ (b′) $\Rightarrow$ (c′) $\Rightarrow$ (d′) $\Rightarrow$ (a′).

(a′) $\Rightarrow$ (b′).
Suppose X is *not* connected. Then there exist disjoint non-empty open sets A, B such that $X = A \cup B$. Then $A^c = B$ and $B^c = A$ are open, hence A, B are closed, and X is the union of the disjoint non-empty closed sets A, B.

(b′) $\Rightarrow$ (c′).
By (b′), $X = A \cup B$ with A, B disjoint, closed, non-empty subsets of X. Therefore A is both open (because $A^c = B$ is closed) and closed, $A \neq \emptyset$, and $A \neq X$ (because $A^c = B \neq \emptyset$). Thus (c′) is valid.

(c′) $\Rightarrow$ (d′).
By (c′), there exists $B \subset X$, both open and closed, $B \neq \emptyset$, and $B \neq X$. Since B is closed, $\overline{B} = B$. Since B is open, $B^\circ = B$ by Exercise 5(a). Hence (cf. Exercise 5(d))

$$\partial B := \overline{B} \setminus B^\circ = B \setminus B = \emptyset.$$

Thus B is a proper subset of X with empty boundary, and (d′) is proved.

(d′) $\Rightarrow$ (a′).
Let B be a proper subset of X with empty boundary. Since $B^\circ \subset B \subset \overline{B}$ and (cf. Exercise 5(d)),

$$\overline{B} \setminus B^\circ := \partial B = \emptyset,$$

we have $B = B^\circ = \overline{B}$. Hence B is both open and closed; therefore $A := B^c$ is both closed and open. Both A, B are non-empty (because B is a *proper* subset of X), and of course $X = A \cup B$. This proves (a′).

9. *The subsets $A, B \subset X$ are separated if*

$$\overline{A} \cap B = A \cap \overline{B} = \emptyset.$$

Prove that $E \subset X$ is connected iff it is not the union of two non-empty separated subsets.

Solution.
We shall apply Exercise 8. We consider $E \subset X$ as a metric space with the metric d restricted to $E \times E$. If $A \subset E$, we denote by $\overline{A}^E$ the closure of A in the metric space

E, while $\overline{A}$ is the closure of A in X. Note that $x \in \overline{A}^E$ iff every ball $B_E(x, r)$ in E meets A. Since $B_E(x, r) = B(x, r) \cap E$, the previous statement is equivalent to the statement: $x \in E$, and for every ball $B(x, r)$ in X, there exists $y \in (B(x, r) \cap A) \cap E$, that is, $x \in \overline{A} \cap E$. Hence

$$\overline{A}^E = \overline{A} \cap E. \tag{A.6}$$

Suppose $E = A \cup B$ with A, B non-empty and separated. By (A.6), since $A \subset \overline{A}$ and $\overline{A} \cap B = \emptyset$, we have

$$\overline{A}^E = \overline{A} \cap (A \cup B) = (\overline{A} \cap A) \cup (\overline{A} \cap B) = A,$$

that is, A is a closed subset of E. Similarly, B is a closed subset of E. Hence E is the union of the non-empty disjoint closed subsets A and B. Therefore E is not connected, by Exercise 8(b).

Conversely, suppose E is not connected. By Exercise 8(b), $E = A \cup B$ with A, B non-empty, disjoint, and closed in E. Hence by (A.6), since A is closed, we have

$$\overline{A} \cap B = \overline{A} \cap (E \cap B) = (\overline{A} \cap E) \cap B = \overline{A}^E \cap B = A \cap B = \emptyset,$$

and by symmetry, $\overline{B} \cap A = \emptyset$. Thus E is the union of the non-empty disjoint *separated* subsets A and B.

Section 1.2.11

1. *Suppose X is compact.*

(a) *Let $F_i \subset X$ $(i \in I)$ be closed sets such that*

$$\bigcap_{j=1}^{n} F_{i_j} \neq \emptyset \tag{A.7}$$

for any $i_j \in I$ and $n \in \mathbb{N}$. Prove that

$$\bigcap_{i \in I} F_i \neq \emptyset. \tag{A.8}$$

Solution.
Suppose

$$\bigcap_{i \in I} F_i = \emptyset.$$

Taking complements, we have by De Morgan's Laws

$$\bigcup_{i \in I} V_i = X,$$

where $V_i := F_i^c$ are open. Therefore $\{V_i;\ i \in I\}$ is an open cover of the compact space X. Let then $\{V_{i_j};\ i_j \in I,\ j = 1, \ldots, n\}$ be a finite subcovering, that is, $X = \bigcup_{j=1}^n V_{i_j}$. Taking complement, we get $\bigcap_{j=1}^n F_{i_j} = \emptyset$, contradicting the hypothesis. Hence

$$\bigcap_{i \in I} F_i \neq \emptyset.$$

(b) *In case $I = \mathbb{N}$, if F_i are non-empty closed subsets of X such that $F_{i+1} \subset F_i$ for all i, then* (A.8) *holds.*
Solution.
For any set $\{i_j;\ j = 1, \ldots, n\} \subset I := \mathbb{N}$, set $i^* := \max_{1 \le j \le n} i_j$. Since $F_{i+1} \subset F_i$ for all $i \in \mathbb{N}$, the non-empty closed subsets F_i satisfy the condition

$$\bigcap_{j=1}^n F_{i_j} = F_{i^*} \neq \emptyset.$$

Therefore (A.8) is valid, by Part (a).

2. *Prove that $E \subset X$ is compact iff every cover of E by open balls has a finite subcover.*
Solution.
Suppose every cover of $E \subset X$ by (open) balls has a finite subcover. Let $\{V_i;\ i \in I\}$ be any open cover of E. By Example 4 in Sect. 1.1.10, each open set V_i is a union of balls, say $V_i = \bigcup_{j \in J_i} B_{ij}$. Hence

$$E \subset \bigcup_{i \in I} \bigcup_{j \in J_i} B_{ij},$$

that is, $\{B_{ij};\ j \in J_i,\ i \in I\}$ is a cover of E by balls. By assumption, there exist indices $i_1, \ldots, i_n \in I$ and appropriate indices $j_1, \ldots, j_m$ such that

$$E \subset \bigcup_{k=1}^n \bigcup_{l=1}^m B_{i_k j_l}.$$

Since $B_{i_k j_l} \subset V_{i_k}$ for all $k = 1, \ldots, n$, we have $E \subset \bigcup_{k=1}^n V_{i_k}$, and we conclude that E is compact. The converse is trivial.

3. *A subset $E \subset X$ is totally bounded if for every $r > 0$, there exist $x_1, \ldots, x_n \in E$ such that*

$$E \subset \bigcup_{i=1}^n B(x_i, r). \tag{A.9}$$

Prove:

(a) *If E is compact, then it is totally bounded.*
Solution.
Suppose E is compact. For any $r > 0$, the family $\{B(x, r);\ x \in E\}$ is an open cover of E, and has therefore a finite subcover $\{B(x_i, r); x_i \in E,\ i = 1, \ldots, n\}$, i.e., E is totally bounded.

(b) *If $\overline{E}$ is compact, then E is totally bounded.*
Solution.
Suppose $\overline{E}$ is compact. By Part (a), $\overline{E}$ is totally bounded. Hence, for every $r > 0$, there exist $x_1, \ldots, x_n \in \overline{E}$ such that

$$\overline{E} \subset \bigcup_{i=1}^{n} B(x_i, \frac{r}{2}). \tag{A.10}$$

Since $x_i \in \overline{E}$, the ball $B(x_i, r/2)$ meets E at some point $y_i \in E$. For any $x \in B(x_i, r/2)$, we have

$$d(x, y_i) \le d(x, x_i) + d(x_i, y_i) < \frac{r}{2} + \frac{r}{2} = r,$$

that is, $B(x_i, r/2) \subset B(y_i, r)$, and it follows from (A.10) that

$$E \subset \overline{E} \subset \bigcup_{i=1}^{n} B(y_i, r).$$

Since $y_i \in E$, this proves that E is totally bounded.

(c) *If $E \subset \mathbb{R}^k$ is totally bounded, then $\overline{E}$ is compact.*
Solution.
Suppose $E \subset \mathbb{R}^k$ is totally bounded, and let $r > 0$. Then there exist $x^j \in E$ $(j = 1, \ldots, n)$ such that $E \subset \bigcup_{j=1}^{n} B(x^j, r)$. Therefore

$$\overline{E} \subset \overline{\bigcup_{j=1}^{n} B(x^j, r)} = \bigcup_{j=1}^{n} \overline{B(x^j, r)} \subset \bigcup_{j=1}^{n} \overline{B}(x^j, r).$$

(Cf. Exercise 5 below.)

The closed balls $\overline{B}(x^j, r)$ are closed bounded sets in $\mathbb{R}^k$, hence compact. Their finite union is then compact, and $\overline{E}$ is therefore compact as a closed subset of a compact set.

Another way to prove Part (c) is to observe that totally bounded sets are bounded, and the closure of a bounded set is bounded (both facts are true in any metric space). If $E \subset \mathbb{R}^k$ is totally bounded, then E is bounded; therefore $\overline{E}$ is bounded and closed, hence compact.

4. *The metric space X is separable if it has a dense countable subset E. Prove that every compact metric space is separable.*
Solution.
See Observation 1, Sect. 3.7.6.

5. *Prove that every bounded set in $\mathbb{R}^k$ has a compact closure.*
Solution.
If X is any metric space and $E \subset X$ is bounded, then $\overline{E}$ is bounded. Indeed, by hypothesis, $E \subset B(p, r)$ for some $p \in X$ and $r > 0$. Since $\overline{B}(p, r)$ is closed and contains $B(p, r)$, we have

$$\overline{B(p, r)} \subset \overline{B}(p, r)$$

by the minimality property of the closure. Hence

$$\overline{E} \subset \overline{B(p, r)} \subset \overline{B}(p, r) \subset B(p, 2r),$$

hence $\overline{E}$ is bounded. It is also closed.

In case $E \subset \mathbb{R}^k$ is bounded, the set $\overline{E}$ is a bounded closed set in $\mathbb{R}^k$. By the Heine-Borel Theorem, $\overline{E}$ is compact.

6. *The metric space X is locally compact if every point $x \in X$ has a compact neighborhood, that is, a compact set K and an open set U such that*

$$x \in U \subset K.$$

Let $Y \subset X$, considered as a metric space with the metric d restricted to $Y \times Y$. Prove that if X is locally compact and Y is either open or closed in X, then Y is locally compact.

Solution.
Observation: if $K \subset Y$ is compact in X, then K is compact in Y.

Indeed, let V_i $(i \in I)$ be open sets in Y such that $K \subset \bigcup_{i \in I} V_i$. There exist U_i open in X such that $V_i = U_i \cap Y \subset U_i$ (cf. Exercise 2, Sect. 1.1.15). Hence $\{U_i;\ i \in I\}$ is an open cover of K in X. By compactness of K in X, there exist $i_1, \ldots, i_n \in I$ such that $K \subset \bigcup_{j=1}^n U_{i_j}$. Since $K \subset Y$, we have

$$K \subset \Big(\bigcup_{j=1}^n U_{i_j}\Big) \cap Y = \bigcup_{j=1}^n U_{i_j} \cap Y = \bigcup_{j=1}^n V_{i_j}.$$

This proves that K is compact in the metric space Y.
Case Y open in X.

Suppose X is locally compact, and let $x \in Y$. Then there exist U open and K compact (both in X) such that $x \in U \subset K$. Since Y is open in X, there exists $r > 0$ such that

$$x \in B(x, r) \subset \overline{B}(x, r) \subset Y.$$

Then

$$x \in U \cap B(x,r) \subset K \cap \overline{B}(x,r) \subset Y. \tag{A.11}$$

The set $U \cap B(x,r)$ is open in X and contained in Y, and is therefore open in Y. The set $K \cap \overline{B}(x,r)$ is compact in X (cf. Corollary 1.2.5) and contained in Y, and is therefore compact in Y by our opening comments. We conclude then from (A.11) that Y is locally compact.
Case Y closed in X.

For any $x \in Y \subset X$, there exist U open and K compact in X such that $x \in U \subset K$. Hence

$$x \in U \cap Y \subset K \cap Y. \tag{A.12}$$

The set $U \cap Y$ is open in Y (cf. Exercise 2, Sect. 1.1.15).

The set $K \cap Y$ is compact in X by Corollary 1.2.5 and contained in Y. Hence $K \cap Y$ is compact as a subset of Y, by the opening observation. We conclude from (A.12) that Y is locally compact.

7. *Let X be a compact metric space, and let $\{V_i;\ u \in I\}$ be an open cover of X. Prove that there exists $r > 0$ such that for every $x \in X$, $B(x,r)$ is contained in some V_i.*
Solution.
Since X is compact, the open cover $\{V_i;\ i \in I\}$ has a finite subcover, say

$$X = \bigcup_{j=1}^{n} V_{i_j},$$

with $i_j \in I$. Thus for each $x \in X$, there exists $j \in \{1,\dots,n\}$ such that $x \in V_{i_j}$. Since V_{i_j} is open, there exists $r_j > 0$ such that $B(x,r_j) \subset V_{i_j}$. Define

$$r = \min_{1 \le j \le n} r_j.$$

Then $r > 0$ and

$$B(x,r) \subset B(x,r_j) \subset V_{i_j},$$

as desired.

Section 1.3.12

1. *Let (X,d) be a metric space.*

(a) *Let $E \subset X$ be compact and $F \subset X$ be closed, both non-empty. Prove that $d(E,F) = 0$ iff $E \cap F \neq \emptyset$.*
Solution.
Suppose

$$d(E,F) := \inf_{x \in E,\, y \in F} d(x,y) = 0. \tag{A.13}$$

Let $\epsilon > 0$. By (A.13), there exist sequences $\{x_n\} \subset E$ and $\{y_n\} \subset F$ such that $d(x_n, y_n) < 1/n$. Since E is compact, the sequence $\{x_n\}$ has a convergent subsequence $\{x_{n_j}\}$: $x_{n_j} \to x \in E$. Thus $d(x_{n_j}, x) < \epsilon/2$ for all $j > j(\epsilon)$ for a suitable index $j(\epsilon)$. Let

$$j^* = \max[j(\epsilon),\ 2/\epsilon].$$

Then for all $j > j^*$,

$$d(y_{n_j}, x) \le d(y_{n_j}, x_{n_j}) + d(x_{n_j}, x) < \frac{1}{n_j} + \frac{\epsilon}{2} \le \frac{1}{j} + \frac{\epsilon}{2} < \epsilon.$$

Hence $y_{n_j} \to x$, and therefore $x \in F$ since F is closed. We conclude that $x \in E \cap F$, so that $E \cap F \ne \emptyset$.

The converse is trivial: if $E \cap F \ne \emptyset$, then for any $x \in E \cap F$, $0 \le d(E, F) \le d(x, x) = 0$, hence $d(E, F) = 0$.

(b) *Part (a) is false if we replace the assumption that E is compact by the assumption that E is merely closed.*
Solution.
Take the non-empty closed sets $E = \{(x, 0);\ x \ge 1\}$ and $F = \{(x, 1/x);\ x \ge 1\}$ in $\mathbb{R}^2$ (cf. Exercise 5(b), Sect. 1.1.15). Then for all $n \in \mathbb{N}$

$$d(E, F) \le d\Big((n, 0),\ (n, \frac{1}{n})\Big) = \frac{1}{n},$$

hence $d(E, F) = 0$. However $E \cap F = \emptyset$.

2. *Let (X, d) be a metric space, and denote by $\mathcal{K}$ the family of all compact subsets of X. Prove*

(a) *$\mathcal{K}$ is closed under finite unions and intersections.*
Solution.
Using induction, it suffices to show that $E \cup F$ and $E \cap F$ are compact for compact sets E, F in X.

Any open cover $\{V_i;\ i \in I\}$ of $E \cup F$ is an open cover of both E and F. Since E, F are compact, there exist indices $i_1, \ldots, i_n \in I$ and $j_1, \ldots, j_m \in I$ such that

$$E \subset \bigcup_{k=1}^{n} V_{i_k}, \quad F \subset \bigcup_{l=1}^{m} V_{j_l}.$$

Then $\{V_{i_1}, \ldots, V_{i_n}, V_{j_1}, \ldots, V_{j_m}\}$ is a finite subcover of the given open cover of $E \cup F$. We conclude that $E \cup F$ is compact.

The compact set F is closed, hence $E \cap F$ is compact as the intersection of a compact set and a closed set (cf. Theorem 1.2.4 and Corollary 1.2.5).

(b) *$\mathcal{K}$ is not closed in general under countable unions.*
Solution.
Let $X = \mathbb{R}$ and $E_n = \{n\}$, $n \in \mathbb{N}$. Then the singletons E_n are compact for all n, but $\bigcup_n E_n = \mathbb{N}$ is not bounded, hence not compact.

(c) *If $E, F \in \mathcal{K}$, does it follow that $E \setminus F \in \mathcal{K}$?*
Solution.
Take $X = \mathbb{R}$, $E = [0, 2]$, and $F = [0, 1]$. Then E, F are compact, but $E \setminus F = (1, 2]$ is not closed, hence not compact.

3. *Let X be a normed space. If $E, F \subset X$, define*

$$E + F := \{x + y;\ x \in E,\ y \in F\}.$$

Prove:

(a) *If E is compact and F is closed, then $E + F$ is closed. Is it true that $E + F$ is compact?*
Solution.
Let p be a limit point of $E + F$. Then there exists a sequence $\{x_n + y_n\} \subset E + F$, $x_n \in E$, $y_n \in F$, such that $p = \lim(x_n + y_n)$. Since E is compact, the sequence $\{x_n\}$ has a subsequence x_{n_j} converging to some point $x \in E$. Then

$$y_{n_j} = (x_{n_j} + y_{n_j}) - x_{n_j} \to p - x.$$

Since F is closed, $p - x \in F$. Hence $p = x + (p - x) \in E + F$. This shows that $E + F$ contains all its limit points, that is, $E + F$ is closed.

In general, $E + F$ is not compact: take $E = \{x\}$ (a singleton) and $F = X$. Then E is compact, F is closed, but $E + F = X$ is not compact, since it is not bounded.

(b) *If both E and F are compact, then $E + F$ is compact.*
Solution.
We show that $E + F$ is compact by using the equivalence of compactness and sequential compactness for metric spaces (cf. Gemignani, Sect. 8.4, Proposition 16). Recall that X is sequentially compact (or has the BWP) if every sequence in X has a convergent subsequence (cf. Sect. 3.7.7). Let $\{x_n + y_n\} \subset E + F$, $x_n \in E$, $y_n \in F$. Since E is compact, there exists a subsequence $x_{n,1}$ of $\{x_n\}$ converging to some point $x \in E$. The subsequence $\{y_{n,1}\}$ (with the same indices) in the compact set F has a subsequence $\{y_{n,2}\}$ converging to some $y \in F$. Then $\{x_{n,2} + y_{n,2}\}$ is a subsequence of $\{x_n + y_n\}$ converging to $x + y \in E + F$. This proves that $E + F$ a sequentially compact metric space (with the restricted metric). Hence $E + F$ is a compact metric space, and is therefore compact in X.

Another proof, which avoids the detour through sequential compactness, uses the fact that $X \times X$ is a metric space for anyone of the equivalent metrics d_p (cf. Exercise 7, Sect. 1.1.15). To fix the ideas, let us use the metric d_1.

As in Exercise 2, Sect. 1.2.11, we show that a set $Q \subset X \times X$ is compact iff every cover of Q by open sets of the form $A_i \times B_i$, with A_i, B_i open in X, has a finite subcover.

It is clear that if $A, B \in X$ are open, then $A \times B$ is open in $X \times X$.

Let E, F be compact subsets of X. We show that $Q := E \times F \subset X \times X$ is compact. Indeed, let $\{V_i;\ i \in I\}$ be an open cover of Q by open sets of the form $V_i = A_i \times B_i$. Then $\{A_i\}$, $\{B_i\}$ are open covers of E, F (respectively). By compactness of E and F, there exist indices $i_1, \ldots, i_n \in I$ and $k_1, \ldots, k_m \in I$ such that $E \subset \bigcup_{s=1}^n A_{i_s}$ and $F \subset \bigcup_{t=1}^m B_{k_t}$. Then

$$\{A_{i_s} \times B_{i_s};\ s = 1, \ldots, n\} \cup \{A_{k_t} \times B_{k_t};\ t = 1, \ldots, m\}$$

is a finite subcover of the given cover $\{A_i \times B_i\}$ of $Q := E \times F$.

We use now the concept of continuity of a map $f : X \to Y$ between metric spaces X, Y, and the fact that for such a continuous map f, $f(Q)$ is compact in Y for Q compact in X (cf. Sect. 1.4). If X is a normed space, the map $f : X \times X \to X$ defined by $f(x, y) = x + y;\ x, y \in X$ is continuous. If E, F are compact subsets of X, then $Q := E \times F$ is compact in $X \times X$, hence $f(Q)$ is compact in X. Since

$$f(Q) = \{f(x, y);\ (x, y) \in Q\} = \{x + y;\ x \in E,\ y \in F\} = E + F,$$

we are done.

4. *Let X be a normed space, and let $\{x_i\}$ be a sequence in X. Prove:*
(a) *If X is complete, then absolute convergence of the series $\sum x_i$ implies its convergence in X.*
Solution.
Suppose $\sum x_i$ converges absolutely. Given $\epsilon > 0$, there exists $n(\epsilon) \in \mathbb{N}$ such that $\sum_{i=n+1}^m ||x_i|| < \epsilon$ for all $m > n > n(\epsilon)$. Then by the triangle inequality for norms,

$$||s_m - s_n|| = ||\sum_{i=n+1}^{m} x_i|| \le \sum_{i=n+1}^{m} ||x_i|| < \epsilon$$

for all $m > n > n(\epsilon)$. This shows that the sequence $\{s_n\}$ is Cauchy, and since X is complete, the sequence converges, that is, the series $\sum_i x_i$ converges.

(b) *If $\{x_n\}$ is Cauchy, it has a subsequence $\{x_{n_k}\}$ such that*

$$||x_{n_{k+1}} - x_{n_k}|| < \frac{1}{2^k} \qquad (k \in \mathbb{N}).$$

Solution.
By Cauchy's condition, for each $k \in \mathbb{N}$, there exists $p_k \in \mathbb{N}$ such that $||x_n - x_m|| < 2^{-k}$ for all $n > m > p_k$. Define

$$n_k = \max(p_1, \ldots, p_k) + k.$$

Then $n_{k+1} > n_k > p_k$, hence $\{x_{n_k}\}$ is a subsequence of $\{x_n\}$, and

$$||x_{n_{k+1}} - x_{n_k}|| < \frac{1}{2^k} \tag{A.14}$$

for all $k \in \mathbb{N}$.

(c) *If absolute converge of any series in* X *implies its convergence in* X, *then* X *is complete.*
Solution.
Suppose X has the property that any absolutely convergent series in X converges in X. Let $\{x_n\}$ be a Cauchy sequence in X. By Part (b), it has a subsequence $\{x_{n_k}\}$ satisfying (A.14). It follows that the series

$$\sum_{k=1}^{\infty} (x_{n_{k+1}} - x_{n_k})$$

converges absolutely, hence converges in X, by hypothesis. Therefore (setting $n_0 := 1$)

$$x_{n_j} = x_1 + \sum_{k=1}^{j} (x_{n_k} - x_{n_{k-1}})$$

converges. Thus $\{x_n\}$ is a Cauchy sequence which has a convergent subsequence. By Lemma 1.3.9, $\{x_n\}$ converges, and we conclude that X is complete.

5. *See notation in* Sect. 1.3.12, *Exercise 5.*
(a) *Prove* $||T|| = \sup_{||x|| \le 1} ||Tx|| = \sup_{||x||<1} ||Tx|| = \sup_{x \ne 0} \frac{||Tx||}{||x||}$.
Solution.
Clearly $||T|| := \sup_{||x||=1} ||Tx|| \le \sup_{||x|| \le 1} ||Tx||$, since $\{||x|| = 1\} \subset \{||x|| \le 1\}$. On the other hand, if $0 < ||x|| \le 1$, then $||tx|| = 1$ for $t := 1/||x||$. Hence

$$||Tx|| = \frac{||T(tx)||}{t} \le \frac{||T||}{t} = ||T||\,||x|| \le ||T||. \tag{A.15}$$

Therefore $\sup_{||x|| \le 1} ||Tx|| \le ||T||$, and equality follows.

The same argument gives

$$\sup_{||x||<1} ||Tx|| \le ||T||.$$

On the other hand, for each x with $||x|| = 1$ and $t \in (0, 1)$, $y := tx$ satisfies $||y|| = t < 1$. Hence $||Tx|| = (1/t)||Ty|| \le (1/t) \sup_{||y||<1} ||Ty||$. Therefore $||T|| \le (1/t) \sup_{||y||<1} ||Ty||$. Letting $t \to 1$, we get the reverse inequality.

For all $x \neq 0$ in X, $y := (1/||x||)x$ satisfies $||y|| = 1$. Hence

$$\frac{||Tx||}{||x||} = ||Ty|| \leq ||T||, \tag{A.16}$$

Hence $\sup_{x \neq 0} \frac{||Tx||}{||x||} \leq ||T||$. The reverse inequality is trivial, since $\{x;\ ||x|| = 1\} \subset \{x;\ x \neq 0\}$.

(b) By (A.16), $||Tx|| \leq ||T||\,||x||$ for all $x \neq 0$. For $x = 0$, we have equality.

(c) The triangle inequality for $T \in B(X, Y) \to ||T||$.
Let $T, S \in B(X, Y)$. For all $x \in X$ with $||x|| = 1$, we have

$$||(T + S)x|| = ||Tx + Sx|| \leq ||Tx|| + ||Sx|| \leq ||T|| + ||S||.$$

Hence $||T + S|| \leq ||T|| + ||S||$ (in particular, $T + S \in B(X, Y)$.

(d) *If* Y *is complete, so is* $B(X, Y)$.
Solution.
Suppose Y is complete, and let $\{T_n\}$ be a Cauchy sequence in $B(X, Y)$. It follows from Part (b) that $\{T_n x\}$ is Cauchy in Y, for each given $x \in X$. Since Y is complete, the latter sequence converges. Define $T : X \to Y$ by $Tx = \lim T_n x$. By the properties of limits, T is clearly linear. Let $x \in X$, $||x|| = 1$. Then for all $n \in \mathbb{N}$,

$$||T_n x|| \leq ||T_n|| \leq M,$$

where $M < \infty$, since $\{T_n\}$ is Cauchy, and Cauchy sequences are bounded (cf. Lemma 1.3.8). Since $||Tx|| = \lim ||T_n x||$, we get $||Tx|| \leq M$, and consequently $||T|| \leq M$. Thus $T \in B(X, Y)$.

Let $\epsilon > 0$. Since $\{T_n\}$ is Cauchy in $B(X, Y)$, there exists $n(\epsilon) \in \mathbb{N}$ such that $||T_n - T_m|| < \epsilon/2$ for all $m > n > n(\epsilon)$. Then for all $x \in X$ with $||x|| = 1$ and $m > n > n(\epsilon)$,

$$||T_n x - T_m x|| = ||(T_n - T_m)x|| \leq ||T_n - T_m|| < \frac{\epsilon}{2}.$$

Letting $m \to \infty$, we get

$$||T_n x - Tx|| \leq \frac{\epsilon}{2},$$

Hence $||T_n - T|| \leq \epsilon/2 < \epsilon$ for all $n > n(\epsilon)$. This proves that $T_n \to T$ in the normed space $B(X, Y)$, that is, $B(X, Y)$ is complete.

(e) *Let* $h \in X^*$ *with* $||h|| = 1$. *For each* $y \in Y$, *define* $T_y x = h(x)y$. *Prove that* $T_y \in B(X, Y)$ *with* $||T_y|| \leq ||y||$. *Conclude that if* $B(X, Y)$ *is complete, so is* Y.
Solution.
T_y is linear because h is linear, and for all $x \in X$ with $||x|| = 1$, $||T_y x|| = |h(x)|\,||y|| \leq ||h||\,||y|| = ||y||$. Hence $||T_y|| \leq ||y||$.

Assume that $B(X, Y)$ is complete.
Let $\{y_n\}$ be a Cauchy sequence in Y. Since

$$||T_{y_n} - T_{y_m}|| = ||T_{y_n - y_m}|| \le ||y_n - y_m||,$$

$\{T_{y_n}\}$ is Cauchy in $B(X, Y)$, hence converges (because $B(X, Y)$ was assumed to be complete). Let $T := \lim T_{y_n}$ (limit in $B(X, Y)$). Then for all $x \in X$,

$$Tx = \lim_n T_{y_n} x = \lim h(x) y_n.$$

Since $||h|| := \sup_{||x||=1} |h(x)| = 1$, there exists $x_0 \in X$ such that $||x_0|| = 1$ and $h(x_0) \neq 0$. Define $y := \frac{1}{h(x_0)} T x_0$. Then

$$||y_n - y|| = \frac{1}{|h(x_0)|} ||h(x_0) y_n - h(x_0) y|| = \frac{1}{|h(x_0)|} ||T_{y_n} x_0 - T x_0|| \to 0$$

as $n \to \infty$, that is, $y_n \to y$, and we conclude that Y is complete.

6. *Let (X, d) be a metric space. Prove that X is complete iff it has property (P)*
(P) If $\{F_i\}$ is a sequence of non-empty closed subsets of X such that $F_{i+1} \subset F_i$ and $\delta(F_i) \to 0$, then $\bigcap_i F_i \neq \emptyset$.
Solution.
Suppose X is complete. Let F_i, $i \in \mathbb{N}$, be non-empty closed sets in X such that $F_{i+1} \subset F_i$ and $\delta(F_i) \to 0$. For each $i \in \mathbb{N}$, we pick $x_i \in F_i$ ($F_i \neq \emptyset$!). If $j \in \mathbb{N}$ and $j > i$, then $x_j \in F_j \subset F_i$, hence $x_i, x_j \in F_i$. Therefore $d(x_i, x_j) \le \delta(F_i) \to 0$ as $i \to \infty$. This shows that $\{x_i\}$ is a Cauchy sequence. Since X is complete, $x := \lim x_i$ exists in X. Since $x_j \in F_i$ for all $j > i$ and F_i is closed, we have $x \in F_i$, for all $i \in \mathbb{N}$, that is, $x \in \bigcap_i F_i$, and X has Property (P).

Conversely, suppose X has Property (P). Let $\{x_n\}$ be a Cauchy sequence in X. Define $E_i := \{x_i, x_{i+1}, \ldots\}$ and $F_i := \overline{E_i}$. Then F_i are non-empty closed sets for all $i \in \mathbb{N}$. Since $E_{i+1} \subset E_i$, we have $F_{i+1} \subset F_i$.

Let $\epsilon > 0$. Since $\{x_n\}$ is Cauchy, there exists $n(\epsilon) \in \mathbb{N}$ such that $d(x_m, x_n) < \epsilon/4$ for all $n, m > n(\epsilon)$. Let $i > n(\epsilon)$. Suppose $x, y \in F_i := \overline{E_i}$. There exist $x', y' \in E_i$ such that $d(x, x') < \epsilon/4$ and $d(y, y') < \epsilon/4$. By definition of E_i, there exist $m, n \ge i$ such that $x' = x_m$ and $y' = x_n$. Hence $d(x', y') = d(x_m, x_n) < \epsilon/4$, because $m, n \ge i > n(\epsilon)$. Therefore

$$d(x, y) \le d(x, x') + d(x', y') + d(y', y) < \frac{3}{4}\epsilon,$$

and we conclude that $\delta(F_i) \le (3/4)\epsilon < \epsilon$ for $i > n(\epsilon)$, that is, $\delta(F_i) \to 0$. By Property (P), it follows that $\bigcap_i F_i \neq \emptyset$. Let $x \in \bigcap_i F_i$. Since $x_i, x \in F_i$ for all i, we have $d(x_i, x) \le \delta(F_i) \to 0$ as $i \to \infty$, that is, the Cauchy sequence $\{x_n\}$ converges (to x). This proves that X is complete.

7. *Let X be a metric space, and let $E \subset X$ be dense in X. Suppose that every Cauchy sequence in E converges in X. Prove that X is complete.*
Solution.
Let $\{x_n\}$ be a Cauchy sequence in X. We must show that $\{x_n\}$ converges.

Let $\epsilon > 0$, and let $n(\epsilon) \in \mathbb{N}$ be such that $d(x_n, x_m) < \epsilon/4$ for all $m > n > n(\epsilon)$.

Since $X = \overline{E}$, there exists $y_n \in E$ such that $d(x_n, y_n) < \epsilon/4$ for each $n \in \mathbb{N}$. Then by the triangle inequality,

$$d(y_n, y_m) \le d(y_n, x_n) + d(x_n, x_m) + d(x_m, y_m) < \frac{3}{4}\epsilon < \epsilon \tag{A.17}$$

for $m > n > n(\epsilon)$. Hence $\{y_n\}$ is Cauchy in E. By hypothesis, $x := \lim y_n$ exists in X. Letting $m \to \infty$ in (A.17) (with $n > n(\epsilon)$ fixed), we obtain from (A.17) that $d(y_n, x) \le (3/4)\epsilon$ for $n > n(\epsilon)$ (cf. Sect. 1.3.1). Hence

$$d(x_n, x) \le d(x_n, y_n) + d(y_n, x) < \frac{\epsilon}{4} + \frac{3}{4}\epsilon = \epsilon$$

for all $n > n(\epsilon)$. This shows that the Cauchy sequence $\{x_n\}$ converges, and we conclude that X is complete.

8. Cf. Sect. 1.3.12 for notation.
(a) *Prove that Y is a normed space, $f_x \in Y$, and $||f_x|| = 1$, where $f_x(y) := d(x, y) - d(x, x_0)$.*
Solution.
Let $f, g \in Y$, and let $h = f + g$. Then for all $x, y \in X$ $(x \neq y)$,

$$\frac{|h(x) - h(y)|}{d(x, y)} \le \frac{|f(x) - f(y)|}{d(x, y)} + \frac{|g(x) - g(y)|}{d(x, y)} \le ||f|| + ||g||,$$

hence $||h|| \le ||f|| + ||g||$. Therefore $h \in Y$ and the triangle inequality is satisfied. The routine verification of the other properties of normed spaces is omitted.

Fix $x \in X$. For $y, z \in X$ $(y \neq z)$, we have by the triangle inequality

$$f_x(y) - f_x(z) = d(x, y) - d(x, z) \le d(y, z).$$

Interchanging y and z, we get

$$-(f_x(y) - f_x(z)) \le d(y, z).$$

Hence

$$|f_x(y) - f_x(z)| \le d(y, z). \tag{A.18}$$

Since $f_x(x_0) = 0$ trivially, we conclude that $f_x \in Y$, and it follows from (A.18) that $||f_x|| \le 1$. On the other hand, if $x \neq x_0$,

$$||f_x|| \geq \frac{|f_x(x) - f_x(x_0)|}{d(x, x_0)} = 1.$$

Hence $||f_x|| = 1$ for $x \neq x_0$.
Similarly, for all $y \neq x_0$,

$$||f_{x_0}|| \geq \frac{|f_{x_0}(y) - f_{x_0}(x_0)|}{d(y, x_0)} = \frac{d(x_0, y)}{d(y, x_0)} = 1,$$

so that $||f_{x_0}|| = 1$. Hence $||f_x|| = 1$ for all $x \in X$.

(b) *Given $x \in X$, define $h_x : Y \to \mathbb{R}$ by $h_x(f) = f(x)$. Prove that $h_x \in Y^*$ with $||h_x|| \leq d(x, x_0)$, etc. (Cf.* Sect. 1.3.12, *Exercise 8.)*
Solution.
The linearity of h_x is trivial. We have for all $f \in Y$ with $||f|| = 1$

$$|h_x(f)| = |f(x)| = |f(x) - f(x_0| \leq ||f||\, d(x, x_0) = d(x, x_0),$$

hence $h_x \in Y^*$ and $||h_x|| \leq d(x, x_0)$.

(c) For all $f \in Y$ and $x, y \in X$,

$$|(h_x - h_y)(f)| = |f(x) - f(y)| \leq d(x, y)\, ||f||, \tag{A.19}$$

hence $||h_x - h_y|| \leq d(x, y)$. On the other hand, since $||f_x|| = 1$ (cf. Part (a)),

$$||h_x - h_y|| \geq |(h_x - h_y)(f_x)| = |f_x(x) - f_x(y)| = d(x, y).$$

Therefore equality holds in (A.19) for $f = f_x$, and $||h_x - h_y|| = d(x, y)$. This means that the map $\pi : x \to h_x$ is an isometry of X into the Banach space Y^* (cf. Exercise 5, Part (d)).

(d) Since $Z := \overline{\pi(X)}$ is a closed subspace of the complete normed space Y^*, Z is complete. Indeed every Cauchy sequence in Z is Cauchy in Y^*, hence converges in Y^*, and the limit is in Z since Z is closed.

10. *Let X be a compact metric space. Prove:*
(a) *X is sequentially compact, that is, every sequence in X has a convergent subsequence.*
Solution.
Let $\{x_n\} \subset X$. If the range of the sequence is finite, one of its values, say x, is repeated infinitely many times. Hence there exist indices $n_1 < n_2 < n_3, \ldots$, such that $x_{n_i} = x$ for all $i \in \mathbb{N}$, and therefore the (constant) subsequence x_{n_i} converges (to x).

If the range of the sequence is infinite, it has a limit point x, because X is compact (cf. Theorem 1.2.2). The ball $B(x, 1)$ contains some x_{n_1}. Assuming we defined $n_1, \ldots, n_k$ such that $n_1 < n_2 < \cdots, n_k$ and $x_{n_i} \in B(x, 1/i)$ for all $i = 1, \ldots, k$, we consider the ball $B(x, 1/(k+1))$. Since x is a limit point of

$\{x_n\}$, $x_n \in B(x, 1/(k+1))$ for infinitely many indices n; hence we can choose $n_{k+1} > n_k$ such that $x_{n_{k+1}} \in B(x, 1/(k+1))$. This construction by induction gives a subsequence $\{x_{n_i}\}$ such that $x_{n_i} \in B(x, 1/i)$, that is, $d(x_{n_i}, x) < 1/i$, for all $i \in \mathbb{N}$, hence $x_{n_i} \to x$.
(b) *X is complete.*
Solution.
Let $\{x_n\} \subset X$ be Cauchy. By Part (a), it has a convergent subsequence, and therefore converges, by Lemma 1.3.9.

11. *Let $E \neq \emptyset$ be a subset of the metric space X. Then $x \in \overline{E}$ iff there exists a sequence $\{x_n\} \subset E$ such that $x_n \to x$.*
Solution.
Let $x \in \overline{E}$. Then $B(x, 1/n) \cap E \neq \emptyset$ for each $n \in \mathbb{N}$. Pick x_n in this intersection. Then $x_n \in E$ and $d(x_n, x) < 1/n$, hence $x_n \to x$.

Conversely, if $x = \lim x_n$ for some sequence $\{x_n\} \subset E$, then for every $\epsilon > 0$, there exists $n(\epsilon) \in \mathbb{N}$ such that $d(x_n, x) < \epsilon$ for all $n > n(\epsilon)$, that is, $B(x, \epsilon)$ meets E at the points x_n for all such indices n. Hence $x \in \overline{E}$.

Section 1.4.21

1. *Let X, Y be metric spaces and $f : X \to Y$. Prove that f is continuous iff $f^{-1}(E)$ is closed in X for every closed subset E of Y.*
Solution.
Let f be continuous. If $E \subset Y$ is closed, E^c is open, and therefore $f^{-1}(E^c)$ is open, by Theorem 1.4.5. Thus $[f^{-1}(E)]^c$ is open, that is, $f^{-1}(E)$ is closed.

The converse is proved the same way, *mutatis mutandis*.

2. *Let X, Y be metric spaces. Prove:*
(a) *If $f : X \to Y$ is continuous, then $f(\overline{E}) \subset \overline{f(E)}$ for all $E \subset X$.*
Solution.
Let $y \in f(\overline{E})$, that is, $y = f(x)$ for some $x \in \overline{E}$. There exists a sequence $\{x_n\} \subset E$ such that $x = \lim x_n$ (cf. Exercise 11, Sect. 1.3.12). Since f is continuous, $y := f(x) = \lim f(x_n)$. Since $\{f(x_n)\} \subset f(E)$, it follows that $y \in \overline{f(E)}$.

(b) *If X is compact and $f : X \to Y$ is continuous, then $f(\overline{E}) = \overline{f(E)}$ for all $E \subset X$.*
Solution.
For all $E \subset X$, the set $\overline{E}$ is a closed subset of the compact metric space X, and is therefore compact by Theorem 1.2.3. It follows that $f(\overline{E})$ is compact, by Theorem 1.4.6, hence closed, by Theorem 1.2.4. Since $E \subset \overline{E}$, we have $f(E) \subset f(\overline{E})$, and therefore $\overline{f(E)} \subset f(\overline{E})$, by the "minimality property" of the closure operation (cf. Example 6, Sect. 1.1.10). Together with Part (a), we get the desired equality.

3. *Let (X, d) be a metric space, and let $\emptyset \neq F \subset G \subset X$, with F closed and G open. Define*

$$h(x) = \frac{d(x, F)}{d(x, F) + d(x, G^c)}.$$

Prove that $h : X \to [0, 1]$ is continuous, $F = h^{-1}(0)$, and $G^c = h^{-1}(1)$.
Solution.
Let $x, y \in X$. For all $z \in F$, we have

$$d(x, F) \le d(x, z) \le d(x, y) + d(y, z).$$

Taking the infimum over all $z \in F$, we obtain

$$d(x, F) \le d(x, y) + d(y, F),$$

that is,

$$d(x, F) - d(y, F) \le d(x, y).$$

Interchanging the roles of x and y, we obtain that the negative of the left hand side is also $\le d(x, y)$, and therefore

$$|d(x, F) - d(y, F)| \le d(x, y).$$

This implies the continuity of the function $d(\cdot, F)$ on X. Similarly, $d(\cdot, G^c)$ is continuous. Hence h is the ratio of two continuous functions. The denominator vanishes iff $d(x, F) = d(x, G^c) = 0$. Since F and G^c are closed, this happens iff both $x \in F$ and $x \in G^c$ (cf. Exercise 5(a), Sect. 1.1.15), which is impossible because $F \subset G$. Thus h is the ratio of two continuous functions with non-vanishing denominator, and is therefore continuous on X.

Clearly $h(x) \in [0, 1]$ for all $x \in X$. We have $h(x) = 0$ iff $d(x, F) = 0$, i.e., iff $x \in F$, because F is closed; hence $h^{-1}(0) = F$. Similarly, we have $h(x) = 1$ iff $d(x, G^c) = 0$, i.e., iff $x \in G^c$, because G^c is closed; hence $h^{-1}(1) = G^c$.

4. *The following real valued functions are defined on $E := \mathbb{R}^2 \setminus \{(0, 0)\}$. For each function, determine its extendability as a continuous function $\tilde{f}$ on $\mathbb{R}^2$.*

(a) $f(x, y) = \frac{x^2 y}{x^2+y^2}$.
Solution.
As a rational function, f is continuous, since its denominator does not vanish on E. Since $|2xy| \le x^2 + y^2$, we have for all $(x, y) \in E$

$$|f(x, y)| \le \frac{|x|\,|2xy|}{2(x^2 + y^2)} \le \frac{|x|}{2} \to 0$$

as $(x, y) \to (0, 0)$. Hence $f(x, y) \to 0$ as $(x, y) \to (0, 0)$. Therefore f is extendable as a continuous function $\tilde{f} : \mathbb{R}^2 \to \mathbb{R}$ by defining $\tilde{f}(0, 0) = 0$.

(b) $f(x, y) = \frac{xy}{\sqrt{x^2+ay^2}}$, $a \geq 1$.
Solution.
The function $\phi(t) = \sqrt{t}$ is continuous on $[0, \infty)$, and the polynomial $x^2 + ay^2 \geq x^2 + y^2$ is continuous and positive in E. Therefore the composite function in the denominator of f is continuous and positive on E. It follows that f is continuous on E as the ratio of two continuous functions with non-vanishing denominator. Also

$$|f(x, y)| \leq \frac{x^2 + y^2}{2\sqrt{x^2 + y^2}} = \frac{1}{2}\sqrt{x^2 + y^2} \to 0$$

as $(x, y) \to (0, 0)$. Hence $f(x, y) \to 0$ as $(x, y) \to (0, 0)$, and f is extendable as a continuous function $\tilde{f} : \mathbb{R}^2 \to \mathbb{R}$ by setting $\tilde{f}(0, 0) = 0$.

(c) $f(x, y) = (x^{2m} + y^{2m}) \log(x^2 + y^2)$, $m \in \mathbb{N}$.
Solution.
Since $\log : (0, \infty) \to \mathbb{R}$ is continuous, the composite function $(x, y) \in E \to \log(x^2 + y^2)$ is continuous, hence f is continuous on E.

By the binomial formula for $(x^2 + y^2)^m$, we have

$$|f(x, y)| \leq (x^2 + y^2)^m |\log(x^2 + y^2)| = 2r^{2m} |\log r| = 2r^{2m-1} |r \log r|,$$

where $r = ||(x, y)||_2 := \sqrt{x^2 + y^2}$. Since $r \log r \to 0$ as $r \to 0$ and $m \geq 1$, it follows that $f(x, y) \to 0$ as $(x, y) \to (0, 0)$. Therefore f is extendable as a continuous function $\tilde{f} : \mathbb{R}^2 \to \mathbb{R}$ by setting $\tilde{f}(0, 0) = 0$.

5. *Let X be a metric space and let $f : X \to \mathbb{R}$ be continuous. Suppose $\gamma_j : [0, 1] \to X$ $(j = 1, 2)$ are paths from p to q and from q to p respectively, where p, q are given points in X. Prove*

(a)–(b) *The functions $F_j := f \circ \gamma_j$ $(j = 1, 2)$ coincide at some point of* $[0, 1]$. *If $f(p) \neq f(q)$, F_j coincide at some point of* $(0, 1)$.
Solution.
The functions $F_j : [0, 1] \to \mathbb{R}$ are continuous as compositions of continuous functions. Consider the continuous function $F := F_1 - F_2 : [0, 1] \to \mathbb{R}$. We have $F(0) = f(p) - f(q)$ and $F(1) = f(q) - f(p) = -F(0)$. If $f(p) = f(q)$, $F(0) = F(1) = 0$, that is, the functions F_j $(j = 1, 2)$ coincide at the endpoints of the interval $[0, 1]$. Otherwise, F assumes non-zero values with opposite signs at the endpoints. By the IVT (for continuous functions of one real variable), there exists $t \in (0, 1)$ such that $F(t) = 0$, i.e., $F_1(t) = F_2(t)$.

6. *Let D be the domain $\{x \in \mathbb{R}^k; 2 < ||x||^2 < 3\}$, where $||\cdot||$ is the Euclidean norm. Let*

$$f(x) = \cos \frac{1}{||x||^2 - 1} \qquad (x \in \mathbb{R}^k,\ ||x|| \neq 1).$$

Prove

(a) *f is uniformly continuous in D.*
Solution.
Consider the set

$$\overline{D} := \{x \in \mathbb{R}^k;\ 2 \leq ||x||^2 \leq 3\}.$$

Since $\overline{D} = \overline{B}(0, \sqrt{3}) \cap B(0, \sqrt{2})^c$, $\overline{D}$ is closed as the intersection of two closed sets. It is also bounded, since it is contained in the ball $B(0, 2)$. Therefore $\overline{D}$ is compact, by the Heine-Borel Theorem. Since $||x||^2 - 1 \geq 1$ on $\overline{D}$, f is continuous on the compact set $\overline{D}$, as the composition of the continuous function $\cos : \mathbb{R} \to [-1, 1]$ and the rational function $x \in \overline{D} \to [||x||^2 - 1]^{-1}$. Therefore, by Theorem 1.4.11, f is uniformly continuous on $\overline{D}$, hence on its subset D.

(b) *Let X, Y be metric spaces, Y complete, and $E \subset X$. Suppose $f : E \to Y$ is uniformly continuous. Then f is extendable to a continuous function on the closure $\overline{E}$ of E.*
Solution.
Denote the metrics on X and Y by d_X and d_Y, respectively. Let $\epsilon > 0$, and let then $\delta > 0$ be such that $d_Y(f(x), f(x')) < \epsilon$ for all $x, x' \in E$ such that $d_X(x, x') < \delta$.

Let $x \in \overline{E}$. There exists a sequence $\{x_n\} \subset E$ such that $x_n \to x$. The sequence is necessarily Cauchy. Therefore there exists $n_1 \in \mathbb{N}$ such that $d_X(x_n, x_m) < \delta$ for all $m > n > n_1$. Hence $d_Y(f(x_n), f(x_m)) < \epsilon$ for all such m, n, that is, $\{f(x_n)\}$ is a Cauchy sequence in Y. Since Y is complete, $y := \lim f(x_n)$ exists. Define $\tilde{f}(x) = y$. If $\{x_n'\} \subset E$ is any sequence converging to x, and $y' := \lim f(x_n')$ as above, it follows from the triangle inequality that $d_X(x_n, x_n') \to 0$. Hence $d_X(x_n, x_n') < \delta$ for $n > n_2 \in \mathbb{N}$, and therefore $d_Y(f(x_n), f(x_n')) < \epsilon$ for all these n. Letting $n \to \infty$, we obtain $d_Y(y, y') \leq \epsilon$, hence $d_Y(y, y') = 0$ and $y = y'$. This shows that $\tilde{f}$ is well-defined at x, its definition being independent on the choice of the sequence $\{x_n\} \subset E$ converging to $x \in \overline{E}$. Clearly $\tilde{f} = f$ on E and $\tilde{f} : \overline{E} \to Y$ is continuous.

(c) *The function f of Part (a) is not uniformly continuous on the unit ball $B(0, 1)$ in $\mathbb{R}^k$.*
Solution.
Assume f is uniformly continuous on $B(0, 1)$. By Part (b), f has a continuous extension $\tilde{f}$ to the closure of the ball $B(0, 1)$, which is the closed ball $\overline{B}(0, 1)$. Consider the points

$$x_n = (\sqrt{1 - \frac{1}{2n\pi}}, 0, \ldots, 0); \quad x_n' = (\sqrt{1 - \frac{1}{2n\pi + \frac{\pi}{2}}}, 0, \ldots, 0)$$

for $n \in \mathbb{N}$ (the indices n do not refer here to components of the points in $\mathbb{R}^k$). The sequences $\{x_n\}$ and $\{x_n'\}$ are contained in $B(0, 1)$ and converge to $e^1 \in \overline{B}(0, 1)$. Since $\tilde{f}$ is continuous on $\overline{B}(0, 1)$, we obtain

$$\tilde{f}(e^1) = \lim \tilde{f}(x_n) = \lim f(x_n) = \lim \cos(-2n\pi) = 1$$

and

$$\tilde{f}(e^1) = \lim \tilde{f}(x_n') = \lim f(x_n') = \lim \cos(-2n\pi - \frac{\pi}{2}) = 0,$$

contradiction! This shows that f is not uniformly continuous on $B(0, 1)$.

7. *Let X, Y be normed spaces, and let $T : X \to Y$ be a linear map. Prove the equivalence of the following statements (a)–(d):*

(a) *T is continuous at $x = 0$.*
(b) *$T \in B(X, Y)$.*
(c) *T is Lipschitz on X.*
(d) *T is uniformly continuous on X.*

Solution.
We prove the implications
(a)⇒ (b) ⇒ (c) ⇒ (d) ⇒ (a).

(a) ⇒ (b). Balls in X are denoted $B(x, r)$ as usual (not to be confused with $B(X, Y)$!)

Take $\epsilon = 1$ in the definition of continuity at $x = 0$. There exists $\delta > 0$ such that

$$||Tu|| = ||Tu - T0|| < 1$$

for all $u \in B(0, \delta)$. If $x \in B(0, 1)$, $u := \delta x \in B(0, \delta)$, and therefore

$$||Tx|| = ||T(\frac{1}{\delta} u)|| = \frac{1}{\delta}||Tu|| < \frac{1}{\delta}.$$

Taking the supremum over all $x \in B(0, 1)$, we obtain $||T|| \leq \frac{1}{\delta} < \infty$. Hence $T \in B(X, Y)$.

(b) ⇒ (c). For all $x, x' \in X$, we have

$$||Tx - Tx'|| = ||T(x - x')|| \leq ||T||\,||x - x'||,$$

and we may take $q = ||T||$ as the Lipschitz constant.

(c) ⇒ (d). If the Lipschitz constant q is zero, $||Tx|| = ||Tx - T0|| \leq 0$, hence $Tx = 0$ for all x, so that T is trivially uniformly continuous. If $q > 0$, then for any given ϵ, choose $\delta = \frac{\epsilon}{q}$. Then for all $x, x' \in X$ such that $||x - x'|| < \delta$, we have

$$||Tx - Tx'|| \leq q\,||x - x'|| < q\delta = \epsilon.$$

(d) ⇒ (a): trivial.

8. *Let S be the unit sphere in $X := (\mathbb{R}^k, ||\cdot||_1)$.*
(a) *Prove that S is compact in X.*
Solution.
$S = \overline{B}(0,1) \cap B(0,1)^c$ is the intersection of two closed sets, hence closed; S is contained in $B(0,2)$, hence bounded. Therefore S is compact, by the Heine-Borel theorem.

(b) *Let $||\cdot||$ be an arbitrary norm on $\mathbb{R}^k$. Denote*

$$K := \max_{1\le i\le k} ||e^i||.$$

Prove that $||x|| \le K\,||x||_1$ for all $x \in \mathbb{R}^k$, and conclude that $||\cdot||$ is continuous on X. etc.
Solution.
We have for all $x \in \mathbb{R}^k$

$$||x|| = ||\sum_{i=1}^k x_i e^i|| \le \sum_i |x_i|||e^i|| \le K \sum_i |x_i| = K\,||x||_1.$$

Obviously $K > 0$ (otherwise $\mathbb{R}^k = \{0\}$!)
Let $\epsilon > 0$. Choose $\delta = \frac{\epsilon}{K}$. For any $x, x' \in X$,

$$\Big|||x|| - ||x'||\Big| \le ||x - x'|| \le K\,||x - x'||_1 < \epsilon$$

if $||x - x'||_1 < \delta$. This proves the (uniform) continuity of the function $||\cdot||$ on X.
The latter (continuous) function assumes its minimum L on the compact set S in X: $L = ||a||$ for some $a \in S$. If $L = 0$, then $a = 0$, but $0 \notin S$. Hence $L > 0$. For all $x \ne 0$, $u := \frac{1}{||x||_1}x \in S$, hence $||x|| = ||x||_1\,||u|| \ge L\,||x||_1$ (also true for $x = 0$, trivially). We proved

$$L\,||x||_1 \le ||x|| \le K\,||x||_1$$

for all $x \in \mathbb{R}^k$, that is, the norm $||\cdot||$ is equivalent to the norm $||\cdot||_1$. It follows that any two norms on $\mathbb{R}^k$ are equivalent.

9. *Suppose E is a compact convex subset of the normed space X, $0 \in E^\circ$. Define*

$$h(x) = \inf\{t > 0;\ t^{-1}x \in E\} \qquad (x \in X).$$

Prove:
$h(x+y) \le h(x) + h(y)$, $h(cx) = c\,h(x)$ for all $x, y \in X$ and $c \in [0,\infty)$, and $h(x) = 0$ iff $x = 0$.

Solution.
Let $x \in E$. Denote

$$L(x) := \{t > 0;\ \frac{x}{t} \in E\}. \tag{A.20}$$

Thus $h(x) := \inf L(x)$.

If $t \in L(x)$ and $t' > t$, then since $0 \in E$ and $\frac{x}{t} \in E$, it follows from the convexity of E that

$$\frac{x}{t'} = (1 - \frac{t}{t'})0 + \frac{t}{t'}\frac{x}{t} \in E,$$

that is, $t' \in L(x)$. This means that $L(x)$ is a ray $[h(x), \infty)$ or $(h(x), \infty)$. It follows that

$$t > h(x) \Rightarrow t \in L(x). \tag{A.21}$$

Let $x, y \in X$, $t > h(x)$, and $s > h(y)$. By (A.21), $\frac{x}{t}$, $\frac{y}{s} \in E$, hence by convexity of E,

$$\frac{x+y}{t+s} = \frac{t}{t+s}\frac{x}{t} + \frac{s}{t+s}\frac{y}{s} \in E.$$

Therefore $t + s \in L(x + y)$, and $h(x + y) := \inf L(x) \le t + s$. Since this is true for all $t > h(x)$ and $s > h(y)$, taking the infimum over all such t and s, we obtain the inequality $h(x + y) \le h(x) + h(y)$.

Since $0 \in E$, $L(0) = (0, \infty)$, and therefore $h(0) = 0$. Hence for $c = 0$, we have for all $x \in \mathbb{R}^k$, $0 = h(0) = h(0x) = 0\,h(x)$. For $c > 0$, we have (writing $ct = s$):

$$cL(x) = \{ct;\ t > 0,\ \frac{x}{t} \in E\} = \{s > 0;\ \frac{cx}{s} \in E\} = L(cx).$$

Taking the infimum, we obtain $h(cx) = c\,h(x)$.

We observed already that $h(0) = 0$. On the other hand, if $h(x) = 0$ for some $x \in X$, then $\frac{x}{t} \in E$ for all $t > 0$. The compact set E is bounded, say $||y|| < K < \infty$ for all $y \in E$. Hence for all $t > 0$,

$$||x|| = t\,||\frac{x}{t}|| < tK,$$

and therefore $x = 0$.

(b) *h is uniformly continuous on X.*
Solution.
Since $0 \in E^\circ$, there exists $r > 0$ such that $B(0, r) \subset E$. Let $0 \ne x \in X$. If $t > 1/r$, we have

$$\frac{x}{t\,||x||} \in B(0, r) \subset E,$$

hence $t\,||x|| \in L(x)$, and therefore $h(x) \le t\,||x||$. Since this is true for all $t > 1/r$, we obtain

$$h(x) \le \frac{||x||}{r} \qquad (x \in X). \tag{A.22}$$

((A.22) is also true for $x = 0$, since $h(0) = 0$.)

For any $y \in X$, we have by Part (a) and (A.22):

$$h(x+y) - h(x) \le h(x) + h(y) - h(x) = h(y) \le \frac{||y||}{r}. \tag{A.23}$$

Replacing y by $-y$, we obtain $h(x-y) - h(x) \le \frac{||y||}{r}$. Set $x' = x - y$; then $h(x') - h(x'+y) \le \frac{||y||}{r}$ for all $x' \in X$. Together with (A.23), this gives the inequality

$$|h(x+y) - h(x)| \le \frac{||y||}{r} < \epsilon$$

for any given $\epsilon > 0$, provided that $||y|| < \delta := r\epsilon$. This proves the uniform continuity of h on X.

(c) $E = h^{-1}([0,1])$.
Solution.
If $x \in E$, then $1 \in L(x)$, hence $h(x) := \inf L(x) \le 1$. This shows that $E \subset h^{-1}([0,1])$. On the other hand, if x belongs to the latter set, we have $h(x) \le 1$, hence $1 + 1/n \in L(x)$ for all $n \in \mathbb{N}$, that is, $\frac{x}{1+1/n} \in E$. Since E is compact, hence closed, $x = \lim_n \frac{x}{1+1/n} \in E$. This shows that $E = h^{-1}([0,1])$.

(d) $E^\circ = h^{-1}([0,1))$.
Solution.
The set $[0,1)$ is open in the metric space $[0,1]$ with the metric of $\mathbb{R}$. Since $h : X \to [0,1]$ is continuous by Part (b), the set $h^{-1}([0,1))$ is open, and contained in E by Part (c). By the maximality property of the operation $E \to E^\circ$, it follows that $h^{-1}([0,1)) \subset E^\circ$. On the other hand, suppose $x \in E^\circ$. If $x = 0$ ($\in E^\circ$ by hypothesis), $h(x) = 0$ (cf. Part (a)), hence $x \in h^{-1}([0,1))$ trivially. Assume that $x \ne 0$. There exists $r > 0$ such that $B(x,r) \subset E$. Let $t > (1 + r/||x||)^{-1}$. Then

$$||\frac{x}{t} - x|| = (\frac{1}{t} - 1)||x|| < r,$$

hence $\frac{x}{t} \in B(x,r) \subset E$, and therefore $h(x) \le t$. Since this is true for all such t, we conclude that

$$h(x) \le \frac{1}{1 + \frac{r}{||x||}} < 1,$$

therefore $E^\circ \subset h^{-1}([0,1))$, and the desired equality follows.

10. *Let E be any non-empty subset of* $\{0\} \times [-1, 1]$ *and let*

$$Y = \{(x, \sin \frac{1}{x});\ x > 0\} \subset \mathbb{R}^2.$$

Prove

(a) $Y \cup E$ *is connected.*
Solution.
The vector valued function

$$f : x \in (0, \infty) \to (x, \sin \frac{1}{x}) \in \mathbb{R}^2$$

is continuous, since both its components are continuous. The set $(0, \infty) \subset \mathbb{R}$ is connected, hence $f((0, \infty)) = Y$ is connected, by Theorem 1.4.12.

We show that $Q := \{0\} \times [-1, 1] \subset Y'$. Let $a \in [-1, 1]$, and let $t \in [-\pi/2, \pi/2]$ be such that $\sin t = a$. Define $x_n = (t + 2n\pi)^{-1}$ $(n \in \mathbb{N})$. Then $p_n := (x_n, \sin(1/x_n)) \in Y$ and

$$p_n = (\frac{1}{t + 2n\pi}, \sin(t + 2n\pi)) = (\frac{1}{t + 2n\pi}, \sin t) \to (0, \sin t) = (0, a) \in Q.$$

Hence $Q \subset Y'$. Therefore

$$Y \subset Y \cup E \subset Y \cup Q \subset Y \cup Y' = \overline{Y}.$$

Since Y is connected, it follows that $Y \cup E$ is connected by Theorem 1.4.20.

(b) $Y \cup \{(0, 0)\}$ *is not pathwise connected.*
Solution.
Suppose that $Y \cup \{(0, 0)\}$ is pathwise connected. We shall reach a contradiction. Fix $c > 0$. There is a path $\gamma \subset Y$ connecting the points $(c, \sin(1/c)) \in Y$ and $(0, 0)$. The range of γ with $(0, 0)$ excluded lies in Y, hence coincides with the graph of $\sin(1/x)$ for $0 < x \le c$, that is, $\gamma(0) = 0$ and $\gamma(x) = \sin(1/x)$ for $0 < x \le c$. Since γ is continuous (by definition of paths!), $\lim_{x\to 0+} \gamma(x) = 0$. Consider however the sequences $x_n = 1/2n\pi$ and $y_n = 1/(2n\pi + \pi/2)$, both in $(0, c)$ for integers $n > 1/(2\pi c)$. Then $\gamma(x_n) = \sin(2n\pi) = 0 \to 0$ and $\gamma(y_n) = \sin(\pi/2+2n\pi) = 1 \to 1$, so that the limit $\lim_{x\to 0+} \gamma(x)$ does not exist, contradiction!

The set $Y \cup \{(0, 0)\}$ is an example of a connected set (Part (a)) which is not pathwise connected (Part (b)).

Section 2.1.15

2. *Let* $f(0, 0) = 0$ *and* $f(x, y) = \frac{x^2 y^2}{x^4+y^4}$ *for* $(x, y) \neq (0, 0)$ *in* $\mathbb{R}^2$.

(a) *Show that for all unit vectors* $u \in \mathbb{R}^2$, $\frac{\partial f}{\partial u}(0,0) = 0$.
Solution.
For $h \neq 0$ real, we have $\frac{f(h,0)-f(0,0)}{h} = 0 \to 0$ as $h \to 0$. Hence $f_x(0,0) = 0$, and by symmetry, $f_y(0,0) = 0$. Hence $\nabla f(0,0) = (0,0)$ and $D_u f(0,0) = u \cdot \nabla f(0,0) = 0$ for all unit vectors $u \in \mathbb{R}^2$.

(b) *Show that* f *is not continuous (hence not differentiable!) at* $(0,0)$.
Solution.
For all $x \neq 0$, we have $f(x,x) = 1/2 \to 1/2$ and $f(x,0) = 0 \to 0$ as $x \to 0$. This shows that $\lim f(x,y)$ as $(x,y) \to (0,0)$ does not exist, and therefore f is not continuous at the point $(0,0)$.

3. *Let* g *be a real valued function defined in a neighborhood of* $a \in \mathbb{R}$, $a \neq 0$, *and differentiable at* a. *Let* $x_0 \in \mathbb{R}^k$ *be such that* $||x_0|| = a$, *and define* $f(x) := g(||x||)$ *for* x *in a neighborhood of* x_0. *Prove*
(a) $\frac{\partial f}{\partial u}(x_0) = \frac{g'(a)}{a}\, u \cdot x_0$.
Solution.
Note first that for $x \neq 0$ and $i = 1, \ldots, k$,

$$\frac{\partial ||x||}{\partial x_i} = \frac{\partial}{\partial x_i}\Big(\sum_j x_j^2\Big)^{1/2} = \Big(\sum_j x_j^2\Big)^{-1/2} x_i = \frac{x_i}{||x||}.$$

Hence $\nabla ||x|| = \frac{x}{||x||}$ is a unit vector.
By the chain rule,

$$f_{x_i} = g'(||x||)\,\frac{\partial ||x||}{\partial x_i} = g'(||x||)\,\frac{x_i}{||x||},$$

hence

$$\nabla f(x) = g'(||x||)\frac{x}{||x||} \tag{A.24}$$

in a neighborhood of x_0. Therefore, in a neighborhood of x_0,

$$\frac{\partial f}{\partial u}(x) = u \cdot \nabla f(x) = \frac{g'(||x||)}{||x||}\, u \cdot x.$$

At $x = x_0$, we obtain

$$D_u f(x_0) = [g'(a)/a]\, u \cdot x_0. \tag{A.25}$$

(b) $\max_u |D_u f(x_0)| = |g'(a)|$, *and the maximum is attained when* $u = x_0/a$.
Solution.
The desired maximum is attained for

$$u^* = \frac{\nabla f}{||\nabla f||}\Big|_{x_0} = \frac{x_0}{a}.$$

(Cf. (A.24).)

We then have by (A.25)

$$\max_u |D_u f(x_0)| = |D_{u^*} f(x_0)| = |\frac{g'(a)}{a} u^* \cdot x_0| = |g'(a)|.$$

6. *Let $f : \mathbb{R}^k \to \mathbb{R}^m$ be defined by*

$$f(x) = \left(\sum x_i, \sum x_i^2, \ldots, \sum x_i^m\right).$$

Given $x \in \mathbb{R}^k$, what is the linear map $df(x) : \mathbb{R}^k \to \mathbb{R}^m$?
Solution.
The Jacobian matrix $\frac{\partial f}{\partial x}\Big|_x$ has the rows $(1, \ldots, 1)$, $2(x_1, \ldots, x_k), \ldots, m(x_1^{m-1}, \ldots, x_k^{m-1})$.

Writing vectors as columns, we have for all $h \in \mathbb{R}^k$

$$df|_x h = \frac{\partial f}{\partial x}\Big|_x h = \left(\sum h_i, 2\sum x_i h_i, \ldots, m \sum x_i^{m-1} h_i\right)^t.$$

7. *Let $f : \mathbb{R}^3 \to \mathbb{R}$ be defined by*

$$f(x, y, z) = \log(x^2 + y^2 + 4z^2) \qquad (x, y, z) \neq (0, 0, 0).$$

Let $g : \mathbb{R}^3 \to \mathbb{R}^3$ be defined by

$$g(r, \phi, z) = (r \cos\phi, r \sin\phi, z) \qquad (r > 0, 0 \leq \phi \leq 2\pi, z \in \mathbb{R}).$$

(a) *Find the partial derivatives of $f \circ g$ with respect to r, ϕ, z for $r^2 + z^2 > 0$ and $0 \leq \phi \leq 2\pi$.*
Solution.
The C^1 condition for the application of the chain rule is satisfied. Hence

$$\frac{\partial(f \circ g)}{\partial r} = \frac{2x}{x^2 + y^2 + 4z^2} \cos\phi + \frac{2y}{x^2 + y^2 + 4z^2} \sin\phi = \frac{2r}{r^2 + 4z^2},$$

$$\frac{\partial(f \circ g)}{\partial \phi} = \frac{2x}{x^2 + y^2 + 4z^2}(-r \sin\phi) + \frac{2y}{x^2 + y^2 + 4z^2}(r \cos\phi) = 0.$$

and

$$\frac{\partial(f \circ g)}{\partial z} = \frac{8z}{r^2 + 4z^2}.$$

(b) (This part is trivial.)

8. *Let* $f : \mathbb{R}^2 \to \mathbb{R}$ *be differentiable at the point* $(x, y) \neq (0, 0)$. *Define* $g(r, \theta) = f(r\cos\theta, r\sin\theta)$. *Prove the formula*

$$||\nabla f\Big|_{(x,y)}||^2 = g_r^2 + r^{-2}g_\theta^2,$$

where $x = r\cos\theta,\ y = r\sin\theta,\ r > 0,\ 0 \le \theta \le 2\pi$.
Solution.
The chain rule is applicable. We have

$$g_r^2 = (f_x\cos\theta + f_y\sin\theta)^2,$$

$$r^{-2}g_\theta^2 = (-f_x\sin\theta + f_y\cos\theta)^2.$$

Therefore

$$g_r^2 + r^{-2}g_\theta^2 = (f_x^2 + f_y^2)(\cos^2\theta + \sin^2\theta) = ||\nabla f\Big|_{(x,y)}||^2,$$

with $(x, y) = (r\cos\theta,\ r\sin\theta)$.

9. *Prove or disprove the differentiability at* $(0, 0)$ *of the following functions* $f : \mathbb{R}^2 \to \mathbb{R}$:
(a) $f(x, y) = \frac{x^2-y^2}{||(x,y)||}$ *for* $(x, y) \neq (0, 0)$ *and* $f(0, 0) = 0$.
Solution.
We have for real $h \neq 0$

$$\frac{f(h, 0) - f(0, 0)}{h} = h^{-1}\frac{h^2}{|h|} = \frac{h}{|h|} := sign\, h,$$

where $sign\, h$ (the sign function) equals 1 for $h > 0$ and -1 for $h < 0$. The limit of the above differential ratio as $h \to 0$ does not exist (since the $\lim_{h\to 0+} = 1 \neq -1 = \lim_{h\to 0-}$), that is, $f_x(0, 0)$ does not exist. Therefore f is not differentiable at $(0, 0)$.

(b) $f(x, y) = (xy)^{2/3}$.
Solution.
For $h \neq 0$, we have $h^{-1}[f(h, 0) - f(0, 0)] = 0 \to 0$, hence $f_x(0, 0) = 0$ and by symmetry, $f_y(0, 0) = 0$. Thus $\nabla f(0, 0) = 0$, and the "candidate" for $df(0, 0)$ is then the zero linear functional $L = 0$ (for all $h = (h_1, h_2) \in \mathbb{R}^2$, $df(0, 0)h = h \cdot \nabla f(0, 0) = 0$ if $df(0, 0)$ exists). Consider the error function $\phi(f)|_{(0,0)} := f(h_1, h_2) - f(0, 0) - Lh = (h_1h_2)^{2/3}$. Since $h_i^2 \le ||h||^2$, we have

$$\frac{\left|\phi(f)|_{(0,0)}\right|}{||h||} \le ||h||^{1/3} \to 0$$

as $h \to 0$. This proves that f is differentiable at $(0, 0)$ and $df(0, 0) = 0$.

10. *Suppose the real valued function f is defined in a neighborhood of $(0, 0)$ in $\mathbb{R}^2$, and satisfies the following conditions:*

(i) *f is differentiable at $(0, 0)$.*
(ii) $\lim_{x\to 0} x^{-1}[f(x, x) - f(x, -x)] = 1$.

Find $f_y(0, 0)$.
Solution.
By (i),

$$f(x, y) - f(0, 0) - xf_x(0, 0) - yf_y(0, 0) = o(||(x, y)||)$$

as $(x, y) \to (0, 0)$. Take in particular the points (x, x) and $(x, -x)$ as $x \to 0$. Then

$$f(x, x) - f(0, 0) - x[f_x(0, 0) + f_y(0, 0)] = o(|x|)$$

and

$$f(x, -x) - f(0, 0) - x[f_x(0, 0) - f_y(0, 0)] = o(|x|).$$

Subtracting these last two relations, dividing by $x \neq 0$, and using (ii), we obtain as $x \to 0$: $1 - 2f_y(0, 0) = 0$, i.e., $f_y(0, 0) = 1/2$.

Section 2.2.17

The exercises are routine applications of the techniques studied in this section. Solutions are given below for some of them.

4. *Let $D \subset \mathbb{R}^k$ be an open star-like set (that is, there exists a point $p \in D$ such that the line segment $\overline{px}$ is contained in D for all $x \in D$). Suppose $f : D \to \mathbb{R}$ is of class C^1 and $||\nabla f|| \leq M$ in D. Prove that $|f(x) - f(p)| \leq M\,||x - p||$ for all $x \in D$.*
Solution.
Let $x(t) := p + t(x - p)$ ($t \in [0, 1]$) be a parametrization of $\overline{px}$ for any given $x \in D$. By the chain rule,

$$f(x(\cdot))'(t) = \nabla f\Big|_{x(t)} \cdot x'(t) = \nabla f\Big|_{x(t)} \cdot (x - p)$$

for all $t \in [0, 1]$. For $t = 1$, we obtain

$$f(x(\cdot))'(1) = \nabla f(x) \cdot (x - p) \qquad (x \in D).$$

By the MVT for functions of one variable, there exists $\theta \in (0, 1)$ such that

$$f(x) - f(p) = f(x(1)) - f(x(0)) = f(x(\cdot))'(\theta)(1 - 0) = \nabla f\Big|_{x(\theta)} \cdot (x - p).$$

Since $x(\theta) \in \overline{px} \subset D$, it follows from Schwarz' inequality and our hypothesis that

$$|f(x) - f(p)| \le ||\nabla f\Big|_{x(\theta)}||\, ||x-p|| \le M\, ||x-p||.$$

5. *Answer.* $x^2 + y^2$.

7. *Let*

$$f(x, y, z) = x^4 + y^4 + z^4 - x^2 - y^2 - z^2 - 2xy - 2xz - 2yz.$$

Find all the critical points of f and determine wether they are local maxima, minima, or neither.
Solution.
We have

$$\nabla f = 4(x^3, y^3, z^3) - 2(x + y + z)(1, 1, 1).$$

The equation $\nabla f = 0$ is then equivalent to the equations

$$2x^3 = 2y^3 = 2z^3 = x + y + z.$$

Equivalently, $x = y = z$, and the common value t satisfies $2t^3 = 3t$, that is, either $t = 0$ or $t = \pm\sqrt{3/2}$. The critical points of f are therefore

$$(0, 0, 0),\ \pm\sqrt{\frac{3}{2}}(1, 1, 1).$$

The Hessian matrix has the rows $2(6x^2 - 1, -1, -1)$, $2(-1, 6x^2 - 1, -1)$, $2(-1, -1, 6x^2 - 1)$. Its principal minors at the points $\pm\sqrt{3/2}(1, 1, 1)$ have the values $M_1 = 16$, $M_2 = 252$, $M_3 = 3888$. Since they are all positive, the points $\pm\sqrt{3/2}(1, 1, 1)$ are local minima.

The principal minors test fails for the critical point $(0, 0, 0)$, because $M_2 = 0$ at this point (also $M_3 = 0$ at this point, but this information is not needed for the above conclusion). We have $f(x, 0, 0) = x^4 - x^2 = x^2(x^2 - 1) < 0 = f(0, 0, 0)$ for $0 < |x| < 1$ and

$$f(x, y, z) = x^4 + y^4 + z^4 - (x + y + z)^2 = x^4 + y^4 + (x + y)^4 > 0 = f(0, 0, 0)$$

for $(x, y, z) \neq (0, 0, 0)$ on the plane through the origin with the equation $x + y + z = 0$. This shows that the point $(0, 0, 0)$ is not a local extremum of f.

8. *Same problem as No. 7 for the following functions.*
(c) $f(x, y) = x^2y^3(2 - x - y)$, $x, y > 0$.
Solution.
We have $f_x = xy^3(4 - 3x - 2y)$, $f_y = x^2y^2(6 - 3x - 4y)$. For $x, y > 0$, $\nabla f = 0$ iff $3x + 2y = 4$ and $3x + 4y = 6$. Hence f has a unique critical point in the given

domain, namely the point (2/3, 1). The Hessian matrix at this point has the rows $-(2/9)(9, 6)$ and $-(2/9)(6, 8)$. Its principal minors are $M_1 = -2$ and $M_2 = 48/27$. Therefore the point (2/3, 1) is a local maximum of f.

(d) $f(x, y) = (x^2 + y^2)e^{-(x^2+y^2)}$ *in the unit disc* $D := \{(x, y);\ x^2 + y^2 < 1\}$.
Solution.
We calculate

$$\nabla f(x, y) = 2e^{-(x^2+y^2)}(1 - x^2 - y^2)(x, y).$$

In the domain D, $\nabla f(x, y) = (0, 0)$ iff $(x, y) = (0, 0)$, that is, the point (0, 0) is the unique critical point of f in D. The Hessian matrix at (0, 0) equals $2I$ (where I stands for the 2×2 identity matrix). Its principal minors are both positive. Therefore (0, 0) is a local minimum of f in D. Since $f \geq 0$ (on $\mathbb{R}^2$), the point (0, 0) is actually the absolute minimum of f (in $\mathbb{R}^2$!)

9. *Let* $f(x, y) = (y - x^2)(y - 3x^2)$. *Let* $g(t) = (a, b)t$ ($t \in \mathbb{R}$; $(a, b) \in \mathbb{R}^2$ *constant,* $(a, b) \neq (0, 0)$) *be any line through* (0, 0). *Prove*
(a) $f \circ g$ *has a local minimum at* $t = 0$.
Solution.
We have

$$(f \circ g)(t) = (bt - a^2t^2)(bt - 3a^2t^2) = b^2t^2 - 4a^2bt^3 + 3a^4t^4.$$

Hence

$$(f \circ g)'(t) = 2t(b^2 - 6a^2bt + 6a^4t^2),$$

so that $t = 0$ is indeed a critical point of $f \circ g$. Recall that $(a, b) \neq (0, 0)$ (so that g indeed defines a line!). If $b \neq 0$, we have $(f \circ g)''(0) = 2b^2 > 0$. If $b = 0$, then $a \neq 0$, and we have $(f \circ g)^{(4)}(0) = 72\,a^4 > 0$, while all lower order derivatives vanish at $t = 0$. In any case, we conclude that $f \circ g$ has a local minimum at $t = 0$.

(b) (0, 0) *is a critical point of* f *but is not a local minimum of* f.
Solution.
Write $f(x, y) = y^2 - 4x^2y + 3x^4$. Then

$$\nabla f\Big|_{(0,0)} = (-8xy + 12x^3,\ 2y - 4x^2)\Big|_{(0,0)} = (0, 0).$$

Hence (0, 0) is a critical point of f.

On the parabolas through (0, 0) with the equations $y = 2x^2$ and $y = 4x^2$, we have for $x \neq 0$ $f(x, 2x^2) = -x^4 < 0 = f(0, 0)$ and $f(x, 4x^2) = 3x^4 > 0 = f(0, 0)$, respectively. This shows that (0, 0) is not a local extremum of f (hence not a local minimum!).

10. *Among all rectangular boxes whose edges have a given total length, find the box with maximal volume.*
Solution.
Denote the length of the edges by x_i $(i = 1, 2, 3)$. Since the volume of the box vanishes if $x_i = 0$ fo some i, and 0 is certainly not the maximal volume, we may assume that $x_i > 0$ for all i. Let c be the given total length of the edges. Then $x_3 = c - x_1 - x_2$, and the volume of the box is given by the function

$$f(x_1, x_2) = x_1x_2(c - x_1 - x_2) = cx_1x_2 - x_1^2x_2 - x_1x_2^2, \qquad x_1, x_2 > 0,\ x_1 + x_2 < c.$$

We have

$$\nabla f(x_1, x_2) = (cx_2 - 2x_1x_2 - x_2^2,\ cx_1 - x_1^2 - 2x_1x_2).$$

Since $x_i > 0$ for $i = 1, 2$, the equation $\nabla f(x_1, x_2) = (0, 0)$ reduces to the equation

$$(c - 2x_1 - x_2,\ c - x_1 - 2x_2) = (0, 0),$$

whose unique solution is $(x_1, x_2) = \frac{c}{3}\,(1, 1)$.

We calculate the Hessian at this point: its rows are $-(c/3)(2, 1)$ and $-(c/3)(1, 2)$. Its principal minors are $M_1 = -2c/3 < 0$ and $M_2 = c^2/3 > 0$. Therefore the point $(c/3)\,(1, 1)$ is a local maximum of f. Actually, this point is the global maximum point of f. To see this, we observe that (x_1, x_2) varies in the open square $(0, c)^2$, because $0 < x_i < x_1 + x_2 + x_3 = c$. The continuous function $f \geq 0$ attains its (global) maximum on the compact set $[0, c]^2$. Since $f = 0$ on the boundary of that closed square, the global maximum is attained in the interior, and is therefore a critical point of f. The uniqueness of the critical point implies that it is the global maximum point of f. We conclude that the box with maximal volume (with total edges length c) is the cube with edges equal to $c/3$.

14. *Among all polygons with a given number $n \geq 3$ of edges inscribed in a circle whose center O is inside the polygon, find the polygon with maximal area.*
Solution.
We partition such polygons into triangles with common vertex at O, with their other vertices on the given circle of radius r. The variables x_i are the angles of the triangles at the vertex O, $i = 1, \ldots, n$. Since $\sum x_i = 2\pi$ and O is assumed to be inside the polygon, we have $0 < x_i < \pi$ for all i, and $x_n = 2\pi - \sum_{i=1}^{n-1} x_i$.

The triangle with the angle x_i at O has the area $2r\cos(x_i/2)\,r\sin(x_i/2) = r^2 \sin x_i$. Therefore the area of the polygon divided by r^2 is given by the function

$$f(x) = \sum_{i=1}^{n-1} \sin x_i + \sin(2\pi - \sum_{i=1}^{n-1} x_i) = \sum_{i=1}^{n-1} \sin x_i - \sin(\sum_{i=1}^{n-1} x_i),$$

where $x := (x_1, \ldots, x_{n-1})$ varies in the domain $D \subset \mathbb{R}^{n-1}$ given by

$$D := \{x \in \mathbb{R}^{n-1}; 0 < x_i < \pi,\ \sum x_i < 2\pi\}.$$

We have

$$\nabla f(x) = (\cos x_1, \ldots, \cos x_{n-1}) - \cos(\sum_{i=1}^{n-1} x_i)(1, \ldots, 1).$$

Therefore $\nabla f(x) = 0$ iff

$$\cos x_j = \cos(\sum_i x_i) = \cos(2\pi - x_n) = \cos x_n$$

for all $j = 1, \ldots, n-1$. Since the cosine function is strictly decreasing in the interval $(0, \pi)$, this happens iff all the angles x_j $(j = 1, \ldots, n)$ are equal (to $2\pi/n$).

The rows of the Hessian $(n-1) \times (n-1)$ matrix are

$$-\sin x_i e^i + \sin(\sum_{i=1}^{n-1} x_i)\,(1, \ldots, 1) = -\sin x_i e^i - \sin x_n\,(1, \ldots, 1).$$

At the critical point, $x_i = 2\pi/n$ for all $i = 1, \ldots, n$. Therefore the Hessian at the critical point, H, is equal to $-\sin(2\pi/n)(I_{n-1} + A_{n-1})$, where I_{n-1} is the $(n-1) \times (n-1)$ identity matrix and the entries of the $(n-1) \times (n-1)$ matrix A_{n-1} are all equal to 1. Let M_j be the j-th principal minor of H. We have for all $j = 1, \ldots, n-1$

$$(-1)^j M_j = \sin^j \frac{2\pi}{n} \det(I_j + A_j).$$

We verify by induction on $j \le n-1$ that $\det(I_j + A_j) = j + 1$. This is trivial for $j = 1$. Assume $\det(I_{j-1} + A_{j-1}) = j$ for some $j \le n-1$. In $\det(I_j + A_j)$, subtract the second row from the first. The resulting first row is $(1, -1, 0, \ldots, 0)$. Expanding the determinant according to its first row, we get

$$\det(I_j + A_j) = Q_1 + Q_2,$$

where Q_1, Q_2 are the minor corresponding to 1 and -1, respectively. Clearly $Q_1 = \det(I_{j-1} + A_{j-1}) = j$, by the induction hypothesis. The rows of Q_2 are

$$(1, \ldots, 1),\ (1, 2, 1, \ldots, 1), \ldots, (1, \ldots, 1, 2).$$

If we subtract the first row from all the other rows, we obtain a determinant with the entry 1 in the diagonal, and the entry 0 under the diagonal. Hence $Q_2 = 1$, and we conclude that $\det(I_j + A_j) = j + 1$ as desired. Hence

$$(-1)^j M_j = \sin^j \frac{2\pi}{n}(j + 1) > 0 \qquad (j = 1, \ldots, n-1)$$

since $0 < 2\pi/n \leq 2\pi/3$, and we conclude that the critical point $(2\pi/n)(1, \ldots, 1)$ is a local maximum of f, hence of the area of the polygon. An argument similar to the one we did in the solution of Exercise 10 shows that the critical point is a global maximum point for the area.

Section 3.2.6

3. (a) *Show that the equation*

$$e^{z-x-y^2} + z^3 + 2z = \cos(x^2 + 4xy - z) \tag{A.26}$$

defines a unique real-valued function $z = \phi(x, y)$ *of class* C^1 *in a neighborhood of the point* $(0, 0)$ *in* $\mathbb{R}^2$, *such that* $\phi(0, 0) = 0$.
Solution.
Let

$$F(x, y, z) := e^{z-x-y^2} + z^3 + 2z - \cos(x^2 + 4xy - z) : \mathbb{R}^3 \to \mathbb{R}.$$

We verify that F satisfies the conditions of the IFT (and even those of its "scholium") in $\mathbb{R}^3$:

(i) F is of class C^1 in $\mathbb{R}^3$;
(ii) $F(0, 0, 0) = 0$;
(iii) $F_z(0, 0, 0) = 3 \neq 0$.

By the said theorem(s), there exists a neighborhood V of $(0, 0)$ in $\mathbb{R}^2$ and a unique function $z = \phi(x, y)$ of class C^1 defined in V, such that $\phi(0, 0) = 0$ and $F(x, y, \phi(x, y)) = 0$ for all $(x, y) \in V$, that is, $z = \phi(x, y)$ solves (A.26) in an $\mathbb{R}^3$-neighborhood of the point $(0, 0, 0)$.

(b) *Calculate* $\nabla\phi(0, 0)$.
Solution.

$$\nabla\phi(0, 0) = -\left(\frac{F_x}{F_z}, \frac{F_y}{F_z}\right)\Big|_{(0,0,0)} = \left(\frac{1}{3}, 0\right).$$

Section 3.4.3

3. *Find the distance from* $(0, 0)$ *to the hyperbola* γ *with the equation* $7x^2 + 8xy + y^2 = 45$.
Solution.
The distance of any point (x, y) to $(0, 0)$ is the Euclidean norm $||(x, y)||_2$. The distance from $(0, 0)$ to the hyperbola γ is the minimum of this norm as (x, y) varies on γ. Equivalently, we need to find the point(s) $(x, y) \in \gamma$ where $f(x, y) := ||(x, y)||_2^2 := x^2 + y^2$ attains its (global) minimum. This is an extremum problem for f with the constraint given by the equation of γ. A necessary condition for such point(s) (x, y) is

$$\nabla[x^2 + y^2 - c(7x^2 + 8xy + y^2 - 45)] = (0, 0), \tag{A.27}$$

where c is a Lagrange multiplier. We get from (A.27)

$$((7c-1)x+4cy,\ 4cx+(c-1)y)=(0,0). \tag{A.28}$$

Since the solution $(x,y)=(0,0)$ of this equation does not satisfy the constraint, the determinant of the homogeneous linear system above must vanish, that is $c(9c-7)=0$. Since $c=0$ implies $(x,y)=(0,0)$, we must have $c\neq 0$, hence $c=7/9$. It follows then from (A.28) that $y=14x$. The constraint implies that $x^2=1/7$, and therefore $y^2=196/7$ and the corresponding distance is $\sqrt{(197/7)}$, attained at the points $(x,y)=\pm(1/\sqrt{7})(1,\ 14)$ on γ.

Let $g(x,y)=7x^2+8xy+y^2-45$. Since g is continuous on $\mathbb{R}^2$, $\gamma=g^{-1}(\{0\})$ is closed (as the inverse image of the closed set $\{0\}$ in $\mathbb{R}$). For any $R>\sqrt{(197/7)}$, the set $E:=\gamma\cap\overline{B}((0,0),R)\subset\mathbb{R}^2$ is then closed (as the intersection of two closed sets) and bounded, hence compact. Therefore the continuous function f attains its global minimum on E, necessarily at an interior point (why?), hence at a critical point of $f-cg$. Since the only critical points are $\pm(1/\sqrt{7})(1,\ 14)$, they are global minimum points of f on γ, and the wanted distance from $(0,0)$ to γ is therefore given by their norm $\sqrt{(197/7)}$.

7. *Find the local extrema of the function $f(x,y,z)=x^2+y^2+z^2$ under the constraint*

(b) $x^4+y^4+z^4=1$.
Solution.
Writing $u=x^2$, $v=y^2$, $w=z^2$, the problem is equivalent to finding the local extrema of $f(u,v,w)=u+v+w$ under the constraint $g(u,v,w)=0$, where $g(u,v,w)=u^2+v^2+w^2-1$. A necessary condition is

$$\nabla(f-cg):=(1-2cu,\ 1-2cv,\ 1-2cw)=(0,0,0).$$

Necessarily $c\neq 0$, and $(u,v,w)=(1/2c)(1,1,1)$. From the constraint, we get $4c^2=3$, hence $c=(\sqrt{3})/2$ (c is positive since at least one of the variables u,v,w is positive). The critical points (x,y,z) are trivially deduced from this.

8. *Find a rectangle with maximal area inscribed in the ellipse $x^2+2y^2=1$.*
Solution.
The vertices of the rectangle are $(\pm x,\pm y)$, $x,y>0$, $x^2+2y^2=1$. We need to maximize the function $f(x,y)=xy$ giving a fourth of the area of the rectangle, under the constraint $g(x,y):=x^2+2y^2-1=0$. A necessary condition is given by $\nabla(f-cg)=0$, where c is a Lagrange multiplier. Thus

$$(y-2cx,\ x-4cy)=(0,0).$$

Since $x,y>0$, the determinant of the homogeneous linear system above vanishes, i.e., $8c^2-1=0$. Since $y=2cx$ and $x,y>0$, we have $c>0$, and therefore $c=1/(2\sqrt{2})$ is the unique solution for c. It follows that $y=x/\sqrt{2}$, hence $1=$

$x^2 + 2y^2 = 2x^2$, and we conclude that $(x, y) = (1/2)(\sqrt{2}, 1)$. The usual argument based on the uniqueness of the critical point and compactness implies that the above critical point is the global maximum point of the area of the inscribed rectangles, and the maximal area is $4(1/4)\sqrt{2} = \sqrt{2}$.

Section 3.5.5

1. (a) *Let M be the quadratic surface in $\mathbb{R}^3$, given by the general equation*

$$xAx^t + 2xb^t = c \qquad x \in \mathbb{R}^3, \tag{A.29}$$

where $b \in \mathbb{R}^3$, $c \in \mathbb{R}$ are given constants and A is a given constant 3×3 symmetric matrix. Let $p \in M$ be permissible, i.e., $Ap^t + b^t$ is not the zero column. Show that the equation of the tangent plane to M at a permissible point $p \in M$ is

$$xAp^t + (x + p)b^t = c. \tag{A.30}$$

Solution.
The surface M has the equation $F(x) = 0$, where

$$F(x) = xAx^t + 2xb^t - c.$$

Since $\frac{\partial x}{\partial x_i} = e^i$, we have

$$F_{x_i}(x) = e^i Ax^t + xA(e^i)^t + 2e^i b^t.$$

However $xA(e^i)^t = [(xA(e^i)^t]^t = e^i Ax^t$ because A is symmetric. Hence

$$F_{x_i}(x) = 2e^i(Ax^t + b^t) = 2(Ax^t + b^t)_i,$$

and

$$\nabla F(x)^t = 2(Ax^t + b^t). \tag{A.31}$$

Therefore the equation of the tangent plane at the point p is

$$(x - p)(Ap^t + b^t) = 0.$$

Note that this is a valid plane equation because $Ap^t + b^t \neq 0^t$ by hypothesis (p permissible!). Expanding we obtain

$$xAp^t + xb^t - (pAp^t + 2pb^t) + pb^t = 0. \tag{A.32}$$

The expression in brackets is equal to c because $p \in M$. Therefore Eq. (A.32) coincides with (A.30).

(b) *Let M be the circular cylinder with radius r and axis along the x_2-axis. Find the equation of the tangent plane to M at the point $p = (r/2)(1, \sqrt{2}, \sqrt{3})$ ($\in M$).*
Solution.
The equation of M is $x_1^2+x_3^2 = r^2$. Replacing the quadratic form by the corresponding bilinear form (as described in the statement of the exercise), we get $(r/2)(x_1+\sqrt{3}x_3 = r^2$, i.e., $x_1 + \sqrt{3}x_3 = 2r$.

(c) *Let M be the circular cone*

$$M : F(x) := x_1^2 + x_2^2 - 2x_3 = 0.$$

Find a point $p \in M$ such that the tangent plane at p is orthogonal to the vector $(1, 1, -2)$.
Solution.
The gradient vector $\nabla F(p)$ is a normal vector to M at p, and is required to be $\lambda(1, 1, -2)$ for some real λ. By (A.31),

$$\nabla F(p)^t = 2(Ap^t + b^t) = 2[(p_1, p_2, 0) + (0, 0, -1)]^t = 2(p_1, p_2, -1)^t.$$

We then have the requirement

$$(p_1, p_2, -1) = c(1, 1, -2),$$

where $c := \lambda/2$. Hence $p_1 = p_2 = c = 1/2$, and since $F(p) = 0$, we have $p_3 = (1/2)[(1/4) + (1/4)] = 1/4$. Thus $p = (1/2,\ 1/2,\ 1/4)$.

2. *Let M be the surface parametrized by*

$$f(s, t) = ((1 + s)\cos t,\ (1 - s)\sin t,\ s) : \mathbb{R}^2 \to \mathbb{R}^3.$$

Find the equations of two tangent planes to M at the point $p = (1, 0, 1) \in M$.

Solution.
By periodicity of the trigonometric functions, we may assume that $t \in [-\pi, \pi]$, while $s \in \mathbb{R}$.

At the point p, $1 = p_3 = s$, and $1 = p_1 = 2\cos t$, hence $t = \pm\pi/3$. We have

$$f_s(1, \pm\pi/3) = (\cos t,\ -\sin t,\ 1)\Big|_{(1,\pm\pi/3)} = (\frac{1}{2},\ \mp\frac{\sqrt{3}}{2},\ 1),$$

and

$$f_t(1, \pm\pi/3) = (-(1 + s)\sin t,\ (1 - s)\cos t,\ 0)\Big|_{(1,\pm\pi/3)} = (\mp\sqrt{3},\ 0,\ 0).$$

The following vectors n are then normal vectors to M at p:

$$n := (f_s \times f_t)(1, \pm\pi/3) = (0, \mp\sqrt{3}, -\frac{3}{2}).$$

The corresponding tangent planes to M at p have the equations $n \cdot (x - p) = 0$ ($x \in \mathbb{R}^3$), that is,

$$\mp\sqrt{3}\,x_2 - \frac{3}{2}(x_3 - 1) = 0.$$

3. *The circle γ with equation $(y - r)^2 + z^2 = a^2$ in the yz-plane ($r > a$) is rotated about the z-axis. The surface M generated in this manner is a torus.*

(a) *Find a parametrization of M.*
Solution.
Use plane polar coordinates (ρ, s) in the yz-plane, with origin at the center $(0, r, 0)$ of γ. The points of γ are then $(0, r + a\cos s, a\sin s)$. The rotation of γ about the z-axis by an angle $\pi/2 - t$ moves the latter point point to

$$((r + a\cos s)\cos t,\ (r + a\cos s)\sin t,\ a\sin s) \in M. \tag{A.33}$$

The function $f(s, t) : [0, 2\pi]^2 \to \mathbb{R}^3$ in (A.33) is a parametrization of M.

(b) *Find the unit normal to M when $r = 2a$ and $s = t = \pi/4$.*
Solution.
We have (when $r = 2a$)

$$f_s(\pi/4, \pi/4) = (-a\sin s\,\cos t,\ -a\sin s\,\sin t,\ a\cos s)\Big|_{(\pi/4,\pi/4)} = -\frac{a}{2}(1, 1, \sqrt{2}),$$

$$f_t(\pi/4, \pi/4) = (r + a\cos s)(-\sin t,\ \cos t,\ 0)\Big|_{(\pi/4,\pi/4)} = a(2 + 1/\sqrt{2})(-1/\sqrt{2},\ 1/\sqrt{2},\ 0).$$

A normal vector to M at the given point is given by

$$n := f_s \times f_t\Big|_{(\pi/4,\pi/4)} = a^2(2 + 1/\sqrt{2})(-1, -1, \sqrt{2}).$$

Since $||n|| = 2a^2(2 + 1/\sqrt{2})$, a unit normal to M at the given point is $\tilde{n} := \frac{n}{||n||} = (-1/2, -1/2, 1/\sqrt{2})$.

Section 3.6.15

The guiding remarks in this section's exercises are very detailed. Consequently, we shall give very few additional details. The lengthy statements are not repeated.

1. (a) By definition, the columns y^j of the fundamental matrix Y are a basis for the solution space $\mathcal{S}(A)$ of the homogeneous system (H). Therefore the general solution

$y \in \mathcal{S}(A)$ is an arbitrary linear combination $y = \sum c_j y^j = Yc$, where c denotes the column with components c_j.

2. (b) Denote $c := [Y(x_0)]^{-1} y^0$, where y^0 is a given constant column. By Part (a), $y := Yc \in \mathcal{S}(A)$, and clearly $y(x_0) = y^0$. Thus y is the unique solution of the Cauchy IVP for (H).

2. (b) Let

$$z(x) := Y(x) \int_{x_0}^{x} Y(t)^{-1} b(t)\, dt.$$

Since $Y' = AY$, we have

$$\begin{aligned} z'(x) &= Y'(x) \int_{x_0}^{x} Y(t)^{-1} b(t)\, dt + Y(x)\, Y(x)^{-1} b(x) \\ &= A\, Y(x) \int_{x_0}^{x} Y(t)^{-1} b(t)\, dt + b(x) = Az(x) + b(x). \end{aligned}$$

Thus z is a particular solution of (NH). Since the general solution of (H) is Yc (cf. Exercise 1(a)), we conclude that the general solution of (NH) is $Y(x)[\int_{x_0}^{x} Y(t)^{-1} b(t)\, dt + c]$, where c is an arbitrary constant column vector.

5. (c) Deriving term-by-term the power series for e^{xA}, we obtain

$$[e^{xA}]' = \sum_{k=1}^{\infty} \frac{x^{k-1}}{(k-1)!} A^k = \sum_{j=0}^{\infty} \frac{x^j}{j!} A^{j+1} = A\, e^{xA} = e^{xA} A.$$

The common factor A of the summands can be factored out to the left or to the right of the series, by the usual distributive law of matrices and the continuity of multiplication in the normed algebra $\mathbb{M}_n$ of $n \times n$ matrices.

(e) By the rule for the derivative of a product of $\mathbb{M}_n$-valued functions,

$$F'(x) = -A\, e^{-xA} e^{x(A+B)} + e^{-xA} (A + B)\, e^{x(A+B)} = e^{-xA} B e^{x(A+B)}.$$

Since B commutes with A, it commutes with A^k for all $k = 0, 1, \ldots$, hence with e^{-xA}, by the distributive law and the continuity of multiplication. Hence

$$F'(x) = B\, F(x).$$

Therefore

$$\left(e^{-xB} F(x)\right)' = -B\, e^{-xB} F(x) + e^{-xB} B\, F(x) = 0.$$

We conclude that $e^{-xB}F(x)$ is a constant matrix. Since this matrix equals I at $x = 0$, it follows that

$$e^{-xB}F(x) = I \tag{A.34}$$

identically. Taking in particular $B = 0$, we get

$$e^{-xA}e^{xA} = I,$$

that is, e^{xA} is non-singular, and its inverse is e^{-xA}. Therefore, for arbitrary commuting matrices $A, B \in \mathbb{M}_n$, (A.34) is equivalent to

$$e^{x(A+B)} = e^{xA}e^{xB}$$

for all real x, which is in turn equivalent to its special case with $x = 1$:

$$e^{A+B} = e^{A}e^{B}.$$

Taking in particular the commuting matrices xA and yA for any real (or complex) x, y, we get the "group relation"

$$e^{(x+y)A} = e^{xA}e^{yA}.$$

6. (b) The matrix A of the system

$$y_1' = 3y_1 + y_2; \quad y_2' = y_1 + 3y_2$$

has the characteristic polynomial $p(\lambda) = (\lambda - 3)^2 - 1$. The eigenvalues of A are therefore $\lambda_1 = 4$ and $\lambda_2 = 2$. Corresponding column eigenvectors are $(1, \pm 1)^t$. A fundamental matrix Y for A has the columns $e^{4x}(1, 1)^t$ and $e^{2x}(1, -1)^t$.

13. (a) We denote by y_j the solution in the interval U of the Cauchy IVP

$$Ly_j = 0, \quad y_j^{(k)}(x_0) = \delta_{k,j-1}, \quad k, j-1 = 0, \ldots, n-1.$$

Let $y = (y_1, \ldots, y_n)$, and let $c \in \mathbb{R}^n$ be such that $c \cdot y = 0$ in U. Evaluating at the point $x_0 \in U$, we get $c_j = 0$ for $j = 1, \ldots, n$. Hence $y_1, \ldots, y_n$ are linearly independent.

(b) Let y be any solution of (H) in U, and define

$$z = \sum_{j=0}^{n-1} y^{(j)}(x_0)y_{j+1}.$$

Then z is a solution of (H), and for all $k = 0, \dots, n-1$,

$$z^{(k)}(x_0) = \sum_j y^{(j)}(x_0) y^{(k)}_{j+1}(x_0) = \sum_j y^{(j)}(x_0)\delta_{k,j} = y^{(k)}(x_0).$$

Thus y and z are solutions of the same Cauchy IVP. By uniqueness, $y = z$ in U, that is, y is a linear combination of $y_1, \dots, y_n$. Together with Part (a), this proves that $y_1, \dots, y_n$ is a basis for the solution space of (H) in U.

(g) Let $y = (y_1, \dots, y_n)$, where y_j are linearly independent solutions of (H) in the interval U, that is, $Ly_j = 0$ in U. The Wronskian matrix $(W) := (W(y))$ is then invertible in U, and its inverse is continuous there. Define $u : U \to \mathbb{R}^n$ by

$$u^t := \int_{x_0}^x b(s)(W(s))^{-1}(e^n)^t ds.$$

Then u is well-defined and of class C^1, and

$$(W)(u^t)' = b(e^n)^t. \tag{A.35}$$

Since the rows of the Wronskian matrix (W) are $y, y', \dots, y^{(n-1)}$, the vector equation (A.35) is equivalent to the n scalar equations

$$y^{(k)}(u^t)' = 0, \quad 0 \le k < n-1; \; y^{(n-1)}(u^t)' = b. \tag{A.36}$$

Define $z := yu^t$. Then by (A.36),

$$z' = y'u^t + y(u^t)' = y'u^t, \dots, z^{(k)} = y^{(k)}u^t \; (k \le n-1),$$

$$z^{(n)} = y^{(n)}u^t + y^{(n-1)}(u^t)' = y^{(n)}u^t + b. \tag{A.37}$$

Since $a_n = 1$ and $Ly = 0$ in U, we obtain from (A.37) that

$$Lz = \sum_{k=0}^n a_k z^{(k)} = \Big[\sum_{k=0}^n a_k y^{(k)}\Big] u^t + b = (Ly)u^t + b = b$$

on U. This proves that $z = yu^t$ is a solution of (NH) in U. The general solution of (NH) in U is then $w = z + yc^t = y(u+c)^t$, where $c \in \mathbb{R}^n$ is an arbitrary constant vector.

Section 4.1.13

1. (a) Leibnitz' rule applies, and F' is continuous (on $\mathbb{R}$). Thus

$$F'(0) = \int_0^1 x^2 e^{x^2 y} dx\Big|_{y=0} = \int_0^1 x^2 dx = \frac{1}{3}.$$

(b) We have

$$2yF'(y) = \int_0^1 2x^2y\, e^{x^2y} dx = \int_0^1 x\, \frac{d}{dx} e^{x^2y} dx$$
$$= x\, e^{x^2y}\Big|_{x=0}^1 - \int_0^1 e^{x^2y} dx = e^y - F(y).$$

3. We use the general Leibnitz rule for $x \neq 0$. After some rearrangement, we get

$$F'(x) = -x \int_0^{x^2} \frac{2y}{x^4 + y^2}\, dy + \frac{\pi x}{2} = -x\, \log(x^4 + y^2)\Big|_{y=0}^{x^2} + \frac{\pi x}{2}$$
$$= x\, \log \frac{x^4}{2x^4} + \frac{\pi x}{2} = \frac{\pi x}{2} - x \log 2.$$

This formula is trivially true for $x = 0$. Since $F(0) = 0$, we get $F(x) = \int_0^x F'(t)\, dt = (\frac{\pi}{4} - \frac{\log 2}{2})x^2$.

4. We have $c \in [\alpha, \beta] \subset (0, \infty)$, and

$$F(y) := \int_0^{\pi/2} \log(y^2 \cos^2 x + c^2 \sin^2 x)\, dx \qquad y \in [\alpha, \beta].$$

Leibnitz rule is applicable in the rectangle $[0, \pi/2] \times [\alpha, \beta]$. Hence for all $y \in [\alpha, \beta]$,

$$F'(y) = \int_0^{\pi/2} \frac{2y \cos^2 x}{y^2 \cos^2 x + c^2 \sin^2 x}\, dx \tag{A.38}$$

$$= \frac{2}{y} \int_0^{\pi/2} \Big[1 - \frac{c^2 \sin^2 x}{y^2 \cos^2 x + c^2 \sin^2 x}\Big] dx = y^{-1}[\pi - 2c^2 G(y)]. \tag{A.39}$$

We have

$$G(y) = (cy)^{-1} \int_0^{\pi/2} \frac{cy^{-1} \tan^2 x}{1 + (cy^{-1} \tan x)^2}\, dx.$$

Writing $\tan^2 x = \sec^2 x - 1$, we obtain from (A.38)

$$G(y) = (cy)^{-1} \arctan[cy^{-1} \tan x]\Big|_{x=0}^{\pi/2} - (2y)^{-1} F'(y)$$
$$= \frac{\pi}{2}\, (cy)^{-1} - (2y)^{-1} F'(y).$$

Therefore, by (A.39),

$$(1 - \frac{c^2}{y^2})\, F'(y) = \frac{\pi}{y}(1 - \frac{c}{y}).$$

For $y \neq c$, we divide this equation by $1 - \frac{c^2}{y^2}$ ($\neq 0$!). Hence $F'(y) = \frac{\pi}{y+c}$ for $y \neq c$. For $y = c$, this formula follows directly from (A.38) by an elementary calculation. We then conclude that

$$F(y) - F(c) = \int_c^y F'(t)\,dt = \pi \int_c^y \frac{dt}{t+c} = \pi \log \frac{y+c}{2c}.$$

Since $F(c) = \pi \log c$ (by the definition of F), we conclude that $F(y) = \pi \log \frac{y+c}{2}$.

6. The function $f(x, y) := x^3 e^{x^2 y}$ is continuous in the square $[0, 1]^2$. Therefore the order of integration can be changed. In the partial integral with respect to y, we make the change of variable $u = x^2 y$ for each fixed $x \in (0, 1]$. We obtain

$$\int_0^1 \Big(\int_0^1 f(x, y)\,dx\Big)\,dy = \int_0^1 x\Big(\int_0^1 e^{x^2 y} x^2 dy\Big)\,dx$$

$$= \int_0^1 x\Big(\int_0^{x^2} e^u du\Big)\,dx = \int_0^1 x\,(e^{x^2} - 1)\,dx = \frac{1}{2}(e^{x^2} - x^2)\Big|_0^1 = \frac{e-2}{2}.$$

8. (a) The function

$$f(x, y) := \frac{x}{(1+x^2)(1+xy)}$$

is continuous in the square $[0, 1]^2$. We may then change the order of integration in the given repeated integral. The latter is equal therefore to

$$\int_0^1 \frac{1}{1+x^2}\Big(\int_0^1 \frac{x\,dy}{1+xy}\Big)\,dx = \int_0^1 \frac{\log(1+x)}{1+x^2}\,dx := A. \tag{A.40}$$

On the other hand, a partial fraction decomposition of f is given by

$$f(x, y) = \frac{1}{1+y^2}\Big[\frac{x}{1+x^2} + \frac{y}{1+x^2} - \frac{y}{1+xy}\Big).$$

Therefore

$$(1+y^2)\int_0^1 f(x, y)\,dx = [\frac{1}{2}\log(1+x^2) + y \arctan x - \log(1+xy)]\Big|_{x=0}^1$$
$$= \frac{\log 2}{2} + \frac{\pi}{4} y - \log(1+y).$$

Hence by (A.40)

$$A = \int_0^1 (\int_0^1 f(x, y)\, dx)\, dy = \frac{\log 2}{2} \arctan y|_0^1 + \frac{\pi}{8} \log(1 + y^2)|_0^1 - A$$

$$= \frac{\pi \log 2}{4} - A.$$

Solving for A, we get $A = \frac{\pi \log 2}{8}$.

(b) We integrate by parts and use Part (a):

$$B := \int_0^1 \frac{\arctan x}{1 + x}\, dx = \int_0^1 \arctan x\, [\log(1 + x)]' dx$$

$$= [\arctan x \ \log(1 + x)]|_0^1 - \int_0^1 (\arctan x)' \log(1 + x)\, dx$$

$$= \frac{\pi \log 2}{4} - A = \frac{\pi \log 2}{8}.$$

Section 4.2.15

1. The given domain

$$\overline{D} := \{(x, y);\ y^2 \le x \le y,\ 0 \le y \le 1\}$$

is normal (of y-type). The function $f(x, y) := \sin(\pi x/y)$ is bounded in $\overline{D}$ (by 1) and continuous in $\overline{D} \setminus \{(0, 0)\}$, hence integrable on $\overline{D}$. However Theorem 4.2.9 is not directly applicable because $f \notin C(\overline{D})$. We give below the details of a procedure to overcome this obstacle.

Let $0 < \epsilon < 1$, and consider the partition $\overline{D} = \overline{D_1} \cup \overline{D_2}$, where

$$\overline{D_1} := \overline{D} \cap \{(x, y);\ 0 \le y \le \epsilon\}$$

$$\overline{D_2} := \overline{D} \cap \{(x, y);\ \epsilon \le y \le 1\}.$$

Then $\overline{D_1} \subset [0, \epsilon]^2$, and since $|f| \le 1$, we have

$$|\int_{\overline{D_1}} f\, dS| \le S([0, \epsilon]^2) = \epsilon^2. \tag{A.41}$$

We may apply Theorem 4.2.9 on the domain $\overline{D_2}$, since $f \in C(\overline{D_2})$. We obtain

$$\int_{\overline{D_2}} f\, dS = \int_{\epsilon}^{1} F(y)\, dy, \tag{A.42}$$

where

$$F(y) := \int_{y^2}^{y} \sin \pi \frac{x}{y}\, dx = \frac{y}{\pi}(\cos \pi y + 1).$$

Hcncc

$$\int_{\overline{D_2}} f\, dS = -\frac{1}{\pi^3}\Big[1 + \pi\epsilon \sin \pi\epsilon + \cos \pi\epsilon\Big] + \frac{1}{2\pi}(1 - \epsilon^2). \tag{A.43}$$

Letting $\epsilon \to 0+$, we conclude that the value of the given integral is $1/2\pi - 2/\pi^3$.

4. The closed domain

$$\overline{D} := \{(x, y);\, 1 \le x^2 + y^2 \le 4,\ \frac{x}{\sqrt{3}} \le y \le x\}$$

corresponds in polar coordinates (r, ϕ) to the closed rectangle

$$\overline{D'} := [1, 2] \times [\frac{\pi}{6}, \frac{\pi}{4}].$$

Therefore

$$\int_{\overline{D}} (x^2 + y^2 - 1)^{1/2} dS = \int_1^2 (\int_{\pi/6}^{\pi/4} d\phi)(r^2 - 1)^{1/2} r\, dr = \frac{\pi\sqrt{3}}{12}.$$

6. The given closed domain $\overline{D}$ is

$$\overline{D} = \{(x, y);\, a \le \frac{y^2}{x} \le b,\ \alpha \le xy \le \beta\}.$$

The map

$$(x, y) \to (u, v) := (\frac{y^2}{x}, xy)$$

associates to $\overline{D}$ the closed rectangle

$$\overline{D'} := \{(u, v);\ a \le u \le b,\ \alpha \le v \le \beta\}.$$

The Jacobian of the map has absolute value $3y^2/x = 3u \neq 0$ in $\overline{D'}$. Hence

$$S(\overline{D}) = \int_a^b (3u)\,du \int_\alpha^\beta dv = \frac{3}{2}(b^2 - a^2)(\beta - \alpha).$$

7. The closed domain $\overline{D'}$ corresponding to the given domain $\overline{D}$ under the map

$$(r, \phi) \to (x, y) = r^4(\cos^4\phi, \sin^4\phi)$$

is the closed rectangle $[0, 1] \times [0, 2\pi]$. We calculate that $|\frac{\partial(x,y)}{\partial(r,\phi)}| = 2r^7|\sin^3\theta|$, where $\theta := 2\phi$. Note that the Jacobian does not vanish in $\overline{D'}$ except on a set of area zero, so that the change of variable formula is applicable. Therefore

$$\int_{\overline{D}} (x^{1/2} + y^{1/2})^3 dS = \int_{r=0}^1 \int_{\theta=0}^{4\pi} r^{13}|\sin^3\theta|\,d\theta\,dr$$

$$= \frac{1}{14}\int_0^{4\pi} |\sin^3\theta|\,d\theta.$$

The function $g(\theta) := |\sin^3\theta|$ is π-periodic, that is, $g(\theta + \pi) = g(\theta)$. Therefore

$$\int_0^{4\pi} g(\theta)\,d\theta = 4\int_0^\pi g(\theta)\,d\theta.$$

Since $\sin\theta \ge 0$ for $\theta \in [0, \pi]$, the wanted integral is equal to

$$\frac{2}{7}\int_0^\pi \sin^3\theta\,d\theta = \frac{8}{21}.$$

8. The continuous function $f(x, y) := |\cos(x + y)|$ over $\overline{D} := [0, \pi]^2$ is equal to $\cos(x + y)$ $(-\cos(x + y))$ over the normal closed subdomains $\overline{D_1}$ and $\overline{D_4}$ ($\overline{D_2}$ and $\overline{D_3}$, respectively), where

$$\overline{D_1} := \{(x, y);\ 0 \le y \le \frac{\pi}{2} - x,\ 0 \le x \le \frac{\pi}{2}\},$$

$$\overline{D_2} := \{(x, y);\ \frac{\pi}{2} - x \le y \le \pi,\ 0 \le x \le \frac{\pi}{2}\},$$

$$\overline{D_3} := \{(x, y);\ 0 \le y \le \frac{3\pi}{2} - x,\ \frac{\pi}{2} \le x \le \pi\},$$

$$\overline{D_4} := \{(x, y);\ \frac{3\pi}{2} - x \le y \le \pi,\ \frac{\pi}{2} \le x \le \pi\}.$$

Since $\{\overline{D_j};\ j = 1, \ldots, 4\}$ is a partition of $\overline{D}$, we have

$$\int_{\overline{D}} |\cos(x+y)|\, dS = \sum_{j=1}^{4} \int_{\overline{D_j}} |\cos(x+y)|\, dS$$

$$= \int_{D_1} \cos(x+y)\, dS - \int_{D_2} \cos(x+y)\, dS - \int_{D_3} \cos(x+y)\, dS + \int_{D_4} \cos(x+y)\, dS$$

$$= \int_0^{\pi/2} \Big(\int_0^{\pi/2-x} \cos(x+y)\, dy - \int_{\pi/2-x}^{\pi} \cos(x+y)\, dy \Big)\, dx$$

$$+ \int_{\pi/2}^{\pi} \Big(-\int_0^{3\pi/2-x} \cos(x+y)\, dy + \int_{3\pi/2-x}^{\pi} \cos(x+y)\, dy \Big)\, dx$$

$$= \int_0^{\pi} 2\, dx = 2\pi.$$

10. The closed disc

$$\overline{D} := \{(x, y);\ x^2 + y^2 \le x\}$$

centred at $(1/2, 0)$ with radius $1/2$ corresponds in polar coordinates to the normal closed domain

$$\overline{D'} := \{(r, \phi);\ 0 \le r \le \cos\phi,\ \phi \in [-\frac{\pi}{2}, \frac{\pi}{2}]\}.$$

Therefore

$$I := \int_{\overline{D}} \sqrt{1 - x^2 - y^2}\, dS = \int_{-\pi/2}^{\pi/2} \Big(\int_0^{\cos\phi} (1 - r^2) r\, dr \Big)\, d\phi.$$

The inner integral is equal to

$$\frac{1}{3}(1 - r^2)^{3/2}\Big|_{\cos\phi}^{0} = \frac{1}{3}(1 - \sin^3\phi).$$

Therefore

$$I = \frac{\pi}{3} - \frac{1}{3}\int_{-\pi/2}^{\pi/2} \sin^3\phi \, d\phi.$$

The integrand of the last integral is an odd function on the symmetric interval $[-\pi/2, \pi/2]$. Consequently, this integral vanishes, and we conclude that $I = \pi/3$.

11. By symmetry with respect to the variables, we have $A_1 = A_2 = A_3$. We calculate A_3 by using spherical coordinates (r, ϕ, θ). The closed domain

$$\overline{D} := \{x \in \mathbb{R}^3;\ x_i \geq 0,\ a \leq ||x|| \leq b\}$$

corresponds to the cell

$$\overline{D'} := [a, b] \times [0, \frac{\pi}{2}] \times [0. \frac{\pi}{2}].$$

Therefore

$$A_3 := \int_{\overline{D}} x_3^q ||x||^p dx = \int_{\overline{D'}} r^{p+q+2} \cos^q\theta \sin\theta \, dr \, d\theta \, d\phi$$

$$= \frac{\pi}{2} \frac{\cos^{q+1}}{q+1}\Big|_{\pi/2}^{0} \int_a^b r^{p+q+2} dr = \frac{\pi}{2(q+1)} \frac{b^{p+q+3} - a^{p+q+3}}{p+q+3}$$

if $p + q + 3 \neq 0$, and $A_3 = \frac{\pi}{2(q+1)} \log \frac{b}{a}$ if $p + q + 3 = 0$.

12. We use the map $(r, \phi, z) \to (ar\cos\phi,\ br\sin\phi,\ z)$. The closed domain $\overline{D}$ corresponds to the normal closed domain

$$\overline{D'} := \{(r, \phi, z);\ 0 \leq z \leq c(1 - r^2),\ (r, \phi) \in [0, 1] \times [0, 2\pi]\}.$$

The Jacobian of the map is equal to abr. Hence

$$vol(\overline{D}) = \int_{\overline{D'}} abr \, dr \, d\phi \, dz$$

$$= 2\pi ab \int_0^1 c(1 - r^2) r \, dr = \frac{\pi}{2} abc.$$

13. We use cylindrical coordinates. The closed domain $\overline{D}$ corresponds to the normal closed domain

$$\overline{D'} := \{(r, \phi, z);\ (1 + z^2)^{1/2} \leq r \leq (4 + z^2)^{1/2},\ (\phi, z) \in [0, 2\pi] \times [0, 1]\}.$$

Hence

$$\int_{\overline{D}} z\,dxdydz = \pi \int_0^1 z[(4+z^2)-(1+z^2)]\,dz = \frac{3\pi}{2}.$$

14. We use spherical coordinates. The closed domain

$$\overline{D} := \{x \in \mathbb{R}^3;\ ||x||^2 \le x_3\}$$

corresponds to the normal closed domain

$$\overline{D'} := \{(r,\theta,\phi);\ 0 \le r \le \cos\theta;\ (\theta.\phi) \in [0,\pi] \times [0,2\pi]\}.$$

Hence

$$\int_{\overline{D}} ||x||\,dx = 2\pi \int_0^{\pi} \Big(\int_0^{\cos\theta} r^3 dr \sin\theta \Big)\,d\theta$$
$$= \frac{\pi}{2} \int_0^{\pi} \cos^4\theta\ \sin\theta\,d\theta = \frac{\pi}{5}.$$

15. The volume is given by the repeated integral

$$\int_0^1 \Big(\int_{x^2}^1 [\int_0^{x^2+y^2} dz]\,dy \Big)\,dx.$$

We omit the routine calculation.

Section 4.3.15

1. (a) The helix γ is parametrized by the C^1 vector function $x(t) = (a\cos t, a\sin t, bt)$, $t \in [0, T]$, where $a, b > 0$ are constants. We have

$$x'(t) = (-a\sin t,\ a\cos t,\ b), \tag{A.44}$$

hence $||x'(t)|| = \sqrt{a^2+b^2}$ and

$$s(t) = \int_0^t ||x'(\tau)||\,d\tau = \sqrt{a^2+b^2}\,t. \tag{A.45}$$

(b) One turn of the helix corresponds to the interval $[0, 2\pi]$ of the parameter t; its length is $s(2\pi) = 2\pi\sqrt{a^2+b^2}$.

(c) By (A.44), we have for one turn γ of the helix

$$\int_\gamma f \cdot dx = \int_0^{2\pi} (-a\sin t,\ a\cos t,\ bt) \cdot (-a\sin t,\ a\cos t,\ b)\,dt$$

$$= \int_0^{2\pi} (a^2 + b^2 t)\, dt = 2\pi(a^2 + \pi b^2).$$

(d) By (A.45), if γ is one turn of the given helix, then

$$\int_\gamma (x^2+y^2+z^2)^{-1/2} ds = (a^2+b^2)^{1/2} \int_0^{2\pi} (a^2+b^2t^2)^{-1/2} dt = \sqrt{1+(\frac{a}{b})^2}\, \sinh^{-1} 2\pi \frac{b}{a}.$$

2. The curve γ is parametrized by the C^1 vector function $x(t) = e^t(\cos t,\ \sin t,\ 1)$, $t \in [0, 1]$. We have

$$x'(t) = e^t(\cos t,\ \sin t,\ 1) + e^t(-\sin t,\ \cos t,\ 0) = e^t(\cos t - \sin t,\ \cos t + \sin t,\ 1).$$

Hence

$$||x'(t)|| = e^t\Big[(\cos t - \sin t)^2 + (\cos t + \sin t)^2 + 1\Big]^{1/2} = \sqrt{3}e^t.$$

Therefore the length of γ is $\int_0^1 ||x'(t)||\, dt = \sqrt{3}(e-1)$.

3. (a) Since $r = g(\theta)$ ($\theta \in [a, b]$) is the equation of the curve γ in polar coordinates (r, θ), a parametrization of γ is given by the C^1 vector function

$$\theta \in [a, b] \to f(\theta) := g(\theta)\,(\cos\theta,\ \sin\theta) \in \mathbb{R}^2.$$

We have

$$f'(\theta) = g'(\theta)\,(\cos\theta,\ \sin\theta) + g(\theta)\,(-\sin\theta,\ \cos\theta).$$

We then calculate

$$||f'(\theta)|| = [g'(\theta)^2 + g(\theta)^2]^{1/2},$$

and (a) follows.

(b) By Part (a),

$$L(\gamma) = \int_0^{2\pi} [2(1-\cos\theta)]^{1/2} d\theta = 2\int_0^{2\pi} \sin\frac{\theta}{2}\, d\theta = 8.$$

5. (a) The given cycloid arc γ parametrized by

$$x(t) = a(t - \sin t,\ 1 - \cos t), \quad t \in [0, 2\pi]$$

has the arc length function

$$s(t) := \int_0^t ||x'(\tau)||\, d\tau = a \int_0^t \Big[(1-\cos\tau)^2 + \sin^2\tau\Big]^{1/2} d\tau$$

$$= 2a \int_0^t \sin\frac{\tau}{2}\, d\tau = 4a(1-\cos\frac{t}{2}).$$

(b) $L(\gamma) = s(2\pi) = 8a$.

(c) The vector field f has components f_i that are rational functions with denominators vanishing at no point of $\mathbb{R}^2$. In particular, f is of class C^1 in $\mathbb{R}^2$. We calculate

$$\frac{\partial f_2}{\partial x} = -\frac{16xy^3}{(1+2x^2+y^4)^2} = \frac{\partial f_1}{\partial y}.$$

Since $\mathbb{R}^2$ is a star-like domain, it follows that the field f is conservative in $\mathbb{R}^2$. Therefore

$$\int_\gamma f \cdot dx = \int_{\gamma'} f \cdot dx,$$

where γ' is the line segment $[0, 2\pi a]$ on the x-axis joining the end points $(0, 0)$ and $(2\pi a,\ 0)$ of γ. A parametrization of γ' is $t \in [0, 2\pi] \to (at, 0) \in \gamma'$. Hence

$$\int_{\gamma'} f \cdot dx = \int_0^{2\pi} f(at, 0) \cdot (a, 0)\, dt$$

$$= \int_0^{2\pi} \frac{4a^2 t}{1+2a^2t^2}\, dt = \log(1+2a^2t^2)\Big|_0^{2\pi} = \log(1+8\pi^2a^2).$$

7. The vector field $f(x, y) = e^x(\sin y, \cos y)$ is of class C^1 in $\mathbb{R}^2$, and

$$\frac{\partial f_2}{\partial x} = e^x \cos y = \frac{\partial f_1}{\partial y}.$$

It follows that f is conservative in $\mathbb{R}^2$. In particular
(a) $\int_\gamma f \cdot dx = 0$ for the given ellipse γ.
(b) For the part γ_1 of γ in the first quadrant, $\int_{\gamma_1} f \cdot dx = \int_\delta f \cdot dx$, where δ joins the end points $(a, 0)$ and $(0, b)$ along any curve. Take for example $\delta = \delta_1 + \delta_2$, where δ_1 is the line segment from $(a, 0)$ to $(0, 0)$ on the x-axis and δ_2 is the line segment from $(0, 0)$ to $(0, b)$ on the y-axis. A parametrization of δ_1 is $t \in [0, a] \to (a-t, 0)$, hence

$$\int_{\delta_1} f \cdot dx = \int_0^a f(a-t, 0) \cdot (-1, 0)\, dt = \int_0^a e^{a-t}(0, 1) \cdot (-1, 0)\, dt = 0.$$

A parametrization of δ_2 is $t \in [0, b] \to (0, t) \in \delta_2$, hence

$$\int_{\delta_2} f \cdot dx = \int_0^b f(0, t) \cdot (0, 1)\, dt = \int_0^b \cos t\, dt = \sin b.$$

We conclude that

$$\int_{\gamma_1} f \cdot dx = \int_{\delta} f \cdot dx = \int_{\delta_1} f \cdot dx + \int_{\delta_2} f \cdot dx = \sin b.$$

(c) Since the field f is conservative in $\mathbb{R}^2$, we have $\int_{\gamma_2} f \cdot dx = \int_{\gamma_1} f \cdot dx = \sin b$ (cf. Part (b)).

Section 4.4.5

In all the exercises of this section, the hypothesis of Corollary 4.4.4 *is satisfied. The routine application of the corollary is presented with the omission of most calculations.*

1. The cardioid γ is parametrized by the C^1 function

$$(x(t),\ y(t)) = (1 - \cos t)\,(\cos t,\ \sin t), \qquad t \in [0, 2\pi].$$

Hence

$$2S(\overline{D}) = \int_0^{2\pi} (-y(t),\ x(t)) \cdot (x'(t),\ y'(t))\, dt$$

$$= \int_0^{2\pi} (1 - \cos t)\Big[(-\sin t,\ \cos t) \cdot (-\sin t + 2\cos t \sin t,\ \cos t + \sin^2 t - \cos^2 t)\, dt$$

$$= \int_0^{2\pi} (1 - \cos t)^2 dt = \int_0^{2\pi} (\frac{3}{2} - 2\cos t + \frac{1}{2}\cos 2t)\, dt$$

$$= 3\pi - 2\sin t\Big|_0^{2\pi} + \frac{1}{4}\sin 2t\Big|_0^{2\pi} = 3\pi.$$

We conclude that $S(\overline{D}) = 3\pi/2$.

2. For all $(x, y) \in \gamma$, $|x/a|$, $|y/b| \le 1$, and therefore we may write $x/a = \cos^3 t$ and $y/b = \sin^3 t$ with t varying in the interval $[0, 2\pi]$. We get the parametrization

$$(x(t),\ y(t)) = (a\cos^3 t,\ b\sin^3 t) \qquad (t \in [0, 2\pi])$$

of the hypocycloid $\gamma = \partial D$. Hence

$$2S(\overline{D}) = \int_0^{2\pi} (-b\sin^3 t,\ a\cos^3 t)\cdot(-3a\cos^2 t\sin t,\ 3b\sin^2 t\cos t)\,dt$$
$$= 3ab\int_0^{2\pi}\sin^2 t\cos^2 t\,dt = \frac{3ab}{8}\int_0^{2\pi}(1-\cos 4t)\,dt = \frac{3}{4}\pi ab.$$

We conclude that $S(\overline{D}) = 3\pi ab/8$.

Section 4.5.7

1. We parametrize the upper hemisphere M as indicated (r, ϕ are the plane polar coordinates in the plane $z = 0$):

$$f(r,\phi) := (r\cos\phi,\ r\sin\phi,\ \sqrt{1-r^2}), \qquad (r,\phi)\in\overline{D} := [0,1]\times[0,2\pi].$$

We calculate

$$n := f_r \times f_\phi = r(1-r^2)^{-1/2}(r\cos\phi,\ r\sin\phi,\ (1-r^2)^{1/2});$$

$$||n|| = r(1-r^2)^{-1/2}.$$

Hence

$$\int_M (x^2+y^2)z\,d\sigma = \int_{\overline{D}} f(r\cos\phi,\ r\sin\phi,\ (1-r^2)^{1/2})\,||n||\,drd\phi$$
$$= \int_0^{2\pi}(\int_0^1 r^3 dr)\,d\phi = \frac{\pi}{2}.$$

2. We use the same parametrization as in Exercise 1, with $(r,\phi)\in\overline{D} := [1/2,1]\times[0,\pi/2]$. We obtain

$$\int_M \arctan\frac{y}{x}\,d\sigma = \int_0^{\pi/2}\Big(\int_{1/2}^1 r(1-r^2)^{-1/2}dr\Big)\phi\,d\phi = \frac{\pi^2}{16}\sqrt{3}.$$

3. (a) We parametrize M by

$$f(x,y) := (x,\ y,\ h(x,y)), \qquad (x,y)\in\overline{D}.$$

We calculate

$$n := f_x \times f_y = (-h_x,\ -h_y,\ 1),$$

$$||n|| = (1 + h_x^2 + h_y^2)^{1/2},$$

$$\int_M g\, d\sigma = \int_{\overline{D}} g(x, y, h(x, y))(1 + h_x^2 + h_y^2)^{1/2} dxdy.$$

(b) Apply Part (a) with $h(x, y) = x^2 + y^2$, $g(x, y, z) = xy(1 + 4z)^{1/2}$, and

$$\overline{D} = \{(x, y);\ x^2 + y^2 \leq 1,\ x, y \geq 0\},$$

and shift afterwards to plane polar coordinates. We obtain:

$$\begin{aligned}\int_M g\, d\sigma &= \int_{\overline{D}} xy[1 + 4(x^2 + y^2)]\, dxdy \\ &= \int_0^{\pi/2} \Big(\int_0^1 (r^3 + 4r^5) dr \Big) \sin\phi \cos\phi\, d\phi = \frac{11}{24}.\end{aligned}$$

4. (a) The divergence of the vector field $F := (xy, 0, -z^2)$ is $y - 2z$. By the Divergence Theorem applied to $V := [0, 1]^3$, we have

$$\Phi_M(F) = \int_V (y - 2z) dV = \int_0^1 \int_0^1 \int_0^1 (y - 2z) dxdydz = -\frac{1}{2}.$$

(b) The outer unit normal $\tilde{n}$ to the faces M_i $(M_i')(i = 1, 2, 3)$ of M lying on the planes $x = 0$, $y = 0$, $x = 0$ (on the planes $x = 1$, $y = 1$, $z = 1$) are $-e^i$ (e^i, respectively). The corresponding values of $F \cdot \tilde{n}$ are $-F_i\Big|_{M_i} = 0$ ($F_i\Big|_{M_i'} = y,\ 0,\ -1$, respectively). Hence

$$\Phi_M(F) = \int_{M_1'} y\, dydz - \int_{M_3'} dxdy = \frac{1}{2} - 1 = -\frac{1}{2}.$$

5. Since $\nabla \cdot F = 1$, the Divergence Theorem implies that $\Phi_M(F)$ is equal to the volume of V, which is πR^3.

The outer unit normal $\tilde{n}$ is $-e^3$ (e^3) on lower (upper) bases M_1 (M_2) of $M := \partial V$. The respective values of $F \cdot \tilde{n}$ are $-F_3\Big|_{M_1} = 0$ ($F_3\Big|_{M_2} = R$, respectively). Let $\overline{D}$ denote the closed disc $x^2 + y^2 \leq R^2$ in the xy-plane. Then

$$\int_{M_1 \cup M_2} F \cdot \tilde{n}\, d\sigma = \int_{\overline{D}} R\, dxdy = \pi R^3.$$

We parametrize the lateral part M_3 of ∂V by

$$f(\phi, z) = (R\cos\phi,\ R\sin\phi,\ z) \qquad (\phi, z) \in [0, 2\pi] \times [0, R].$$

We calculate

$$n = f_\phi \times f_z = R(\cos\phi,\ \sin\phi,\ 0),$$

$$\tilde{n} = \frac{n}{||n||} = (\cos\phi,\ \sin\phi,\ 0),$$

$$\begin{aligned}\int_{M_3} F \cdot \tilde{n}\, d\sigma &= \int_0^{2\pi} \int_0^R (R^2 \sin^2\phi,\ R^3\cos^3\phi, z) \cdot (\cos\phi, \sin\phi, 0)\, d\phi\, dz \\ &= R^3 \int_0^{2\pi} (\sin^2\phi\cos\phi + R\cos^3\phi\sin\phi)\, d\phi = 0.\end{aligned}$$

We conclude that $\int_M F \cdot \tilde{n}\, d\sigma = \int_{M_1 \cup M_2} \dots d\sigma = \pi R^3$.

6. (a) When the curve γ is revolved by the angle $\pi/2 - \phi$ from its position in the yz-plane, the point $(0, y(t), z(t)) \in \gamma$ moves to the point $(y(t)\cos\phi,\ y(t)\sin\phi,\ z(t)) \in M$. Therefore a parametrization of M is given by

$$f(t, \phi) = (y(t)\cos\phi,\ y(t)\sin\phi,\ z(t)), \qquad (t, \phi) \in [a, b] \times [0, 2\pi].$$

We calculate

$$n := f_t \times f_\phi = y(t)\,(-z'(t)\cos\phi,\ -z'(t)\sin\phi,\ y'(t)),$$

$$||n|| = |y(t)|\,[y'(t)^2 + z'(t)^2]^{1/2},$$

and consequently

$$\sigma(M) = \int_0^{2\pi} \int_a^b ||n||\, dt\, d\phi = 2\pi \int_a^b |y(t)|\,[y'(t)^2 + z'(t)^2]^{1/2} dt.$$

(b) A parametrization of the circle γ centred at $(0, R, 0)$ with radius $r < R$ is given by

$$t \in [0, 2\pi] \to (0,\ R + r\cos t,\ r\sin t).$$

By Part (a), the area of the torus M obtained by revolving γ about the z-axis is given by

$$\begin{aligned}\sigma(M) &= 2\pi \int_0^{2\pi} |R + r\cos t|\,[(-r\sin t)^2 + (r\cos t)^2]^{1/2} dt \\ &= 2\pi \int_0^{2\pi} (Rr + r^2\cos t)\, dt = 4\pi^2 Rr.\end{aligned}$$

7. (a) We parametrize M by

$$f(x, y) = (x,\ y,\ x^2 - y^2), \qquad (x, y) \in [-1, 1]^2.$$

We calculate

$$n := f_x \times f_y = (-2x,\ -2y,\ 1),$$

$$\nabla \times F = (0, 2z, 0).$$

Hence

$$\begin{aligned}\Phi_M(\nabla \times F) &= \int_{[-1,1]^2} [(\nabla \times F) \circ f] \cdot n\, dxdy \\ &= 4 \int_{-1}^{1} \int_{-1}^{1} y(y^2 - x^2)\, dxdy = 0.\end{aligned}$$

Therefore, by Stokes' formula, $\int_\gamma F \cdot dx = 0$.

(b) By 2π-periodicity of the trigonometric functions, we may parametrize the curve γ by the vector valued function

$$x(t) := (\cos t,\ \sin t,\ \cos 2t), \qquad t \in [-\pi, \pi].$$

(Note the interval of the parameter is chosen as $[-\pi, \pi]$.) We calculate

$$\int_\gamma F \cdot dx := \int_0^{2\pi} F(x(t)) \cdot x'(t)\, dt$$

$$= \int_{-\pi}^{\pi} (\cos^2 t + \cos^2 2t,\ \sin t,\ \cos 2t) \cdot (-\sin t,\ \cos t,\ -2\sin 2t)\, dt$$

$$= \int_{-\pi}^{\pi} [(\cos^2 t + \cos^2 2t)(-\sin t) + \sin t \cos t - \sin 4t]\, dt.$$

The last integrand is an odd (continuous) function of the variable t. Therefore its integral over the symmetric interval $[-\pi, \pi]$ is equal to zero.

References

T.M. Apostol, Mathematical Analysis, 2nd edn. (Addison-Wesley, 1974)
C.R. Buck, Advanced Calculus (McGraw-Hill, New York, 1978) (reprinted 2002)
J.C. Burkill, H. Burkill, A Second Course on Mathematical Analysis (Cambridge University Press, 1970) (reprinted 2002)
J.J. Callahan, *Advanced Calculus* (A Geometric View, Undergraduate Texts in Mathematics (Springer, 2010)
L.J. Corwin, *Multivariable Calculus* (M. Dekker, New York, 1982)
R. Courant, *Differential and Integral Calculus*, vol. II (Interscience, London, 1937)
A. Devinatz, *Advanced Calculus* (Holt, Rinehart and Winston, New York, 1968)
C.H. Edwards, *Advanced Calculus of Several Variables* (Dover, New York, 1994)
H.H. Edwards, *Advanced Calculus, a Differential Forms Approach* (Birkhauser, Boston, 1994)
P.M. Fitzpatrick, Advanced Calculus, 2nd edn., The Sally Series, Pure and Applied Undergraduate Texts 5 (Amer. Math. Soc., Providence, R.I., 2006)
H. Flanders, *A Second Course in Calculus* (Academic Press, New York, 1974)
L. Flatto, *Advanced Calculus* (William and Wilkins, Baltimore, 1976)
W.H. Fleming, *Functions of Several Variables* (Springer, New York, 1977)
G.B. Folland, Advanced Calculus (2002)
A. Friedman, *Advanced Calculus* (Holt, Rinehart and Winston, New York, 1968)
W. Fulks, *Advanced Calculus: An Introduction to Analysis* (Wiley, New York, 1978)
M.C. Gemignani, Elementary Topology, 2nd edn. (Addison-Wesley, Reading, 1972)
J.H. Hubbard, *Vector Calculus, Linear Algebra, and Differential Forms: A Unified Approach* (Prentice-Hall, Upper Saddle River, 1999)
W. Kaplan, Advanced Calculus (Addison-Wesley, Reading, 1984)
R.G. Kuller, Topics in Modern Analysis (Prentice-Hall, 1969)
S. Lang, Calculus of Several Variables (Addison-Wesley, Reading, 1973)
L.H. Loomis, S. Sternberg, Advanced Calculus (Addison-Wesley, Reading, 1968)
J.E. Marsden, *Basic Multivariable Calculus* (Freeman, New York, 1993)
E.K. McLachlan, *Calculus of Several Variables* (Books/Cole Publ. Co., Belmont, 1968)
M. Moskowitz, F. Palogiannis, *Functions of Several Real Variables* (World Scientific Publ. Co., Imperial College Press, London, 2011)
M.E. Munroe, Modern Multidimensional Calculus (Addison-Wesley, Reading, 1963)
A.M. Ostrowski, *Differential and Integral Calculus* (Scott, Foresman, Glenview, Ill, 1968)
G.B. Price, *Multivariable Analysis* (Springer, New York, 1984)
H. Sagan, *Advanced Calculus* (Houghton Mifflin, Boston, 1974)

S. Kantorovitz, *Several Real Variables*, Springer Undergraduate Mathematics Series, DOI 10.1007/978-3-319-27956-5

R.T. Seeley, *Calculus of Several Variables* (Scott, Foresman, Glenview, Ill, 1970)
R. Sikorski, *Advanced Calculus: Functions of Several Variables* (Polish Scientific Publ, Warsaw, 1969)
A.E. Taylor, W.R. Mane, *Advanced Calculus* (Wiley, New York, 1983)
D.V. Widder, Advanced Calculus, 2nd edn. (Dover, 1989)
R.E. Williamson, *Multivariable Mathematics* (Prentice-Hall, Englewood Cliffs, 1979)

Index

S. Kantorovitz, *Several Real Variables*, Springer Undergraduate Mathematics Series, DOI 10.1007/978-3-319-27956-5

Printed in the United States
By Bookmasters